Methods in
Molecular Biology

Volume 4

New
Nucleic Acid
Techniques

Edited by

John M. Walker

The Hatfield Polytechnic, Hatfield, Hertfordshire, UK

Humana Press • Clifton, New Jersey

© 1988 The Humana Press Inc.
Crescent Manor
PO Box 2148
Clifton, NJ 07015

Printed in the United States of America.

Library of Congress Cataloging in Publication Data
Main entry under title:

Methods in molecular biology.

(Biological methods)
Includes bibliographies and indexes.
Contents: v. 1. Proteins—v. 2. Nucleic acids—v. 3. New protein techniques—v. 4. New nucleic acid techniques.

1. Molecular biology—Technique. I. Walker, John M., 1948–   .   II. Series.

QH506.M45   1984   574.8'8'078   84-15696

ISBN 0-89603-062-8 (v. 1)
ISBN 0-89603-064-4 (v. 2)
ISBN 0-89603-126-8 (v. 3)
ISBN 0-89603-127-6 (v. 4)
ISBN 0-89603-150-0 (v. 5)

Preface

In recent years there has been a tremendous increase in our understanding of the functioning of the cell at the molecular level. This has been achieved in the main by the invention and development of new methodology, particularly in that area generally referred to as "genetic engineering."

Although this revolution has been taking place in the field of nucleic acids research, the protein chemist has at the same time developed fresh methodology to keep pace with the requirements of present day molecular biology. Today's molecular biologists can no longer be content with being experts in one particular area alone. They need to be equally competent in the laboratory at handling DNA, RNA, and proteins, moving from one area to another as required by the problem that is being solved. Although many of the new techniques in molecular biology are relatively easy to master, it is often difficult for a researcher to obtain all the relevant information necessary for setting up and successfully applying a new technique. Information is of course available in the research literature, but this often lacks the depth of description that the new user requires. This requirement for in-depth practical details has become apparent by the considerable demand for places on our Molecular Biology Workshops held at Hatfield each summer.

Volume 2 of this series described practical procedures for a range of nucleic acid techniques frequently used by research workers in the field of molecular biology. Because of the limitations on length necessarily inherent in producing any book, one obviously had to be selective in the choice of titles for Volume 2. The production of Volume 4 therefore allows the development of the theme initiated in Volume 2. This volume contains a further selection of detailed protocols for a range of analytical and preparative nucleic acid techniques and should be seen as a continuation of Volume 2. In particular we have introduced protocols for the rapidly developing area of plant molecular biology. Companion Volumes 1 and 3 provide protocols for protein methodology.

Each method is described by an author who has regularly used the technique in his or her own laboratory. Not all the techniques described necessarily represent the state of the art. They are, however, dependable methods that achieve the desired result.

Each chapter starts with a description of the basic theory behind the method being described. The main aim of this book, however, is to describe the practical steps necessary for carrying out the method successfully. The Methods section, therefore, contains a detailed step-by-step description of a protocol that will result in the successful execution of the method. The Notes section complements the Methods section by indicating any major problems or faults that can occur with the technique, and any possible modifications or alterations.

This book should be particularly useful to those with no previous experience of a technique, and, as such, should appeal to undergraduates (especially project students), postgraduates, and research workers who wish to try a technique for the first time.

John M. Walker

Contents

Contributors

ALAN N. BATESON • MRC Group in Human Genetic Diseases, Department of Biochemistry, King's College, University of London, London, UK

HANS E. N. BERGMANS • Department of Molecular Cell Biology, State University of Utrecht, Utrecht, The Netherlands

ALBERT BORONAT • Department of Biochemistry, University of Barcelona, Barcelona, Spain

MICHAEL M. BURRELL • Twyford Plant Laboratories, Baltons Borough, Somerset, UK

WILLIAM G. CHANEY • Department of Cell Biology, Albert Einstein College of Medicine, Bronx, New York

MICHAEL J. CLEMENS • Cancer Research Campaign, Department of Biochemistry, St. George's Hospital Medical School, London, UK

WILLIAM J. DONNELLY • The Agricultural Institute, Fermoy, Ireland

KEITH DUDLEY • Department of Biochemistry, King's College, University of London, London, UK

RACHEL FALLON • Department of Biochemistry, Charing Cross and Westminster Medical School, London, UK

HERMIA FIGUEIREDO • Department of Biochemistry, Charing Cross and Westminster Medical School, London, UK

NEIL FISH • Department of Biochemistry, Rothamsted Experimental Station, Hertfordshire, UK

ANTHONY C. FORSTER • Commonwealth Special Research for Gene Technology, Department of Biochemistry, University of Adelaide, South Australia, Australia

WIM GAASTRA • Department of Infectious Diseases, State University of Utrecht, Utrecht, The Netherlands

DOMINGO GALLARDO • Department of Biochemistry, University of Barcelona, Barcelona, Spain

KEVAN M. A. GARTLAND • School of Life Sciences, Leicester Polytechnic, Leicester, UK

GRAHAM H. GOODWIN • Institute of Cancer Research, Chester Beatty Laboratories, London, UK

ALEXANDER GRAHAM • Biotechnology Division, Inveresk Research International Ltd., Musselburgh, Scotland

PETTER GUSTAFSSON • Institute of Cell and Molecular Biology, Umea University, Umea, Sweden

JOHN D. HALEY • Ludwig Institute for Cell and Molecular Biology, Umea University, Umea, Sweden

ROBERT HARR • Institute of Cell and Molecular Biology, Umea University, Umea, Sweden

HARALD HAYMERE • Department of Biotechnology, Sandoz AG, Basel, Switzerland

RICHARD D. HENFREY • Division of Biology and Environmental Sciences, The Hatfield Polytechnic, Hertfordshire, UK

JOACHIM HERZ • European Molecular Biology Laboratory, Heidelberg, Federal Republic of Germany

EILEEN G. HOAL • MRC Unit for Molecular and Cellular Cardiology, Department of Medical Physiology and Biochemistry, University of Stellenbosch Medical School, Tygerberg, South Africa

DANIEL R. HOWARD • Department of Cell Biology, Albert Einstein College of Medicine, Bronx, New York

MATTHEW CLEMENT JONES • The Biotechnology Centre, University of Cambridge, Cambridge, UK

MICHAEL G. K. JONES • Department of Biochemistry, Rothamsted Experimental Station, Hertfordshire, UK

JANE A. LANGDALE • Department of Biochemistry, Charing Cross and Westminster Medical School, London, UK

KEITH LINDSEY • Department of Biochemistry, Rothamsted Experimental Station, Hertfordshire, UK

ALAN D. B. MALCOLM • Department of Biochemistry, Charing Cross and Westminster Medical School, London, UK

ROBERT MCGOOKIN • Inveresk Research International Ltd., Musselburgh, Scotland

JAMES L. MCINNIS • Commonwealth Special Research for Gene Technology, Department of Biochemistry, University of Adelaide, South Australia, Australia

D. MCKECHNIE • Biotechnology Division, Inveresk Research International Ltd., Musselburgh, Scotland

DANIEL M. O'CALLAGHAN • The Agricultural Institute, Fermoy, Ireland

MARTIN J. PAGE • Department of Molecular Biology, Wellcome Biotechnology Ltd., Kent, UK

CAROLINE R. PERRY • Department of Microbiology, King's College, University of London, London, UK

MARK A. PLUMB • Beatson Institute for Cancer Research, Glasgow, UK

JEFFREY W. POLLARD • MRC Group in Human Genetic Diseases, Department of Biochemistry, King's College, University of London, London, UK

SANDRA SALLUSTIO • Department of Cell Biology, Albert Einstein College of Medicine, Bronx, New York

NIGEL W. SCOTT • School of Life Sciences, Leicester Polytechnic, Leicester, UK

ADRIAN SLATER • School of Life Sciences, Leicester Polytechnic, Leicester, UK

ROBERT J. SLATER • Division of Biological and Environmental Sciences, The Hatfield Polytechnic, Hertfordshire, UK

KEITH K. STANLEY • European Molecular Biology Laboratory, Heidelberg, Federal Republic of Germany

PAMELA STANLEY • Department of Cell Biology, Albert Einstein College of Medicine, Bronx, New York

JOHN STEVEN • Biotechnology Division, Inveresk Research International Ltd., Musselburgh, Scotland

ROBERT H. SYMONS • Commonwealth Special Research for Gene Technology, Department of Biochemistry, University of Adelaide, South Australia, Australia

CHRISTOPHER F. THURSTON • Department of Microbiology, King's College, University of London, London, UK

PAUL D. VAN HELDEN • MRC Unit for Molecular and Cellular Cardiology, Department of Medical Physiology and Biochemistry, University of Stellenbosch Medical School, Tygerberg, South Africa

JOHN M. WALKER • Biological Sciences, The Hatfield Polytechnic, Hatfield, Hertfordshire, UK

H. A. WHITE • Department of Biochemistry, University College London, London, UK

J. LESLEY WOODHEAD• Department of Biochemistry, Charing Cross and Westminster Medical School, London, UK

Chapter 1

Electrophoresis of RNA Denatured with Glyoxal or Formaldehyde

Christopher F. Thurston, Caroline R. Perry, and Jeffrey W. Pollard

1. Introduction

The first successful method for electrophoretic analysis of the full size range of cellular RNA molecules was described by Loening (1), and its introduction allowed for major advances, most particularly in the molecular biology of eukaryotic organisms. The method had, nevertheless, two significant disadvantages in that the gels (composed of acrylamide at very low concentrations) were mechanically fragile, and the migration of RNA molecules did not necessarily reflect their size because RNA secondary structure was not disrupted.

More modern methods have consequently sought to incorporate conditions under which RNA is fully denatured and that avoid the use of very fragile gels. Fragility of the gel was overcome by use of agarose either in combination with, or in place of, acrylamide. The problem of RNA denaturation is, however, more complex. It is necessary both for analysis of the integrity of RNA samples, because secondary structure can mask strand breaks to a significant extent, and because both intramolecular and intermolecular interactions must be avoided if the size of RNA molecules is to be deduced from their rate of migration during electrophoresis. In our view, glyoxal (2) and formaldehyde (3) are the preferable denaturing reagents. The use of formamide (4) or urea (5) involves more complicated procedures without giving more thorough denaturation. Methyl mercuric hydroxide is commonly cited as the most efficient denaturant of RNA (6,7), but it is so toxic that its use cannot be justified, except when no other method gives adequate denaturation. Indeed, we have not seen an example in the literature in which an RNA was denatured successfully by methyl mercuric hydroxide and not by glyoxal or formaldehyde.

In this chapter we describe the separation of RNA on flatbed gels using only agarose (because the elimination of acrylamide makes the procedure simpler and less hazardous and has no significant disadvantages) and either glyoxal or formaldehyde as denaturing agents. Both these methods provide RNA separation that is suitable for detection of a specific sequence in a complex mixture by hybridization after blotting onto filters (Northern blots), for which a method is described in detail in Chapter 2.

2. Materials

2.1. Glyoxal Gel Method

1. 10x Electrophoresis buffer A: 100 mM sodium phosphate, pH 6.7. The concentrated buffer should be sterilized by autoclaving and diluted for use (to 10 mM) with sterile water.
2. Dimethyl sulfoxide (DMSO).
3. Glyoxal. This can be obtained as a 40% (w/v) aqueous solution or as a solid. Before use, remove oxidation products by treatment with a mixed-bed ion-exchange resin

(Amberlite MB-3, Biorad AG-501-X8, or equivalent). One gram of resin is maintained in suspension in 10 mL of the 40% solution by slow magnetic stirring for 1 h at room temperature. The deionized glyoxal solution is recovered by filtration or by decanting after the resin beads have been allowed to settle and may be stored in small amounts in securely capped tubes at –70°C .

4. Electrophoresis grade agarose.
5. Denaturation mixture: DMSO/40% glyoxal/buffer A, 10/3/2 by vol.
6. Glycerol mix: glycerol/0.2% bromophenol blue in buffer A, 1/1 by vol (*see* Note 4 in section 4).

2.2. Formaldehyde Gel Method

1. 5x Electrophoresis buffer B: 0.2M sodium morpholino-propane-sulfonate (MOPS), 50 mM EDTA, pH 7.0. Sterilize as for step 1 in section 2.1.
2. Formaldehyde: This is usually supplied as a 37% (w/v) solution. The molecular weight of formaldehyde is 30.03, so the concentration of the solution is 12.3M. The pH of this solution should be greater than 4.0.
3. Formamide: This should be deionized as for glyoxal in step 3 in section 2.1.
4. Denaturation mix: 50% (v/v) formamide, 2.2M formaldehyde, in 1 x electrophoresis buffer B (7).
5. Agarose bead loading buffer: Prepare by melting agarose (0.2%) in 10 mM Tris-HCl, 20 mM EDTA, 10% (v/v) glycerol, 0.2% bromophenol blue, pH 7.5. When the mixture has solidified, it is forced several times through a 21-gage needle with a syringe to give a fine slurry. Store at 4°C (*see* Note 4 in section 4).

2.3. Staining of RNA after Electrophoresis

1. 1% Toluidine blue-0 in 1% acetic acid.
2. Destaining solution: 1% acetic acid.

3.   1 µg/mL Ethidium bromide in distilled water.
4.   0.1M Ammonium acetate.
5.   1 µg/mL Ethidium bromide in 0.1M ammonium acetate.

3. Method

3.1. Glyoxal Gel Electrophoresis (see Fig. 1).

1.   Cast 1.1% (w/v) agarose (*see* Note 1 in section 4) in electro-phoresis buffer A as a 3-mm deep gel in an apparatus that allows the gel to be submerged in buffer during electro-phoresis and is equipped for recirculation of buffer between the electrode compartments. Amounts for this and subsequent steps are given in Table 1 for typical gel apparatus of two different sizes.
2.   Add RNA samples (*see* Note 2 in section 4) to 3 vol of denaturation mix in microfuge tubes and incubate capped at 50°C for 1 h. Cool to room temperature in ice/water.
3.   Add 0.25 vol of glycerol mix to the denatured RNA samples. Use a positive displacement dispenser, since this is viscous.
4.   Fill the electrophoresis apparatus with buffer A so that the gel is covered by a layer 3–5 mm deep.
5.   Load the RNA samples and connect the power supply such that the sample wells are at the cathode end of the gel (*see* Notes 3 and 4 in section 4).
6.   Electrophoresis is performed with constant voltage to give up to 4 V/cm (with respect to the distance between the electrodes, not the length of the gel). Allow 10 min for migration of RNA into the agarose and start the buffer recirculation pump (*see* Table 1). The bromophenol blue marker dye migrates about 2.5 cm/h, and electrophoresis should be stopped when the dye has migrated about 80% of the distance from the sample wells to the end of the gel, since tRNA migrates ahead of the dye.

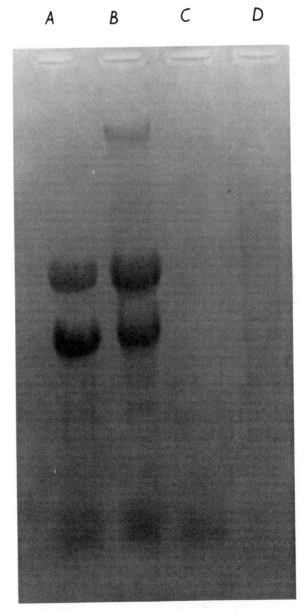

Fig. 1. Electrophoresis of glyoxal-denatured RNA, stained with toluidine blue. Tracks contain (A) 10 µg total RNA from *Agaricus bisporus*, (B) 10 µg total RNA from *Chlorella fusca* (C), 1 µg tRNA from *Escherichia coli*, (D) 3 µg l-phage DNA digested with the restriction enzyme Hind lll.

Table 1
Composition of Gel and Sample Mixtures for Glyoxal Gels
of Two Common Sizes

Dimensions of the gel (mm)	100 x 64 x 3	200 x 200 x 3
Total volume of buffer A required	300 mL	2200 mL
Agarose (in buffer A)	0.33 g in 30 mL	1.32 g in 120 mL
Denaturation mix:		
DMSO	50 µL	100 µL
40% (w/v) glyoxal	15 µL	30 µL
Buffer A	10 µL	20 µL
Glycerol mix:		
Glycerol	10 µL	20 µL
Buffer A	10 µL	20 µL
(+/- Bromophenol blue, *see* note 3 in section 4)		
RNA in sterile deionized water (*see* note 2 in section 4)	2 µL	4 µL
Denaturation mix	6 µL	12 µL
Glycerol mix	2 µL	4 µL
Volume of sample loaded	10 µL	20 µL
Voltage	65 V	120 V
Buffer recirculation	200 mL/h	500 mL/h
Duration of electrophoresis (approx.)	2 h	3 h

3.2. Formaldehyde Gel Electrophoresis

1. First melt the agarose in distilled water, and when it has cooled to 60°C, add 5x buffer B and formaldehyde. The amounts for a minigel are: 0.3 g of agarose in 18.6 mL of water, 6.0 mL of buffer (B), and 5.4 mL of formaldehyde (this gives a 1% agarose gel, total volume, 30 mL; scale up in proportion for larger gels).

2. Cast the gel (*see* Note 1 in section 4).
3. Place the gel in the electrophoresis apparatus and submerge in 1x buffer.
4. Add 4.5 µL of RNA solution (*see* Note 2 in section 4) to 2 µL of 5x buffer B, 3.5 µL formaldehyde, and 10 µL formamide. Incubate in a capped tube for 15 min at 55°C. Cool in ice/ water.
5. Add 4 µL of agarose bead loading buffer and 2 µL of a 1 mg/1 mL ethidium bromide stock (*see* Note 5 in section 4) to the denatured RNA sample and load into a sample well.
6. Electrophoresis is performed with constant current at 40V. After 15 min start the buffer recirculation. Electrophoresis is typically run overnight.

3.3. Staining RNA Bands after Electrophoresis

1. Slide the gel from the plate (on which it was cast) into a tray containing about 1 cm depth of 1% toluidine blue solution. Stain for 15–60 min at room temperature, preferably on a reciprocating shaker (10–30 strokes/min).
2. Drain off the staining solution and destain with several changes of destain solution until the background of the gel is completely clear (otherwise faintly stained bands will not show up). Store the gel in destain solution.

3.4. Fluorescent Staining

1. Glyoxal gels may be stained directly in aqueous ethidium bromide, and RNA bands may be visualized on a UV illuminator after about 30 min.
2. Formaldehyde gels must first be washed with distilled water for 2 h, using four or five changes, in order to remove the formaldehyde (*see* Notes 5 and 6 in section 4).
3. After washing, soak the gel in 0.1M ammonium acetate twice for 1 h.
4. Stain for 1 h with ethidium bromide in ammonium acetate.

5. Destain for 45 min in ammonium acetate and visualize on a UV illuminator.

4. Notes

1. When preparing the agarose gel, it is essential to completely melt the agarose, which can be done successfully in a microwave oven, in a steamer, or with careful direct heating over a Bunsen burner. When the solution has gone clear because the bulk of the agarose has dissolved, there will still be a small proportion of swollen agarose beads undissolved. These can be seen if the flask is held up to the light. Continue heating until they have dissolved, since they interfere with the RNA separation. At the same time vigorous boiling must be avoided, since it can lead to significant loss of water altering the agarose concentration. Allow the agarose to cool to about 50°C (not uncomfortably hot to touch) before pouring. If there are persistent bubbles on the surface of the agarose after pouring, they can be collapsed by briefly flaming the surface with a Bunsen burner.

   The protocols described give agarose concentrations suitable for the separation of a wide range of molecular weight species. When the marker bromophenol blue has run 80% of the length of the gel, tRNA has not run off, and 16–18S ribosomal RNAs are approximately half way. Other agarose concentrations may be more appropriate for some specialized applications

2. Both of the methods described require relatively high concentrations of RNA because the samples are diluted with denaturing reagents. Generally 5–50 µg of RNA are run on the gel. If running total RNA either to assess integrity or to transfer to a nitrocellulose filter for hybridization, 50 µg might be a suitable amount. If separating mRNA for transfer and hybridization, however, 5 µg would be sufficient and would give superior resolution.

3. The loading buffers described all contain bromophenol blue, but it is generally better to omit the tracker dye from the samples and run the outside tracks of the gel with dye but no sample. Most particularly do not include bromophenol blue in molecular weight marker or sample tracks if it is intended to visualize the RNA with ethidium bromide: the dye can mask important RNA bands.

4. The incorporation of macerated agarose in sample mixtures improves the resolution (by reducing sample tailing) in the formaldehyde system, as it does for electrophoresis of DNA, but we have not been able to show that it has any effect in the glyoxal system. In both systems great care must be taken in sample application. The volume of sample loaded must not fill the well above the surface of the gel. This is possible because surface tension draws up the agarose around the well-forming comb. If the sample occupies that part of the well that is above the main body of the gel surface, it streams at the gel–buffer interface. It is equally important that the well-forming comb is clear of the base plate so that the sample cannot leak out along the underside of the gel.

5. For many purposes, ribosomal and tRNA markers are adequate. The values for their molecular weights are shown in Table 2. For blotting experiments (Chapter 2), ribosomal markers run in flanking tracks are cut off and stained with toluidine blue. This is convenient because the intensity of staining does not diminish significantly for several weeks, enabling direct comparison of the marker bands with labeled bands visualized by autoradiography. Alternatively ethiduim bromide can be incorporated directly into the sample and the rRNA bands visualized on a UV transilluminator either on the gel or on the filter.

    Use of DNA markers: Wild type λ-phage DNA cut with Hind lll treated with glyoxal under the conditions described for RNA migrate with the same relative mobility as RNA (*8*). Under these conditions the DNA is single strand-

Table 2
Molecular Weight of RNA and DNA Markers

Ribosomal (and transfer) RNA		$10^{-6} \times M_r$	Approximate no. of nucleotides
Mammalian			
(mouse)	28S	1.70	4700
	18S	0.71	1900
Chlorella	25S	1.18	3600
	23S	1.00	3000
	18S	0.68	1900
	16S	0.57	1800
	5.8S	0.050	150
	5S	0.037	110
	tRNA	0.023	70
Escherichia coli	23S	1.07	2000
	16S	0.56	1600
	tRNA	0.025	75
λ-Phage Hind III			
restriction fragment DNA		7.13	23000
		2.91	9400
		2.05	6600
		1.36	4380
		0.71	2300
		0.62	2000
		0.174	560
		0.039	125

ed, and consequently about 10 µg are required to give clear staining. In the conditions of the formaldehyde gels, DNA migrates more slowly than RNA of the same size (9). Within the range of 0.5–7 million, RNA molecular weight can be derived by adding 0.56 million to the molecular weight of DNA, which has the same relative mobility.

6. If formaldehyde gels are processed for Northern blotting (Chapter 2, Note 1), they have been sufficiently depleted of

formaldehyde for ethidium bromide staining prior to the wash in 20x SSC.

References

1. Loening, V. E. (1967) The fractionation of high-molecular weight ribonucleic acid by polyacrylamide-gel electrophoresis. *Biochem. J.* **102**, 251–257.
2. McMaster, G. K. and Carmichael, G. G. (1977) Analysis of single and double-stranded nucleic acids on polyacrylamide gels by using glyoxal and acridine orange. *Proc. Natl. Acad. Sci. USA* **74**, 4835–4838.
3. Lehrach, H., Diamond, D., Wozney, J. M., and Boedtker, H. (1977) RNA molecular weight determinations by gel electrophoresis under denaturing conditions, a critical re-examination. *Biochemistry* **16**, 4743–4751.
4. Staynov, D. Z., Pinder, J. C., and Gratzer, W. B. (1972) Molecular weight determination of nucleic acids by gel electrophoresis in non-aqueous solution. *Nature New Biol.* **235**, 108–110.
5. Reijnders, L., Sloof, P., Sival, J., and Borst, P. (1973) Gel electrophoresis of RNA under denaturing conditions. *Biochem. Biophys. Acta* **324**, 320–333.
6. Bailey, J. M. and Davidson, N. (1976) Methylmercury as a reversible denaturing agent for agarose gel electrophoresis. *Anal. Biochem.* **70**, 75–80.
7. Maniatis, T., Fritsch, E. F., and Sambrook, J. (1982) *Molecular cloning. A Laboratory Manual* Cold Spring Harbor, New York.
8. Carmichael, G. G. and McMaster, G. K. (1980) The analysis of nucleic acids in gels using glyoxal and acridine orange. *Meth. Enzymol.* **65**, 380–391.
9. Wicks, R. J. (1986) RNA molecular weight determination by agarose gel electrophoresis using formaldehyde as denaturant: Comparison of DNA and RNA molecular weight markers. *Int. J. Biochem.* **18**, 277–278.

Chapter 2

Northern Blotting

Jeffrey W. Pollard, Caroline R. Perry, and Christopher F. Thurston

1. Introduction

Northern blotting is a way of detecting specific RNA sequences, homologous to a hybridization probe, within a complex RNA sample. In this procedure RNA is first separated by electrophoresis under denaturing conditions (*see* Chapter 1), followed by blotting onto a suitable filter. Specific sequences are then detected by hybridization. The RNA species are both immobilized on the filter and denatured so that when the filter is immersed in a solution containing a labeled nucleic acid probe, the probe binds to RNA of complementary base sequence (hybridizes), specifically labeling the position on the filter that this RNA sequence has been blotted to from the gel. This defines the size of the RNA molecule complementary to the probe and can be used to compare relative amounts of the RNA species in different samples.

In earlier experiments of this type, because it was thought that RNA did not bind tightly to nitrocellulose filters, diazobenzyloxymethyl (DBM) paper was used, which binds RNA covalently (1). In the presence of high salt concentration, however, RNA binds perfectly well to both nitrocellulose and nylon filter membranes (2). We describe here the technique developed by Thomas (2) for transfer of RNA to either nitrocellulose or nylon filters and its subsequent hybridization. A typical result is shown in Fig. 1.

In this and many other procedures it is necessary to label the nucleic acid being used as a hybridization probe, and we therefore include a protocol for nick translation of DNA (3). It should be noted that there are now a number of alternatives to this procedure (*see* Note 6 in section 4). The probe must of necessity be highly labeled, and consequently the elimination of nonspecific binding of probe to filter is essential. This is achieved by the prehybridization step. When the probe has been allowed to hybridize, a rigorous washing procedure is employed before autoradiographic analysis, since most probes will bind to a range of RNA species from totally complementary sequences to those that show only very limited homology. The washing denatures the relatively weak binding of species with little homology—the greater the homology of hybrid-forming species, the more stringent the washing conditions they will withstand. When a probe is not completely complementary (homologous) to the RNA species sought, the stringency of washing may have to be reduced.

2. Materials

2.1. Transfer of RNA from Gel to Filter and Hybridization

1. Nitrocellulose sheets (Schleicher and Schuell BA 83, 0.2 μm) or nylon membrane sheets (Gene-screen from New England Nuclear, Hybond from Amersham, or Biodyne from PALL Ultrafine Filtration Corp.).
2. Sterile distilled water.
3. 20x SSC: 3*M* NaCl, 0.3*M* sodium citrate, pH 7.6. Autoclave and store at room temperature.
4. 3x SSC: dilute 20x SSC with sterile distilled water 1 + 5.67 by vol.
5. 100x Denhardt's solution: 2% (w/v) bovine serum album-

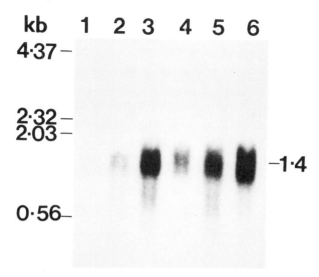

Fig. 1. Analysis of the expression of c-*ras*[H] in relationship to cellular proliferation in the mouse uterine luminal epithelium. Autoradiogram of a Northern blot of total uterine luminal epithelial RNA separated on a formaldehyde gel and probed with a v-*ras*[H] genomic clone, showing a single hybridizing mRNA of 1.4 kb. Each lane had 50 μg of total RNA from mice treated to a variety of combinations of estradiol and progesterone, as detailed in ref. 7 (figure reproduced with permission.)

    in, 2% (w/v) polyvinylpyrrolidone, 2% (w/v) Ficoll (Pharmacia) 0.2% sodium azide. Store at –20°C.

6. Formamide. Deionize before use, as described in Chapter 1.

7. Denatured carrier DNA: Salmon sperm (or other nonhomologous DNA) at 10 mg/mL in sterile distilled water, heated for 10 min at 95°C, and cooled by immersion in ice/salt or ice/ethanol. After cooling, the solution is expelled five times through an 18-gage syringe needle to shear the DNA.

8. 10% (w/v) sodium dodecyl sulfate (SDS).

9. Prehybridization mix: 30% (v/v) formamide; 5x SSC; 2x Denhardt's solution; 0.1% SDS; 100 μg/mL denatured carrier DNA; 25 μg/mL poly A; 0.05 *M* sodium phosphate, pH 7.0.

10. Hybridization mix: all the components of the prehybrid-
    ization mix, except the poly A, but in addition, labeled
    probe nucleic acid (typically 200 ng in 600 µL).
11. Wash buffers: (A) 2x SSC, 0.1% SDS, 0.05% sodium pyro-
    phosphate. (B) 1x SSC, 0.1% SDS, 0.05% sodium pyro-
    phosphate. (C) 0.1x SSC, 0.1% SDS, 0.05% sodium pyro-
    phosphate.
12. Containers for hybridization: either plastic bags (auto-
    clavable disposal bags are suitable) and a bag sealer, or
    rectangular plastic containers that are shallow (preferably
    less than 1 cm deep) and as similar as possible to the
    dimensions of the filter to be treated. Bag sealers supplied
    for use in preparation of food for storage in domestic deep
    freezers are suitable and cheap.

2.2. Materials for Nick Translation of DNA and Autoradiography

1. 10x nick-translation buffer: 0.5M Tris-HCl, pH 7.6, 0.1M
    $MgSO_4$, 10 mM dithiothreitol, 500 µg/mL bovine plasma
    albumin. Filter sterilize and store at –20°C.
2. 10x dXTPs: make separate solutions in sterile distilled
    water of dATP, dCTP, dGTP, and dTTP each at 0.2 mM.
    Adjust the pH to between 7.0 and 7.5 with sodium hydrox-
    ide immediately after dissolving the nucleotide, using
    narrow-range pH papers if the volume of solution is
    insufficient to cover a pH electrode. Keep these solutions
    on ice while in use and store at –20°C.
3. Labeled nucleotide: 367 MBq/mL α^{32}P-dATP or dCTP,
    specific activity 10^{13}–10^{14} Bq/mmol (10 mCi/mL at a spe-
    cific activity of 400–3000 Ci/mmol). Note that the ^{35}S-sub-
    stituted analog of dATP may be used as an alternative.
4. Probe DNA: The DNA to be radiolabeled should be
    0.1–0.5 mg/mL.
5. DNA polymerase 1 from *Escherichia coli*.
6. DNAse 1: dilute to 50 pg/µL in 50 mM Tris-HCl, pH 7.5,

10 mM $MgSO_4$, 1 mM dithiothreitol, and 50 μg/mL bovine serum albumin immediately before use.

7. Stop buffer: 10 mM sodium EDTA, pH 7.5, 0.1% SDS.
8. TE buffer: 10 mM Tris-HCl, pH 7.5, 1 mM EDTA.
9. Sephadex G-50: Equilibrate G-50 medium in TE buffer and pour a column in a Pasteur pipet or the barrel of a 2-mL disposable syringe on the day of use.
10. Photographic film for autoradiography: film designed for use in X-ray cameras such as Kodak XOmat AR or Fuji RX are most suitable and are most easily handled in cassettes specifically designed for autoradiography, particularly if calcium tungstate intensifying screens are used (*see* Note 9 in section 4).

3. Method

3.1. Transfer of RNA

Glyoxylated RNA may be transferred immediately after electrophoresis, but formaldehyde gels require pretreatment, as described in Note 1 (in section 4).

1. A glass plate must be supported in a glass or plastic tray such that the surface of the plate is within a few mm of the top of the tray, as shown in Fig. 2. A 20 x 20-cm plate supported on Petri dishes in a glass penicillin assay plate works well.
2. Pour 20x SSC into the tray to give at least 1 cm depth, but not enough so that the glass plate is submerged (500 mL or more).
3. Cut a sheet of Whatman 3 mm (or two layers of Whatman No. 1) filter paper to give about 2 cm overlap of the glass plate on all sides. Cut out the corners of the overlapping paper so that the excess can be folded under to reach the bottom of the tray when the plate is placed in it. Expel air bubbles trapped between the paper and the glass plate

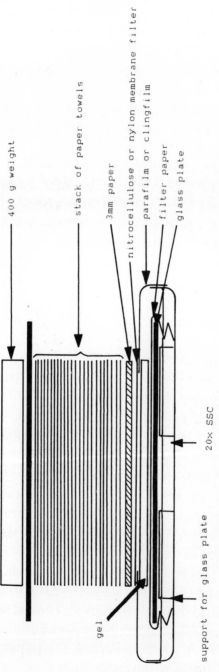

Fig. 2. Diagram of the arrangement of gel and membrane filter during blotting (not drawn to scale).

when the filter paper has become thoroughly wetted with 20x SSC by rolling toward the corners with a sterile 10-mL pipet.

4. Cut a piece of cellulose nitrate or nylon membrane filter to cover the area of gel from which RNA tracks are to be transferred and wet by immersion in sterile distilled water (*see* Note 2 in section 4). Soak the wetted membrane filter in 20x SSC for 5 min.

5. When electrophoresis is finished, remove the gel and briefly drain off electrophoresis buffer. Invert the gel and carefully lower it onto the paper-covered glass plate, removing any air bubbles by rolling as described in step 3 above (*see* Note 1 in section 4).

6. Lay the cellulose nitrate or nylon membrane filter on top of the gel and again remove any air bubbles by rolling with a pipet.

7. Surround the membrane filter, overlapping its edges by 2–3 mm with strips of parafilm or sandwich wrap, which should extend out over the sides of the tray. This serves to minimize the extent to which fluid can be drawn round the membrane filter rather than through it.

8. Cover the membrane filter with Whatman 3 mm paper or two or three layers of Whatman No. 1 paper cut to the size of the gel and well soaked in 20x SSC.

9. Place on top at least 10-cm thickness of paper towels or other absorbent paper, covering the area of the gel, and press down by means of a glass plate and about 500 g weight (such as 400 mL water in a 1-L glass beaker or a second penicillin assay plate).

10. As fluid is absorbed by the paper towels, the 20x SSC drawn through the gel transfers the RNA to the membrane filter. This process is allowed to proceed at room temperature overnight (*see* Note 3 in section 4).

11. When transfer has been effected, the paper on top of the membrane filter is removed.

12. Before the membrane filter is separated from the gel, the

position of the sample wells is marked on the membrane filter with a soft pencil or a ballpoint pen.

13. The membrane filter is carefully peeled off the gel, washed in 2x SSC and allowed to dry on clean Whatman No. 1 paper for 30–60 min.

14. Bake the dry membrane filter at 80°C for 2 h. After baking, the membrane filter may be stored desiccated for many months prior to hybridization. Baking fixes the adsorbed nucleic acid to the membrane filter.

3.2. Labeling of Probe DNA by Nick Translation

1. Make up the following reaction mixture in a sterile microfuge tube: 2.5 µL of 10x nick-translation buffer, 2.5 µL of each 0.2 mM dATP, dCTP, dGTP, and dTTP, 5 µL of α^{32}P-dCTP (1.83 MBq, 50 µCi; *see* Note 4 in section 4), 5 U of DNA polymerase 1, 0.5 µL of DNAse 1, 0.1–0.5 µg of substrate DNA, and sterile distilled water to a total volume of 25 µL.

2. Incubate at 15°C for 2 h.

3. Add 25 µL of stop buffer.

4. Separate the labeled DNA from unincorporated nucleotides by passage through the G-50 column. If the column is allowed to drain before the reaction mixture is applied, addition of 150-µL amounts of TE buffer result in consistent delivery of three-drop fractions that may be collected in microfuge tubes. Collect 15 such fractions (*see* Note 5 in section 4).

5. Count radioactivity in the fractions as Cerenkov radiation from the closed microfuge tubes in a scintillation counter (using the energy window set for counting tritium if optimum settings for Cerenkov radiation from ^{32}P are unknown).

6. Pool the fractions containing the excluded peak and keep on ice until added to the hybridization mix.

7. Calculate the specific activity of the labeled DNA as total count in the excluded peak, corrected for the efficiency of counting, divided by the weight of DNA added to the reac-

tion (hence the units are dpm/µg DNA; *see* Note 4 in section 4).

3.3. Hybridization

1. Wet the membrane filter by flotation on 3x SSC.
2. Place the wet filter in a plastic bag and seal around three sides of the filter as close as possible without trapping the edge of the filter.
3. Pipet prehybridization mix into the bag, using 0.2 mL/cm² of filter, and work all air bubbles up to the open end of the bag.
4. Seal the open end of the bag about 5 mm from the end of the filter (this end has to be opened and resealed), excluding air bubbles.
5. Incubate at 42°C for 2–4 h (*see* Note 7 in section 4).
6. After prehybridization, cut the corner off the bag and squeeze out the prehybridization mix.
7. Pipet hybridization mix into the bag (about 0.1mL/cm² filter) and reseal after removing air bubbles. Nick translation can be conveniently carried out while the filter is incubating in prehybridization mix, but regardless of whether the labeled probe is freshly prepared or is being reused, it should be denatured before addition to the filter. Denaturation is achieved by heating to 95°C for 5 min followed by rapid cooling in ice/water (*see* Note 8 in section 4).
8. Incubate at 42°C for 16–20 h to allow for hybridization (*see* Note 8 in section 4).
9. After hybridization, cut a corner off the bag and carefully express the hybridization mix into a sealable container. The mix can be reused, but should be carefully collected anyway to avoid dissemination of radioactivity.
10. Wash the filter in a 250 mL vol of wash buffers as follows: buffer A, 2x 20 min at room temperature; buffer A, 30 min at 55°C (prewarm wash buffers to temperature of use); buffer B, 30 min at 55°C. For maximum stringency when a homologous probe is being used, a further wash in buffer

C, 30 min at 55°C, may be included.
11. Allow the filter to dry on a sheet of Whatman filter paper, at room temperature.

3.4. Autoradiography

1. Immobilize the filter (by tape on the extreme edge) on a sheet of Whatman filter paper cut to fit an autoradiography cassette and cover the filter with sandwich wrap, folding it around the supporting filter paper sheet in order to remove all creases from the area over the membrane filter.
2. Place in a cassette and load a sheet of X-ray film and an intensifying screen on top. Expose at –70°C for 1–4 d (*see* Note 9 in section 4).
3. Remove the X-ray film after the cassette has warmed to room temperature and process as recommended by the manufacturer. Considerable care is required when separating the X-ray film from the sandwich wrap covering the membrane filter, since if this is done too quickly it is common to get a discharge of static electricity that fogs the photographic emulsion.
4. For reuse of filters, *see* Note 10 in section 4.

4. Notes

1. Formaldehyde gels may require pretreatment before blottting as follows: Soak the gel in distilled water for 2 h with at least two changes to remove formaldehyde. Transfer the gel to 50 mM NaOH, 10 mM NaCl for 30 min, to 0.1M Tris-HCl, pH 7.5, for 30 min and, finally, to 20x SSC for 45 min.
2. If nitrocellulose membrane filters do not wet within 2 or 3 min, they should not be used, since transfer to unevenly wetted nitrocellulose is unreliable. Nitrocellulose that has been handled with greasy fingers will never wet!
3. The rate of transfer of RNA from gel to membrane filter is

a function of its size and the porosity of the gel. Small RNA molecules (<1 kb) transfer from 1.1% agarose in 2–3 h, but high moleular weight RNA may take 10–15 h.

4. Addition of 1.83 MBq (50 μCi) α^{32}P-dCTP at a specific activity of 14.7 TBq/mmol (400 Ci/mmol) should result in the probe being labeled to a specific activity of 2×10^8 dpm/μg. For further information, *see* ref. 4 and Vol. 2 of this series.

5. The labeled DNA elutes in the void volume. If blue dextran is added to the reaction mix [10 μL of a 1% solution in TE buffer (20)], the blue eluate collected as a single fraction will contain the labeled DNA.

6. Alternative labeled probes may be produced in several different ways. Random oligonucleotide-primed synthesis of DNA is now often used as an alternative to nick translation. Kits from commercial sources are becoming available for derivatizing probe DNA so that it can be detected on a filter with an enzyme-tagged monoclonal antibody. Similarly, biotinylated DNA detected by avidin–enzyme conjugate binding is another way of monitoring hybridization while avoiding the use of radioactive labeling (*see* Chapters 20 and 33). None of these methods have as yet the reliability and sensitivity of radiolabeling. RNA probes are increasingly being used and are essential for some procedures. RNA transcripts can provide very highly labeled probes (5) and Bogorad's procedure (6) for end-labeling RNA with T4 polynucleotide kinase is also very useful.

7. The purpose of prehybridization is to minimize nonspecific binding of probe nucleic acid to the membrane filter. This step can be run overnight for convenience.

8. The inclusion of 30% formamide allows the hybridization to be carried out at 37–42°C. In its absence the temperature would have to be at least 65°C, which is undesirable both because it is less easy to set up and because degradation of RNA would be significantly increased. Inclusion of 10% (w/v) dextran sulfate in the hybridization mix increases

the rate of hybridization, possibly as much as 10-fold. Because rate of hybridization is not commonly the limiting factor in Northern blots, inclusion of dextran sulfate rarely gives improved results, since it has the disadvantages of being difficult to handle (the necessarily concentrated solutions are very viscous) and possibly leads to increased nonspecific binding of probe to the membrane filter.

9.   If the probe-specific activity given in Note 4 (this section) is achieved, and the RNA species detected is of average abundance, with the loadings for electrophoresis described in Chapter 1 the bound radioactivity will give a visible band after 2 days of autoradiography at –70°C with two intensifying screens. Autoradiography at room temperature without intensifying screens is about 10-fold less sensitive.

10.   Filters washed as described below may be reused, e.g., with a different probe. First, wash for 1 h at 65°C in 5 mM Tris-HCl, pH 8, 0.2 mM EDTA, 0.1x Denhardt's solution, and 0.05% sodium pyrophosphate. Second, wash in the above mixture diluted with an equal vol of sterile distilled water, for 1 h at 65°C. Third, wash in sterile distilled water for 1 h at 65°C. Bake the filter as described in step 14 in section 3.1.

References

1.   Alwine, J. C., Kemp, D. J., and Stark, G. R. (1977) Method for detection of specific RNAs in agarose gels by transfer to diazobenzyloxymethyl-paper and hybridization with DNA probes. *Proc. Natl. Acad. Sci. USA* **74,** 5350–5354.

2.   Thomas, P.S. (1980) Hybridization of denatured RNA and small DNA fragments transferred to nitrocellulose. *Proc. Natl. Acad. Sci. USA* **77,** 5201–5205.

3.   Rigby, P. W. J., Dieckmann, M., Rhodes, C., and Berg, P. (1977) Labelling deoxyribonucleic acid to high specific activity *in vitro* by nick translation with DNA polymerase 1. *J. Mol. Biol.* **113,** 237–251.

4.   Technical Bulletin 80/3 (1980) Labelling of DNA with 32p by nick translation. Amersham International, Amersham.

5. Melton, D. A., Krieg, P. A., Rebagliati, M. R., Maniatis, T., Zinn, K., and Green, M. R. (1984) Efficient *in vitro* synthesis of biologically active RNA and RNA hybridization probes from plasmids containing a bacteriophage SP6 promoter. *Nucleic Acid. Res.* **12**, 7035–7056.

6. Bogorad, L., Gubbins, E. J., Krebbers, E., Larrinva, I. M., Mulligan, B., Muskavitch, K. M. T., Orr, E. A., Rodermal, S. R., Schantz, R., Steinmetz, A. A., De Vos, G., and Ye, Y. K. (1983) Cloning and physical mapping of maize plastid genes. *Meth. Enzymol.* **97**, 524–554.

7. Cheng, S. V. Y. and Pollard, J. W. (1986) c-*ras*[H] and ornithine decarboxylase are induced by oestradiol-17β in the mouse uterine luminal epithelium independently of the proliferative status of the cell. *FEBS Lett.* **196**, 309–314.

Chapter 3

Hybrid-Release Translation

Adrian Slater

1. Introduction

Many situations arise in recombinant DNA research in which it is necessary to identify the mRNA encoded by a particular cloned DNA sequence. Cloned DNA fragments can be characterized by hybridization to mRNA, the complementary mRNA then being identified by translation in vitro. There are two different approaches to this procedure; hybrid-arrest translation (1) and hybrid-release translation (2). In the former, hybridization of cloned DNA to an mRNA population in solution can be used to identify the complementary mRNA, since the mRNA–DNA hybrid will not be translated in vitro (*see* Chapter 4). In the latter, described here, cloned DNA bound to a solid support is used to isolate the complementary mRNA, which can then be eluted and translated in vitro.

Hybrid-arrest translation has a number of limitations. It is necessary for the mRNA to be sufficiently abundant for the translation product to be identifiable by gel electrophoresis. The technique also requires complete hybridization of the mRNA to obtain a convincing inhibition of translation. In contrast, hybrid-release translation can be used to identify cloned DNA complementary to rare mRNAs (3) and does not require complete hybridization to obtain an unambiguous result. Furthermore, because the technique involves isolation of one specific mRNA, it can be extended to further characterize the translation product by peptide mapping (4) or immunoprecipitation (5). It can also be used to study posttranslational processing in vitro (6).

There are a number of variations in the method for hybrid-release translation that depend on the choice of solid support for the DNA. For example, diazobenzyloxymethyl (DBM) cellulose (7), DBM paper (2), nitrocellulose (1,3), and nylon filters can be used for this purpose. The method described in this chapter uses DBM paper. Although the initial preparation of the DBM filters is time-consuming (unlike nitrocellulose filters), the DBM papers can be reused several times because the DNA is irreversibly bound to the DBM paper.

The method has been divided into four sections. The first section describes a procedure for synthesis of *m*-aminobenzyloxymethyl (ABM) paper, a relatively stable precursor of DBM paper. The next two stages cover the preparation of denatured DNA, the activation of ABM paper to DBM paper, and the covalent bonding of the single-stranded DNA to the DBM paper. The final section describes conditions for hybridization of RNA to the DNA paper, elution of the hybridized RNA, and subsequent washing prior to translation in vitro.

2. Materials

2.1. Chemicals and Equipment

1. *N*-(3-Nitrobenzyloxymethyl) pyridinium chloride (NBPC).
2. *N,N*-dimethylformamide (DMF).
3. Sodium dithionite.
4. Sodium nitrite.
5. Whatman 540 paper, 9 cm diameter.
6. Amberlite MB-3. Monobed mixed-ion exchange resin.
7. Piperazine-*N,N'bis* (2-ethanesulfonic acid) (PIPES).

8. Calf liver tRNA.
9. Pancreatic RNAse.

2.2. Solutions

1. 10% (w/v) Sodium dithionite in 50 mM sodium hydroxide. Prepare fresh before use.
2. 5% (w/v) sodium nitrite. Prepare fresh, and cool on ice before use.
3. 1M HCl.
4. 1M Sodium acetate, pH 4.0.
5. Deionized formamide. Deionize AR-grade formamide with 1 g/10 mL Amberlite MB-3 mixed-bed resin by stirring for 1 h at room temperature. Filter through Whatman No. 1 filter paper and store at –20°C.
6. 1M NaOH.
7. 5M Acetic acid.
8. 100 mM Sodium acetate, 20 mM EDTA, pH 6.0.
9. 400 mM PIPES, 20 mM EDTA, pH 6.4.
10. 4M NaCl.
11. 10% Sodium dodecyl sulfate (SDS).
12. Hybridization buffer: This is prepared from solutions in steps 5, 9, 10, and 11 to give final concentration of 20 mM PIPES, 1 mM EDTA, 0.6M NaCl, 0.2% (w/v) SDS, and 50% (v/v) formamide.
13. TE buffer: 10 mM Tris-HCl, 1 mM EDTA, pH 7.4 and 8.0.
14. 3M Sodium acetate, pH 6.0.
15. 0.5M Potassium acetate. Autoclave this stock solution, then mix 1 vol with 4 vol of ethanol.
16. 10 μg/μL calf liver tRNA in sterile distilled water.
17. 1 mg/mL pancreatic RNAse in 30 mM EDTA.

Since the DBM paper will eventually be used for hybridization with RNA, precautions against ribonuclease contamination should be applied throughout. Thus, all solutions should be autoclaved when possible or prepared in sterile distilled water. Particular attention should be paid to wearing gloves whenever the filters are handled.

3. Method

3.1. Preparation of ABM Paper

This method is based substantially on an improved method for synthesis of DBM paper described by Christophe et al. (*8*). All the volumes given are to prepare one 9-cm diameter ABM filter and should be adjusted according to the size and number of filters prepared.

1. Dissolve 280 mg of NBPC in 1.2 mL of DMF.
2. Lay a 9-cm diameter Whatman 540 filter in the bottom of a large beaker, and soak it with the NBPC solution.
3. Place the beaker in a 130°C oven for 30 min to form nitrobenzyloxymethyl (NBM) paper. Beware of pyridine fumes.
4. Allow to cool to room temperature, then wash twice with 50 mL acetone, swirling rapidly for 2 min with each wash.
5. Wash twice with 100 mL of distilled water in the same way.
6. Add 100 mL of freshly prepared 10% (w/v) sodium dithionite in 50 mM NaOH and agitate for 2 h at room temperature in a fume hood to eliminate SO_2. This step reduces the NBM paper, to give ABM paper.
7. Wash four times with 100 mL of water for 2 min each wash, with rapid swirling.
8. Wash twice with 50 mL of acetone.
9. Dry for 10 min at 37°C.
10. Seal in a polythene bag and store at –20°C (*see* Note 1 in section 4).

3.2. Preparation of DNA

1. Prepare a solution of purified DNA in TE buffer, pH 8.0, at a concentration of 0.2 mg/mL (*see* Note 2 in section 4).

2. Add 25 µL of 1M HCl to 10 µL of DNA solution in a 1.5-mL microfuge tube and incubate for 5 min at room temperature (*see* Note 2 in section 4).
3. Add 175 µL of 1M NaOH and leave at room temperature for 1 h.
4. Neutralize by adding 33 µL of 5M acetic acid, then 667 µL ethanol. Cool to –20°C, and collect the DNA by centrifugation in a microfuge for 15 min.
5. Dry the pellet in a vacuum desiccator and resuspend in 20 µL of 1M sodium acetate, pH 4.0, to give a final concentration of 1 mg/mL.

3.3. Binding of DNA to DBM Paper

1. Freshly prepare 5% (w/v) sodium nitrite solution and cool on ice. Also cool 1M HCl and 1M sodium acetate, pH 4.0, solutions.
2. Cut the required number of 1-cm squares from a circle of ABM paper. All the volumes given below are for a single 1-cm square of paper.
3. Soak each square in 1 mL of 1M HCl containing 22 µL of 5% (w/v) sodium nitrite on ice for 30 min. Nitrous acid converts the ABM paper to DBM paper.
4. Wash the paper three times with 1 mL of cold 1M sodium acetate, pH 4.0, for 2 min each wash.
5. Remove the paper squares and blot dry between two clean filter papers (e.g., Whatman No. 1).
6. Place each square in a separate, sterile, siliconized bijou (5 mL screwtop) bottle, and immediately apply 20 µg of denatured DNA in 20 µL of 1M sodium acetate, pH 4.0. Stopper tightly and incubate overnight at 4°C to allow reaction between the single-stranded DNA and the diazonium groups of the DBM paper to occur.
7. Wash each square three times with 1 mL of distilled water at room temperature to remove unreacted DNA.
8. Wash twice with 0.5 mL of deionized formamide at room temperature and once at 68°C for 2 min to remove any

DNA that has renatured during the reaction. Successful diazotization is indicated by development of a deep orange color during these washes.

9. Wash three times with 1 mL of distilled water at room temperature.
10. Store in 1 mL of 100 mM sodium acetate, pH 6.0, 20 mM EDTA, at 4°C (*see* Note 3 in section 4).

3.4. RNA Hybridization to DNA Paper

1. Prepare fresh hybridization buffer and warm to 42°C. Wash each 1-cm square DNA paper four times with 0.5 mL of hybridization buffer at 42°C.
2. Blot dry on a clean filter paper, and place the DNA papers in a sterile, siliconized bijou bottle.
3. Prepare a solution of RNA in hybridization buffer. Use a 20 mg/mL stock solution of total RNA or 0.2 mg/mL poly (A)$^+$ RNA in sterile distilled water (*see* Note 4 in section 4). To 125 µL of RNA solution, add, in order, 15 µL of H$_2$O, 25 µL of 400 mM PIPES, pH 6.4, 20 mM EDTA, 10 µL of 10% (w/v) SDS, 250 µL of deionized formamide, and 75 µL of 4M NaCl, to give a final volume of 500 µL (*see* Note 5 in section 4).
4. Apply the RNA solution to the DNA papers, seal the bottle tightly, and incubate at 42°C for 16 h (*see* Note 6 in section 4).
5. Decant off the RNA solution, and wash the filters twice with 2.5 mL of hybridization solution at 45°C. Pool the RNA solution and washes (*see* Note 7 in section 4).
6. Separate different DNA papers into individual bijou bottles, and wash each paper eight times with 2.5 mL of hybridization solution at 45°C.
7. Preheat TE buffer, pH 7.4, to 70°C. Elute the hybridized mRNA from the DNA paper by washing three times with 250 µL of TE buffer at 70°C, for 2.5 min each wash. Pool the three washes into a 1.5-mL microcentrifuge tube on ice.

8. To the 750 μL of eluted RNA, add 50 μL 3M sodium acetate, pH 6.0, and 1 μL of 10 μg/μL calf liver tRNA. Mix, and transfer 400 μL to a separate tube. Add 1.0 mL of ethanol to each tube and precipitate the RNA at −20°C. One tube can then be stored at −20°C, whereas the other is processed for translation in vitro. Centrifuge the precipitated RNA in a microcentrifuge for 30 min at 4°C (*see* Note 8 in section 4), and remove the ethanol with a Pasteur pipet.

9. Wash the pellet twice with 500 μL of 0.1M potassium acetate, pH 7.0, in 80% ethanol. Disperse the pellet twice with ethanol in the same way. The RNA can be stored in the last ethanol wash at −20°C (*see* Note 9 in section 4).

10. Dry the pellet under vacuum in a desiccator. The dried pellet can be dissolved in a small volume of distilled water or directly resuspended into the cell-free translation system of choice, and the selected translation product identified by SDS polyacrylamide gel electrophoresis followed by fluorography (Fig. 1) (*see* Notes 10 and 11 in section 4).

4. Notes

1. Each filter can be sealed into a separate compartment of a polythene bag with a heat sealer and stored for many months at −20°C.

2. For example, tomato fruit cDNA cloned into pAT153 has been characterized by this method (9). The plasmid DNA was isolated by the cleared lysate method and purified on ethidium bromide/CsCl gradients (10). It is possible to use the entire recombinant plasmid for hybrid selection, using the vector DNA bound to DBM paper as a control. The circular plasmid is denatured by this method because the HCl treatment partially depurinates the DNA, and the subsequent alkali treatment cleaves the DNA at the depurination sites and denatures the DNA. The DNA is estimated to become fragmented to an average size of 1.5 kb under these conditions.

3.   The filters can be stored in this solution at 4°C for many months and reused several times. To regenerate the paper after each use, wash with 0.1M NaOH for 20 min at room temperature and then six times with distilled water before equilibrating with hybridization buffer.

4.   It is possible to use total (i.e., cellular or cytoplasmic) RNA, and this has a number of advantages compared to using poly(A)[+] RNA. The large proportion of ribosomal RNA in the sample acts as a carrier, minimizing nonspecific binding of mRNA to the filter and reducing the effect of ribonuclease contamination. Furthermore, the time required to isolate sufficient RNA for this procedure may itself be an important consideration. The order of addition of hybridization buffer constituents to total RNA is critical in that a precipitate may form if the 4M NaCl is added before the other components.

5.   It is posssible to hybridize a number of different DNA papers in the same RNA solution, provided they contain unrelated sequences. Up to five 1-cm square papers can be immersed in 0.5 mL RNA solution in the bottom of a bijou bottle. The filters should be identified by a pattern of nicked edges and corners, rather than using pencil, which may fade.

6.   These hybridization conditions have been used successfully to select some, relatively abundant, tomato mRNAs (*9,11*; Fig. 1). Other tomato fruit cDNA clones have not consistently hybridized sufficient RNA under these conditions to be detectable by translation in vitro. Therefore, it may be necessary to optimize the hybridization conditions; e.g., adjusting the formamide concentration or temperature will alter the stringency, whereas altering the incubation time will determine the extent of hybridization.

7.   The pooled nonbound RNA and washes can be precipitated with 2.5 vol of ethanol and could conceivably be reused for hybridization to other, unrelated DNA papers.

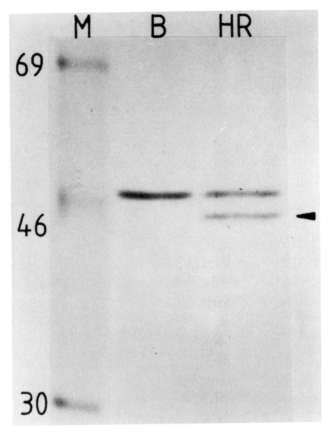

Fig. 1. Hybrid-release translation of tomato fruit mRNA complementary to a ripening-specific cDNA clone. Ripe tomato fruit total RNA (14) was hybridized to ripening-specific cDNA clone pTOM 5 bound to DBM paper (9). The hybridized RNA was eluted and translated in a rabbit reticulocyte lysate containing [^{35}S]-methionine (5). The translation products were separated by SDS-polyacrylamide gel electrophoresis and visualized by fluorography. (B) Translation blank with no added mRNA, showing the major rabbit reticulocyte endogenous band (15). (HR) pTOM 5 hybrid-released mRNA, showing one specific translation product (arrow). (M) ^{14}C-Labeled molecular weight markers ($M_r \times 10^{-3}$).

Alternatively, the RNA can be washed with $0.1M$ potassium acetate in 80% (v/v) ethanol prior to translation in vitro. This is a good internal check for ribonuclease degradation of the RNA during hybridization.

8.   Centrifugation time, rather than the time and temperature
      of precipitation, appears to be the critical factor in recov-
      ering small amounts of nucleic acid by ethanol precipita-
      tion (12).

9.   The potassium acetate wash replaces sodium with potas-
      sium ions prior to translation in vitro. The extensive wash-
      ing of the RNA with this solution and then ethanol also en-
      sures complete removal of formamide. The pellet formed
      with the tRNA carrier should be just visible. The pellet
      tends to be rather loose after the ethanol washes, so care
      must be taken when removing the supernatant. Leave a
      few drops of ethanol in the tube, if necessary. To avoid los-
      ing the pellet when drying under vacuum, prick a few
      holes in the lid of the microfuge tube, or use a pierced lid
      cut from a separate tube.

10.   The tRNA carrier increases the formation of labeled pep-
      tidyl-tRNAs, which may appear on SDS polyacrylamide
      gels as a number of bands with an apparent molecular
      weight of about 30,000 (Fig. 2). These can be removed by
      ribonuclease treatment (13). After translation, add 0.2 vol
      of (1 mg/mL) pancreatic RNAse in 30 mM EDTA and in-
      cubate at 37°C for 30 min before addition of SDS sample
      buffer.

11.   It is often not possible to detect any stimulation of protein
      synthesis by the hybrid-released mRNA, as determined
      by incorporation of radioactivity into acid-precipitable
      material, even when a strong band can be seen after gel
      electrophoresis and fluorography. (Note: Since the tRNA
      carrier stimulates incorporation of radioactivity into acid-
      precipitable material, tRNA should also be added to the
      translation blank.) It is therefore useful to include a known
      RNA sample to check the activity of the translation system
      and to analyze all hybrid-released translation products by
      gel electrophoresis and fluorography regardless of their
      apparent acid precipitable radioactivity.

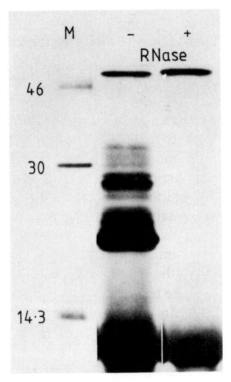

Fig. 2. Removal of peptidyl-tRNAs by ribonuclease treatment. Rabbit reticulocyte lysate containing [^{35}S]methionine was incubated with 1 µg/µL of calf liver tRNA for 1 h at 30°C. The [^{35}S] methionine-labeled products were separated by SDS-polyacrylamide gel electrophoresis and visualized by fluorography. One sample was treated with pancreatic ribonuclease (+RNAse) as described (note 10) and compared with an untreated sample (–RNAse). Ribonuclease treatment removes the prominent peptidyl-tRNA bands between 20,000 and 30,000 molecular weight, but not the endogenous band of about 48,000 molecular weight.

References

1.   Paterson, B. M., Roberts, B. E., and Kuff, E. L. (1977) Structural gene identification and mapping by DNA-mRNA hybrid-arrested cell-free translation. *Proc. Natl. Acad. Sci. USA* **74**, 4370–4374.

2.   Goldberg, M. L., Lefton, R. P., Stark, G. R., and Williams, J. C. (1979) Isolation of specific RNAs using DNA covalently linked to diazobenzyloxymethyl-cellulose on paper. *Meth. Enzymol.* **68**, 206–220.

3.   Parnes, J. R., Velan, B., Felsenfeld, A., Ramanathan, L., Ferrini, V., Appella, E., and Seidman, J. G. (1981) Mouse β_2 microglobulin cDNA clones: A screening procedure for cDNA clones corresponding to rare mRNAs. *Proc. Natl. Acad. Sci. USA* **78**, 2253–2257.

4.   Fischer, S. G. (1983) Peptide mapping in gels. *Meth. Enzymol.* **100**, 424–430.

5.   Grierson, D., Slater, A., Speirs, J., and Tucker, G. A. (1985) The appearance of polygalacturonase mRNA in tomatoes: One of a series of changes in gene expression during development and ripening. *Planta* **163**, 263–271.

6.   Jackson, R. C. and Blobel, G. (1977) Post-translational cleavage of presecretory proteins with an extract of rough microsomes from dog pancreas containing signal peptidase activity. *Proc. Natl. Acad. Sci. USA* **74**, 5598–5602.

7.   Noyes, B. E. and Stark, G. R. (1975) Nucleic acid hybridisation using DNA covalently coupled to cellulose. *Cell* **5**, 301–310.

8.   Christophe, D., Brocas, H., and Vassart, G. (1982) Improved synthesis of DBM paper. *Anal. Biochem.* **120**, 259–261.

9.   Slater, A., Maunders, M. J., Edwards, K., Schuch, W., and Grierson, D. (1985) Isolation and characterisation of cDNA clones for tomato polygalacturonase and other ripening-related proteins. *Plant Mol. Biol.* **5**, 137–147.

10. Clewell, D. B. and Helinski, D. R. (1970). Properties of a supercoiled deoxyribonucleic acid–protein complex and strain specificity of the relaxation event. *Biochemistry* **9**, 4428–4440.

11. Smith, C. J .S., Slater, A., and Grierson, D. (1986) Rapid appearance of an mRNA correlated with ethylene synthesis encoding a protein of molecular weight 35000. *Planta* **168**, 94–100.

12. Zeugin, J.A. and Hartley, J.L. (1985) Ethanol precipitation of DNA. *Focus* **7**, 1-2.

13. (1980) *In vitro* translation of eukaryotic mRNAs. *Focus* **2**, 1-6.

14. Slater, A. (1988) Extraction of RNA from Plants, in *Methods in Molecular Biology*, Vol. 4 (Walker, J., ed.) Humana, Clifton, New Jersey.

15. Morch, M. D. and Benicourt, C. (1980) Post-translational proteolytic cleavage of an *in vitro*-synthesised turnip yellow mosaic virus RNA-coded high-molecular-weight protein. *J. Virol.* **34**, 85–94.

Chapter 4

Hybrid-Arrested Translation

Keith Dudley

1. Introduction

Hybrid arrested translation (1) is a means of identifying recombinant DNA clones by their ability to hybridize to, and thus prevent the translation of, a specific messenger RNA in a cell-free system. In this respect the approach is similar to "hybrid selection," also known as "hybrid release" (*see* Chapter 3), in which messenger RNAs homologous to cDNA clones are specifically trapped as RNA/DNA hybrids on nitrocelloulose filters. By washing the filters at specific temperatures and salt concentrations, RNA molecules bound nonspecifically are removed leaving only the RNA homologous to the DNA bound. This RNA can then be released by boiling the filter, precipitated with ethanol, and translated in vitro. The fundamental difference between the two approaches is that, whereas in hybrid selection translation reveals one or a few polypeptides, hybrid-arrested translation identifies a gene product by its disappearance from the pattern of translation products.

1.1. Principles of Hybridization of Nucleic Acids in Solution

Hybrid-arrested translation takes place in two distinct phases; formation of a DNA/RNA hybrid, followed by translation in vitro of the unhybridized material. The rate at which the first two of these happens is dependent upon the incubation temperature, the salt concentration, and the presence or absence of formamide. Formamide acts to reduce the temperature at which stable hybrids form and also acts to promote the formation of DNA/RNA hybrids as opposed to DNA/DNA hybrids that are less stable in formamide, by 10–30°C (2). The rate constant of hybridization between RNA and DNA is, however, reduced several-fold under these conditions. The great advantage of using formamide for hybridization prior to in vitro translation is that the hybridization temperature can be kept relatively low. This is important to maintain the integrity of the RNA sample. It is no use demonstrating that by incubating a cDNA clone with an RNA preparation a specific translation product disappears if RNA has been degraded during the hybridization period; the translation products of the RNA prior to and subsequent to hybridization will bear no comparison.

The other major consideration is the quality of the formamide to be used. Commercially available formamide needs to be deionized before use since it contains, among other things, formate and heavy metal ions. The heavy metal ions are of particular concern since they can catalyze the nonenzymatic breakdown of nucleic acids. Recrystallization of formamide is therefore detailed in this chapter.

1.2. Preparation of DNA Samples

Hybrid-arrested translation does not depend upon having a full-length cDNA clone. A cDNA insert in a plasmid vector of as little as 400 bp is more than sufficient to form a stable hybrid with the messenger RNA. It is important, however, that the cDNA clone corresponds to a region of the mRNA that is translated; if it is all 3 prime untranslated, the mRNA may still be available for translation. It is also not necessary to purify the cDNA insert away from the plasmid sequences, as long as a control hybridization is set up with just plasmid sequences alone. The recombinant plasmid does need to be linearized, however. A site should be chosen within the vector that is unique to the recombinant and produces after digestion a linear molecule with the cDNA insert uncut (3). Generally the hybridization is done in DNA excess in terms of specific sequence, but usually in order to have enough of the specific RNA in the reaction, the RNA concentration has to be above that of the DNA.

1.3. RNA Samples

Methods of preparing RNA have been dealt with elsewhere (4). The amount of RNA added to a hybrid arrested translation experiment is dictated by the amount of RNA it is necessary to translate in order to see the relevant band on a 1-D gel (or spot on a 2-D gel). If the identity of the gene product corresponding to the cDNA under study is not known, this poses a problem. Generally, however, the amount of RNA needed to detect the translation product on a gel is dictated more by the efficiency of the translation system than by the absolute amount of RNA used, and this is particularly true when two-dimensional gels are needed to reveal the product of interest. If the experiment is designed to show that a particular cDNA clone codes for a particular translation product, deciding the amount of RNA is not a problem. It is not normally necessary to purify poly (A)$^+$ RNA for hybrid arrested translation. In cases in which the expression of a messenger RNA is particularly low, this may be desirable, but I have found that by using total cellular RNA the messenger RNA is protected from degradation by any ribonuclease that may have entered the system. The basis for this is uncertain, but it may be that the huge excess of ribosomal RNA saturates any ribonuclease. A compromise that may be useful under some conditions is to use sucrose gradient-enriched total RNA fractions, but this demands a previous knowledge of the size of the translation product.

As a good starting point, 10 µg total RNA has proved sufficient to demonstrate convincingly that a polypeptide of 66,000 molecular weight was coded for by the cauliflower mosaic virus genome (5) and that a clone designated B1.4 coded for a t-locus protein Tcp-1 (6).

2. Materials

The following items and reagents are necessary for hybridization prior to the translation. When appropriate, all equipment and solutions should be sterilized.

1.   Deionized or preferably recrystallized formamide (*see* section 3), stored in small aliquots at – 70°C.
2.   Hybridization buffer: this is prepared as a 10x strength stock being 4M NaCl, 100 mM PIPES (pH 6.6). This buffer is stored at –20°C. It does not freeze at this temperature because of depression of the freezing point by the NaCl.
3.   Sterile water.
4.   Eppendorf 1.5-mL reaction tubes.
5.   Variable Gilson pipets with sterile tips.
6.   RNA (10 µg or as desired) as a pellet washed twice with 70% ethanol and dried.
7.   DNA (1–2 µg) linearized with a restriction endonuclease (7) phenol extracted, ethanol precipitated, and dried.

For the stages following the hybridization:

8.   2M Sodium acetate and 95% ethanol for precipitation of DNA–RNA hybrids.
9.   70% Ethanol to wash RNA pellets.
10.   Reticulocyte lysate messenger RNA-dependent translation system containing all the ingredients for in vitro translation (8,9).
11.   A suitable radioactive amino acid, usually *l*-(^{35}S)-methionine, although this is not present in all polypeptides.

12. Equipment for analysis of polypeptides on gels, including materials for autoradiography or fluorography (*10*).

2.1. Preparation of Hybridization Grade Formamide

Several companies sell reagents described as molecular biology grade (Sigma, BRL). Formamide in this category is usually of a higher purity than the other commercially available makes, and can be further deionized by stirring with a mixed-bed resin for 1–2 h. A superior alternative to this is to recrystallize the formamide.

1. The formamide can be recrystallized by stirring it vigorously in a beaker set in an ice bath in a fume cupboard. A 500-mL beaker is placed in the ice bath, and the ice is packed around the sides to maximize contact.
2. Formamide (100–200 mL) is added to the beaker, and a stirring bar is added. The formamide is stirred vigorously; failure to do this will prevent recrystallization.
3. Solid sodium chloride is added to the ice bath to reduce the temperature, taking care that during its addition no salt gets into the formamide.
4. Over a period of 30 min the temperature of the formamide drops until it reaches 2–3°C when crystals are formed. These can be collected on a sterile sintered glass funnel on a buchner flask with vacuum applied, and transferred to sterile universals.
5. The formamide should be stored in small (500-µL) aliquots at –70°C.

2.2. Hybrid-Arrested Translation

1. The DNA in an Eppendorf tube is boiled for 30 s and then quick chilled to render the DNA single stranded and prevent "snap-back".
2. The dried RNA pellet is taken up in 5 µL of H_2O. Recrystallized formamide (40 µL) are added, as well as 5 µL of

10x hybridization buffer. This is then added to the DNA to give a final volume of 50 µL.

3. The reaction is incubated at 50°C for 2–4 h.
4. Cold sterile water (200 µL) is added, and the nucleic acids are precipitated by the addition of 2 vol of 95% ethanol and sodium acetate to a concentration of 0.2*M*. This is left at –20°C overnight.
5. The precipitated nucleic acids are collected by centrifugation in a microcentrifuge and washed twice with 200 µL of ice-cold 70% ethanol.
6. The pellet is dried under vacuum and dissolved in 10 µL of sterile distilled water.
7. Half the sample, 5 µL, is removed and heated to 100°C for 1 min and quick chilled. This is an important control since any translation product absent after the hybridization should reappear when the sample is boiled. This demonstrates that its absence is actually caused by the hybridization of the mRNA to the DNA, and not some specific degradation process.
8. Each sample is added to a 20 µL in vitro translation mix containing a radioactively labeled amino acid, translated, and analyzed by polyacrylamide gel electrophoresis and autoradiography (*see* Vol. 2, Chapters 20 and 21, in this series).

3. Notes

1. If the DNA used in the hybrid arrest is a linearized recombinant containing plasmid sequences, it is essential that a control experiment using just the linearized plasmid is performed, to be certain that it is the insert that has homology to the messenger RNA. It is also a sound policy to carry out a positive control with each experiment using, for example, an actin clone to demonstrate that the system is functioning normally. The final order of samples on the polyacrylamide gel might be:

A. No RNA added to the translation system.
B. Translation products of RNA after hybrid arrest.
C. Translation products of RNA after hybrid arrest and heating to 100°C to melt the hybrid.
D. Translation products of RNA with no hybrid arrest.
E. Translation products of RNA after hybrid arrest with a control DNA sequence such as actin.
F. Molecular weight markers.
G. Translation products of RNA after hybrid arrest using just plasmid sequences.

References

1. Paterson, B. M., Roberts, B. E., and Kuff, E. L., (1977) Structural gene identification and mapping by DNA.mRNA hybrid-arrested cell free translation. *Proc. Natl. Acad. Sci USA* **74**, 4370–4374.
2. Anderson, M. L. M. and Young, B. D. (1985) Quantitative Filter Hybridization, in *Nucleic Acid Hybridization—A Practical Approach* (Hames, B. D. and Higgins, S. J., eds.) IRL, Oxford.
3. Forde, B. G. (1983) Synthesis of cDNA for Molecular Cloning, in *Techniques in Molecular Biology* (Walker, J.M. and Gaastra W.M., eds.) MacMillan, New York.
4. Slater, R. J. (1983) The Extraction and Fractionation of RNA, in *Techniques in Molecular Biology* (Walker, J.M. and Gaastra, W., eds.) MacMillan, New York.
5. Odell, J. T. and Howell, S. H. (1980) The identification, mapping and characterisation of mRNA for p66 a cauliflower mosaic virus encoded protein. *Virology* **102**, 349–359.
6. Willison, R., Dudley, K. and Potter, J. (1986) Molecular cloning and sequence analysis of a haploid expressed gene encoding t-complex polypeptide 1. *Cell* **44**, 727–738.
7. Mooi, F. R. and Gaastra, W. (1983) The Use of Restriction Endonucleases and T4 DNA Ligase, in *Techniques in Molecular Biology* (Walker, J. M. and Gaastra, W. , eds.) Macmillan, New York.
8. Grierson, D. and Spiers, J. (1983) Protein Synthesis In Vitro, in *Techniques in Molecular Biology* (Walker, J.M. and Gaastra, W., eds.) MacMillan, New York.
9, Pelham, H. R. B. and Jackson, R. J. (1976) An efficient mRNA-dependent translation system from reticulocyte lysates. *Eur. J. Biochem.* **67**, 247–256.
10. Laskey, R. A. and Mills, A. D. (1975) Quantitative film detection of ^{3}H and ^{14}C in polyacrylamide gels by fluorography. *Eur. J. Biochem.* **56**, 335–341.

Chapter 5

In Vitro Translation and Analysis of Early Events in Protein Synthesis Initiation in Nonreticulocyte Mammalian Cells

Jeffrey W. Pollard and Michael J. Clemens

1. Introduction

The most commonly used systems to translate mRNAs in vitro are the rabbit reticulocyte lysate and the wheat germ extract (*1,2* and *see* Vol. 2 in this series). Both these systems have a number of advantages, including the ease and cheapness of preparation, the relative absence of RNAse, the high level of re-initiation of ribosomes onto mRNA, the high fidelity of translation, and the maintenance of activity upon long-term storage. Both of them have also been shown, upon addition of dog pancreatic membranes, to support processing of translated peptides (*1*). Furthermore, the reticulocyte has been the major source of information concerning the mechanism and control of protein synthesis in mammalian cells (2). For the investigation of many of the regulatory mechanisms in mammalian cells, however, neither the reticulocyte lysate nor the wheat germ system will be suitable. One, after all, is derived from plant cells and the other from a very specialized mammalian cell. It is, therefore, unwise to assume that control mechanisms discovered in either of these are applicable to mammalian cells in general. Furthermore, it is impossible to study the regulation of protein synthesis by growth factors and other physiological stimuli or the responses of cells to viral infection using these systems. For these purposes it is necessary to have cell-free systems from a variety of nonreticulocyte sources, and indeed such systems have been prepared from Ehrlich ascites tumor, CHO, and L cells, as well as from muscle and liver (2-4). Caution does need to be exercised even in these cases, however, because the cell lines used are often neoplastically transformed and may not respond normally to growth signals. Also, even with normal tissues the cell-free systems derived have a low activity and reinitiate rather poorly (1). Nevertheless, they do offer the advantages of being able to permit translation of homologous mRNAs, approximate the controls found in whole cells (1,3), and facilitate the analysis of the partial reactions of protein synthesis in cells that have mutations that affect protein synthetic regulation (5).

In this chapter we will describe the preparation of cell-free protein-synthesizing systems from cultured cells and methods for analyzing some of the early steps in protein synthesis initiation involving the initiation factor eIF-2 that have been shown to be points of regulation. These include the initial binding of [Met]-tRNA$_i$ to the binary complex of eIF-2 and GTP to give the ternary complex eIF-2–GTP–[Met]-tRNA$_i$; the binding of this complex to the small ribosomal subunit to give the 43S pre-initiation complex; and guanine nucleotide exchange on eIF-2, necessary for recycling of the eIF-2–GDP complex, which is ejected from the ribosome at the end of initiation, in order for the cycle of initiation to commence again (2).

2. Materials

1. Growth medium: a rich medium to grow cells should be used. We use minimal essential medium supplemented with 8% (v/v) fetal calf serum.
2. Wash buffer: 10 mM N-2-hydroxyethylpiperazine-N'-2 ethane sulfonic acid (Hepes), 140 mM KCl adjusted to pH 7.6 with KOH.
3. Lysis buffer: 10 mM K$^+$–Hepes, 10 mM KCl, 1.5 mM magnesium acetate, 7 mM β-mercaptoethanol, pH 7.6.
4. 10% (v/v) Nonidet P40 (NP40) in sterile distilled water.
5. 5X buffer E: 100 mM K$^+$–Hepes, 550 mM KCl, 2.5 mM spermidine, 3.5 mM β-mercaptoethanol, 50% (v/v) glycerol, pH 7.6.
6. Sephadex G25 equilibrated in 90 mM KCl, 2 mM magnesium acetate, 7 mM β-mercaptoethanol, 20 mM K$^+$–Hepes, pH 7.6.
7. 0.2M CaCl$_2$.
8. *Staphylococcus aureus* nuclease (micrococcal nuclease).
9. 0.1M Ethylene glycol *bis*-(β-aminoethyl ether)-N, N'-tetraacetate (EGTA).
10. 1M Hepes, adjusted to pH 7.6 with KOH.
11. 2M KCl.
12. 0.1M Magnesium acetate.
13. 0.1M ATP in 0.5M K$^+$–Hepes, pH 7.6.
14. 10 mM GTP in 50 mM K$^+$–Hepes, pH 7.6.
15. 0.5M Spermidine. This is very deliquescent, and it is advisable to make up the complete stock bottle in one step.
16. 0.5M Creatine phosphate stock.
17. 18 mg/mL Creatine phosphokinase (EC 2.7.3.2).
18. 0.5M β-Mercaptoethanol stock.
19. 20% (w/v) Trichloroacetic acid (TCA).
20. Buffer A: 10 mM MgCl$_2$, 10% (v/v) glycerol, 10 mM Tris-HCl, pH 8.0.

21. Buffer B: 20 mM β-mercaptoethanol, 1 mM MgCl$_2$, 10% (v/v) glycerol, 20 mM KH$_2$PO$_4$, pH 7.5.
22. Buffer C: As in solution 21, but at pH 6.5 and KH$_2$PO$_4$ at 0.25M.
23. Buffer D: 40 mM β-mercaptoethanol, 10% (v/v) glycerol, 15% (v/v) polyethylene glycol 6000, 2 mM KH$_2$PO$_4$, pH 7.0.
24. 0.1M Dithiothreitol.
25. 0.1M CTP in 0.5M K$^+$–Hepes, pH 7.6.
26. 2M Sodium acetate, pH 5.0.
27. Water-saturated phenol. Phenol causes severe burns and should only be handled with gloves and eye protection.
28. 50 mM Sodium acetate, 5 mM magnesium acetate, pH 5.0.
29. 0.1M EDTA, pH 7.0.
30. 0.3M Phosphoenol pyruvate.
31. 18 mg/mL Pyruvate kinase (EC 2.7.1.40).
32. Buffer F: 20 mM K$^+$-Hepes, pH 7.6, 100 mM KCl, 2 mM magnesium acetate.
33. Buffer H: 0.1 mM Sodium cacodylate, 100 mM KCl, 5 mM magnesium acetate, pH 6.6. Remember, this solution contains an arsenate and is potentially toxic.
34. 2% (w/v) Cetyl trimethyl ammonium bromide (CTAB).
35. Buffer G: 20 mM K$^+$–Hepes, 90 mM KCl, 1 mM magnesium acetate, pH 7.6.

All solutions should be made up in autoclaved H$_2$O and the pH adjusted at room temperature. Suitable precautions to avoid ribonuclease should be taken, such as use of baked glassware (250°C for 4 h) and autoclaved plastic pipet tips. Solutions 13, 14, 15, 24, and 25 are reasonably unstable and should be aliquoted and stored at –20°C. Solutions 8, 16, 17, 30, and 31 should be aliquoted, snap frozen, and stored at –70°C. The solution 5 should be stored at –20°C and is stable for a month. The remainder, except solutions 1, 18, 19, 27, 33, and 34, can be autoclaved and stored at 4°C, but remember to add the β-mercaptoethanol after autoclaving.

3. Methods

3.1. Preparation of Postmitochondrial Supernatants (S10) and Postribosomal Supernatants (PRS) (ref. 5)

1. Cells are seeded at 0.5×10^5 cells/mL in 1–3 L of growth medium 2 d prior to the preparation of the S10s. These spinner cultures are stirred in temperature-regulated water baths (or in a warm room at the appropriate temperature, usually 37°C).

2. On the day of the experiment when the cells are fully in the exponential phase of growth ($2–4 \times 10^5$ cells/mL; *see* Note 1 in section 4), the cells are harvested by pouring them directly onto an equal volume of crushed ice in 1-L centrifuge bottles. The cells are recovered by centrifugation at $700g$ for 10 min at 4°C.

3. The supernatant is gently poured off and the cell pellet resuspended in 200 mL of ice-cold wash buffer and recentrifuged at $700g$.

4. The supernatant is again removed and the pellet resuspended in 15 mL of ice-cold wash buffer and centrifuged at $3000g$ for 10 min in 15-mL graduated centrifuge tubes.

5. The supernatant is carefully removed, the tubes wiped dry with a cotton tip, and the packed cell volume measured (approx. 0.6 mL/L of cell culture).

6. 2 Vol of cold lysis buffer are then added and the cells resuspended by vigorous vortexing.

7. To this 10% NP-40 is added to give a final concentration of 0.25% (v/v) and the cells lysed by vigorous vortexing. *Immediately* add 0.4x the packed cell volume of 5x buffer E and vortex.

8. Transfer the tube contents to a 5-mL centrifuge tube and centrifuge at $10,000g$ for 10 min at 4°C.

9. Remove the postmitochondrial supernatant (S10), being careful not to disturb the pellet and avoiding as much as

possible the lipid layer at the surface, and snap freeze it in 100–200 µL aliquots.

10. Store these in capped tubes in liquid nitrogen. The S10 preparations should contain approximately 60–100 A_{260} U/mL, and the yield is between 3.5 and 4.0 mL/10^9 cells.

11. The S10 preparations can be gel filtered at 4°C (*see* Note 2 in section 4) prior to freezing by passing through a Sephadex G25 column using a 5-mL bed volume per mL of sample. Wash the column through with the buffer used to equilibrate the G-25.

12. Postribosomal supernatants (PRS) are prepared from these S10s by centrifugation at 130,000g for 3 h or at 356,000g for 50 min (Beckman TL100 centrifuge) at 4°C.

13. These PRS preparations are snap frozen in 100-µL aliquots and stored in liquid nitrogen (*see* Note 3 in section 4).

3.2. Preparation of a Messenger-Dependent Lysate

1. Prior to freezing of the S10, add 0.2M CaCl$_2$ to give a final concentration of 1 mM.

2. Add micrococcal nuclease at 25–100 U/mL (the amount needed to inhibit endogenous protein synthesis completely should be determined experimentally), then incubate at 20°C for 15 min.

3. Add 0.1M EGTA to give a final concentration of 2 mM, and place on ice.

4. The lysates can be used immediately (*see* note 3 in section 4) and will translate up to 40 µg/mL of added mRNA (*see* Note 4 in section 4), or they can be aliquoted and stored in liquid nitrogen.

3.3 In Vitro Protein Synthesis (see Note 4 in section 4)

1. A 100-µL reaction mix is prepared on ice that contains 60% (v/v) S10. The S10 is estimated to contain 20 mM K$^+$–Hepes, pH 7.6, 80 mM KCl, 1 mM Mg acetate, 4.7 mM

β-mercaptoethanol, and 0.33 mM spermidine. These ions must be allowed for in calculating the final conditions. The final concentrations of the components (taking into account the components in the S10) are: 25 mM K$^+$–Hepes, pH 7.6, 110 mM KCl, 1 mM magnesium acetate, 1 mM ATP, 0.25 mM GTP, 0.4 mM spermidine, 4% (v/v) glycerol (provided by the S10), 5 mM creatine phosphate, 180 μg/ mL creatine kinase, 6 mM β-mercaptoethanol, and radioactively labeled amino acids; usually ^{35}S-methionine at 50 μCi/mL or ^{3}H-leucine up to 500 μCi/mL.

2. Gently mix and incubate the reaction mix at 30°C for appropriate times.
3. To measure the rate of protein synthesis, 5-μL aliquots are withdrawn 2, 5, 10, 15, 20, and 30 min after the start of the incubation and added to 1 mL of water containing 10 μg/ mL bovine serum albumin.
4. Add to this an equal volume of ice-cold 20% TCA and incubate it on ice for 10 min.
5. Boil this for 5 min to hydrolyze aminoacyl-tRNA bonds. Cool it and collect the precipitate on GF-C glass fiber filters using an appropriate filtering device.
6. Each filter is washed five times with 8 mL of ice-cold 5% (w/v) TCA, once with 95% ethanol, and then dried and prepared for liquid scintillation counting (*see* Note 5 in section 4).

3.4. Preparation of (^{35}S)-[Met]-tRNA$_i^{Met}$

3.4.1. Preparation of Crude Escherichia coli Aminoacyl-tRNA Synthetases

1. 5 g of lyophilized *E. coli* B are ground to a paste with 15 g of alumina (type 305) in sufficient buffer A to give a thick paste.

2.   Add an equal volume of buffer A and centrifuge at 800*g* for 5 min to remove the alumina.
3.   Centrifuge the resulting supernatant at 100,000*g* for 2 h.
4.   After centrifugation, load the supernatant onto a DEAE column (1 x 9 cm) pre-equilibrated with buffer B. Wash the column thoroughly with buffer B, then elute the crude synthetase mix with buffer C.
5.   Measure the eluate at 280 nm and pool the peak protein fractions.
6.   Dialyze these against three changes of buffer D.
7.   After dialysis, add an equal volume of glycerol and store the crude synthetase mix at –20°C, at which the activity is stable for many months.

3.4.2. Preparation of (^{35}S)-[Met]-tRNA$_i$^{Met}(see Note 6 in section 4)

1.   Make up a 1-mL reaction mix containing 50 mM KCl, 8 mM magnesium acetate, 2 mM DTT, 4 mM ATP, 1 mM CTP, 100 µCi (^{35}S)-methionine (highest specific activity available), 50 mM K$^+$-Hepes (pH 7.5), 1 mg of deacylated calf liver tRNA (commercially obtained), and 80 µL *E. coli* aminoacyl-tRNA synthetases prepared as described above.
2.   Incubate at 30°C for 30 min.
3.   After incubation, rapidly cool on ice and add 0.1 mL of 2M sodium acetate (pH 5.0) and 2 mL of water-saturated phenol.
4.   Shake vigorously for 5 min at 4°C and then separate the phases by centrifugation at 1000*g* (or greater) for 10 min.
5.   Collect the aqueous (upper) layer and re-extract the phenol with 0.8 mL of 50 mM sodium acetate and 5 mM magnesium acetate and recentrifuge as in step 4 above.
6.   Pool the aqueous phases and precipitate the tRNA by adding 2 vol of 95% ethanol at –20°C. (*See* Note 7 in section 4).
7.   Collect the precipitate by centrifugation at 10,000*g* for 15 min.

8. Wash the precipitate with 70% ethanol, regain it by centrifugation, and dry it in a Speed Vac or desiccator.
9. Dissolve the pellet in 20 mM sodium acetate pH 5.0 and determine the radioactivity of a small aliquot by liquid scintillation counting.
10. Store the remainder in 0.1 mL aliquots at −70°C until use.

3.5. Analysis of the Rate of elF-2–GTP–[Met]-tRNA$_i^{Met}$ Ternary Complex Formation

1. A 0.2-mL reaction mix consisting of 20% (v/v) postribosomal supernatant, 25 mM K$^+$-Hepes, pH 7.6, 100 mM KCl, 1 mM magnesium acetate, 1 mM ATP, 0.25 mM GTP, 0.5 mM dithiothreitol, 0.2 mM EDTA, 3 mM phosphoenol pyruvate, and 18 μg/mL pyruvate kinase is prepared.
2. To this mix, (^{35}S)-[Met]-tRNA$_i^{Met}$ (approximately 5 x 10^6 cpm/mL, *see* above) is added, and incubation is performed at 30°C.
3. At various times (usually 0, 2, 5, 10, 15, 20, and 30 min) 20-μL aliquots are removed and immediately placed in 1 mL of ice-cold buffer F.
4. The radioactivity bound to eIF-2 is immediately collected on 2.5-cm cellulose nitrate filters (0.45 μm) by filtration.
5. Each filter is washed twice with buffer F and dried, and the radioactivity is determined in a toluene-based scintillation fluid by liquid scintillation counting.

3.6. Assay of 43S Preinitiation Complexes

1. A 100-μL reaction mix containing 60% (v/v) S10 postmitochondrial supernatant, 25 mM K$^+$-Hepes, pH 7.6, 110 mM KCl, 1 mM magnesium acetate, 6 mMβ-mercaptoethanol, 4% glycerol, 0.4 mM spermidine, 1 mM ATP, 0.25 mM GTP, 5 mM creatine phosphate, and 180 μg/mL creatine kinase is prepared on ice.
2. (^{35}S)-[Met]-tRNA$_i^{Met}$ (prepared as described above) is

added at approximately 5×10^6 cpm/mL and the reaction mix incubated for 2 min at 30°C (*see* Note 8 in section 4).

3. The whole reaction mix, after being rapidly cooled in a –10°C alcohol bath, is layered over a 5-mL 20–50% (w/v) sucrose gradient made up in 0.1 mM sodium cacodylate buffer.

4. The gradients are centrifuged for 3 h at 200,000g.

5. After centrifugation the gradients are fractionated and the optical density at 260 nm monitored using an appropriate gradient analyzer.

6. Fractions are precipitated with 1 mL of 2% (w/v) CTAB in the presence of 1 mL of 0.5M sodium acetate, pH 5.0, containing 0.5 mg of carrier tRNA (commercially obtained).

7. Precipitates are collected on glass fiber filters (Whatman GF/C), washed with 5 mL of water, and dried.

8. Radioactivity on the filters is counted using a toluene-based scintillation fluid.

3.7. Assay of Guanine Nucleotide Exchange Activity of eIF-2B (Guanine Nucleotide Exchange Factor, GEF) (refs. 6,7)

1. Incubate 15 μg of pure initiation factor eIF-2 (refs. 6,7) with 2 μM (^{3}H)-GDP or (^{32}P)-GDP for 10 min at 30°C in a 220-μL volume containing 20 mM K$^+$-Hepes, pH 7.6, 60 mM KCl, 1 mM dithiothreitol, and 100 μg/mL creatine kinase.

2. After incubation the complexes are stabilized by addition of magnesium acetate to a final concentration of 2 mM.

3. To measure displacement of GDP by GTP from the eIF-2–GDP complex, 50 μL of the eIF-2: (^{3}H) (or ^{32}P) GDP is added to a 350-μL volume containing 150 μL S10 postmitochondrial supernatant, 20 mM K$^+$-Hepes, pH 7.6, 90 mM KCl, 1 mM Mg acetate, 0.1 mM dithiothreitol, 180 μg/mL creatine kinase, 5 mM creatine phosphate, and 0.5 mM GTP. The mix is incubated at 30°C.

4. At 0, 2, 5, 10, and 15 min, 60-μL aliquots are removed, immediately diluted with 20 mM K$^+$-Hepes, pH 7.6, 90

mM KCl, and 1 mM Mg acetate, and rapidly filtered through 2.5-cm cellulose nitrate filters (0.45 µm).

5. The filters are washed three times with 5 mL of buffer G and dried, and the bound radioactivity is determined in a toluene-based scintillation fluid by liquid scintillation counting.

4. Notes

1. It is essential that cells are in the exponential phase of growth when they are harvested because the rate of protein synthesis rapidly drops either as the cells enter stationary phase at high density or as the result of nutrient deprivation.

2. The gel-filtered lysate is sometimes used in order to have more defined components and higher specific activities of radioactive amino acids. The procedure results, however, in a 2–3-fold dilution and a substantial loss of activity. Endogenous energy sources are also removed by this procedure and may need to be compensated for.

3. Although the S10s are reasonably stable in liquid nitrogen for periods up to a year, the PRS preparations appear to lose activity fairly rapidly. It may therefore be better to prepare small volumes on the day of use; this can be achieved either by centrifugation in the Beckman benchtop TL-100 ultracentrifuge or in an airfuge. The micrococcal nuclease-treated lysates may also be gel filtered, but in most cases this is unnecessary.

4. Each mRNA has its own Mg^{2+} and K^+ optimum, and this should be determined for the cell-free system used. It is also advisable to examine the products of translation by gel electrophoresis and autoradiography to ensure that premature termination is not occurring. The advantages of this system are that processing of the products, including proteolytic cleavage of precursor proteins, can occur. In vitro protein synthesis may also be used to analyze the

effects of adding back specific protein synthetic factors
and in mRNA competition studies.

5. To determine the absolute rate of incorporation of a particular isotope into protein, unless the gel filtered lysate is
used, the endogenous amino acid pool needs to be measured. This is performed by adding the ice-cold lysate
sulfosalicyclic acid to a final concentration of 3% (w/v).
The precipitate is removed and the concentration of each
amino acid in the supernatant is determined with an
amino acid analyzer.

6. The crude *E. coli* synthetase mix only charges the mammalian initiator tRNA$_i^{Met}$ enabling a total tRNA mix to be used
for preparation of (^{35}S)-[Met]-tRNA$_i^{met}$.

7. If it is important to remove all traces of ATP and CTP, the
pooled aqueous phases can be dialyzed against 0.5M
NaCl, 50 mM sodium acetate (pH 5.0) for 6 h, followed by
20 mM sodium acetate (pH 5.0) for 17 h. Then proceed as
above.

8. As an alternative to (^{35}S)-[Met]-tRNA$_i^{Met}$, free ^{35}S-Met may
be used in preinitiation complex formation assays because
endogenous aminoacyl-tRNA synthetases charge both
the endogenous initiator and elongation species of
tRNA^{Met}. In this case gradient fractionation separates the
methionine used in elongation of polysome-bound nascent chains from that associated with the 40S ribosomal
subunit.

9. Although precautions are taken to prevent ribonuclease
contamination, this may still be a problem for translation
of mRNAs in vitro. Addition of RNAsin (Jagus, personal
communication) has been reported to act as a good ribonuclease inhibitor that does not affect protein synthesis. The
use of diethyl pyrocarbonate-treated water or vanadyl–ribonucleoside complex should be avoided, since trace
contamination by these reagents can affect the translational machinery.

10. Protein synthesis systems derived from nonreticulocyte

mammalian cells are very inefficient. In fact many hardly reinitiate at all, but simply display elongation (run-off) of nascent polypeptide chains on ribosomes. It is possible to determine the extent of reinitiation by incubating the extract with an inhibitor of initiation and determining the residual protein synthesis. It is essential that this inhibitor does not affect either elongation or termination. Usually edeine at 4 μM is the inhibitor of choice. It is not certain why these in vitro systems are so poor. Part of it is undoubtedly because of dilution during the preparation of the extracts. There is also a substantial loss of protein synthetic factors during the preparation of the S10. In fact, using immunoblotting, we have shown that about 50% of the eIF-2 can be found in the pellet. Furthermore, the protein synthetic machinery may require an intact cytoskeleton for optimal translation, and this structure is, of course, lost during the preparation of the extracts.

It may be that much greater activities can be achieved for the translation of added mRNAs by the readdition of purified factors such as eIF-2, eIF-4F, and, particularly, eIF-2B (GEF). Recently, it has also been shown that the addition of a reticulocyte postribosomal supernatant greatly stimulates translation in these systems (*8*). Care must, of course, be exercised in using these factors during the analysis of any regulatory system in vitro.

Acknowledgments

This chapter was prepared while the authors' research was supported by the Medical Research Council, UK (JWP), the Gunnar Nilsson Cancer Research Trust Fund (MJC), and the Cancer Research Campaign, UK. MJC was a recipient of a Career Development Award from the Cancer Research Campaign.

References

1.   Clemens, M. J. (1984) Translation of Eukaryotic Messenger RNA in Cell-Free Extracts, in *Transcription and Translation: A Practical Approach* (Hames, B.D. and Higgins, S. J., eds.) IRL Press, Oxford and Washington.

2.   Pain, V. M. (1986) Initiation of protein synthesis in mammalian cells. *Biochem. J.* **235,** 625–637.

3.   Pain, V. M., Lewis, J. A., Huvos, P., Henshaw, E. C., and Clemens, M. J. (1980) The effects of amino acid starvation on regulation of polypeptide chain initiation in Ehrlich ascites tumor cells. *J. Biol. Chem.* **255,** 1486–1491.

4.   Morley, S. J. and Jackson, R. J. (1985) Preparation and properties of an improved cell-free protein synthesis system from mammalian liver. *Biochim. Biophys. Acta* **825,** 45–56.

5.   Austin, S. A., Pollard, J. W., Jagus, R., and Clemens, M. J. (1986) Regulation of polypeptide chain initiation and activity of initiation factor eIF-2 in Chinese-hamster ovary cell mutants containing temperature-sensitive aminoacyl-tRNA synthetases. *Eur. J. Biochem.* **157,** 39–47.

6.   Matts, R. L. and London, I. M. (1984) The regulation of initiation of protein synthesis by phosphorylation of eIF-2(α) and the role of reversing factor in the recycling of eIF-2. *J. Biol. Chem.* **259,** 6708–6711.

7.   Clemens, M. J., Galpine, A., Austin, S. A., Panniers, R., Henshaw, E. C., Duncan, R., Hershey, J. W. B., and Pollard, J. W. (1987) Regulation of polypeptide chain initiation in CHO cells with a temperature-sensitive leucyl-tRNA synthetase: Changes in phosphorylation of initiation factor eIF-2 and in the activity of the guanine nucleotide exchange factor, GEF. *J. Biol. Chem.* **262,** 767–771.

8.   Morley, S. J., Buhl, W-J., and Jackson, R. J. (1985) A rabbit reticulocyte factor which stimulates protein synthesis in several mammalian cell-free systems. *Biochim. Biophys. Acta* **825,** 57–69.

Chapter 6

Use of Elutips to Purify DNA

Paul D. Van Helden

1. Introduction

Elutips are essentially minicolumns of reverse-phase resin that offer an extremely rapid method for DNA purification and may often replace conventional techniques for manipulation of DNA. The Elutip-d columns commercially available are manufactured by Schleicher and Schuell. Each column consists of a pipet tip, such as may be used on a 0–200 µL automatic pipet, that is filled with an insoluble matrix.

Use of an Elutip offers the researcher an alternative to routine, time-consuming DNA recovery and purification steps, which usually involve enzymatic treatment (such as RNAse), phenol and chloroform extractions, and possibly column chromatography or adsorption to solid supports. Elutips may also be used successfully for concentrating DNA samples from dilute solutions (e.g., 1 µg/10 mL), since the final elution volume is approximately 400 µL.

Possibly the most useful application for Elutips lies in the recovery of DNA after gel electrophoresis. Recovery of DNA fragments from gels (either polyacrylamide or agarose) involves many hours of work and frequently results in samples that are contaminated by pieces of gel or inhibitors of enzymes such as ligases or restriction endonucleases that will be used later to manipulate DNA. Such inhibitors are often extracted from agarose gels (in particular) and are not necessarily removable by phenol or chloroform extractions. Contaminating pieces of gel or inhibitors will be removed by passage of DNA over an Elutip, and this procedure will restore and enhance the reactivity of DNA to enzymes.

Elutip treatment of DNA will also remove undesirable UV-absorbing molecules, such as nucleotides and traces of phenol, which will result in the qualitative improvement of DNA and allow for more accurate quantitation of DNA.

Extraction and purification of DNA from gels are possible in one step with Elutips. This procedure takes approximately 1 h, requires the use of low-melting-temperature agarose, and removes both agarose and any enzyme inhibitors from DNA by passing molten gel through the tip.

The minicolumns can be used to manipulate DNA ranging in size from at least 16,000–50,000 bp, with recovery rates between 40 and 95% in amounts varying from 10 to 100 µg of DNA.

2. Materials

1. Elutip™-d (Schleicher and Schuell).
2. Application and wash buffer: Buffer 1: 0.2M NaCl, 20 mM Tris-HCl (pH 7.5), 1.0 mM EDTA. Buffer 2: as for 1, but 20 mM NaCl. Buffer 3: 0.2M ammonium acetate, 10 mM magnesium acetate, 1.0 mM EDTA, 0.1% SDS. The above three buffers may be used for different applications, as described, and are collectively known as low-salt buffers.
3. Elution buffer: 1.0M NaCl, 20 mM Tris-HCl (pH 7.5), 1.0 mM EDTA, also known as the high-salt buffer.
4. Disposable syringes: 1 and 5 mL.
5. Disposable cellulose acetate filters (e.g., Schleicher and Schuell, FP027/2).

3. Method

3.1. Method 1: Standard DNA Purification

This method may be used to purify dsDNA (e.g., plasmid DNA) from RNA or contaminants such as noncovalently bound drugs or for recovery of dsDNA after elution from acrylamide or agarose gels (after electro-elution, crushing, diffusion, or gel dissolution; *see* Chapter 10 in Vol. 2 of this series).

1. Prepare the DNA sample in low-salt buffer 1, kept to a minimum (Note 1 in section 4).
2. Prepare a syringe with 1–2 mL of high-salt solution (Note 2 in section 4).
3. Cut off the tip of an Elutip column immediately below the matrix, remove the cap attached to the column head, and attach the column to the syringe.
4. Slowly wash high-salt buffer through the tip.
5. Remove the tip from the syringe, load the second syringe with 5 mL of low-salt buffer (buffer 1), and slowly wash buffer through the tip.
6. Remove the column from the syringe (Note 3 in section 4).
7. Attach a disposable filter to the syringe (Note 4 in section 4), attach the column to the outlet of the filter, and load the syringe with the DNA sample, which should be pre-equilibrated in low-salt buffer (Note 1 in section 4).
8. Fit the plunger to the syringe and slowly force the DNA sample through the column (recommended flow rate: 1 drop eluted/5 s). Slow flow is essential for adsorption of DNA to the column matrix.
9. Remove the column from the filter and syringe and wash the column using the same syringe with 2–3 mL of low-salt buffer.
10. Load the first syringe with 0.4–0.5 mL of high-salt buffer, attach the column, and slowly elute the DNA into a small tube (e.g., Eppendorf).

11. The DNA may be recovered by ethanol precipitation—
    add 2 vol of ethanol (–20°C), allow the DNA to precipitate
    (–20°C overnight or 20 min on dry ice), and then centrifuge
    to obtain the DNA as a pellet.

3.2. Method 2: Nick-Translation Application

This method has been described for the separation of
radiolabeled dsDNA from unincorporated labeled nucleotides
(*see* Schleicher and Schuell procedure, pamphlet no. 208 and
Chapter 38, vol. 2, of this series for a description of nick translation).

1. Prepare the DNA sample by adding high-salt solution to
   adjust the final concentration to approximately $0.35M$
   NaCl.
2. Prepare the Elutip and load the DNA as described for
   Method 1 in section 3.1, steps 2–8.
3. Wash the Elutip with 1 mL of $0.4M$ NaCl, 20 mM Tris-HCl
   (pH 7.5), 1.0 mM EDTA (Note 5 in section 4).
4. Elute the DNA with 400 µL of high-salt solution, as described in section 3.1. (Note 6 in section 4).

3.3. Method 3: Alternative Buffers—Ammonium Acetate

DNA may be eluted from polyacrylamide gels in a $0.5M$
ammonium acetate buffer (*1*; Note 7 in section 4). These samples may be applied to an Elutip column without prior buffer
exchange to NH_4Ac from NaCl.

1. Dilute the DNA sample in $0.5M$ NH_4Ac solution (*1*) with
   water to a final concentration of $0.2M$ NH_4Ac.
2. Follow the protocol described for Method 1 in section 3.1.,
   but replace low-salt buffer 1 by low-salt buffer 3, i.e.,
   NH_4Ac, in all relevant steps.

3.4. Method 4: Oligonucleotide Purification

The Elutip columns are most frequently used for the purification of DNA fragments of 100–20,000 bp in size. It is possible to use them to purify smaller fragments (<50 bp) (2). For this application, it is essential to use a lower ionic strength low-salt buffer.

1. Prepare DNA in low-salt buffer 2.
2. Follow the protocol described for Method 1 in section 3.1., but replace the low-salt buffer 1 by low-salt buffer 2 in all relevant steps.

3.5. Method 5: Direct Recovery of DNA from Low-Gelling Temperature Agarose

Recovery of DNA from gels is a problem that has to be overcome by many researchers. One useful option is the use of low melting temperature agarose. Many researchers using this agarose use a hot phenolic ex-traction step to recover DNA. In the method described below, melted agarose is simply passed through the column, which retains the DNA that may be eluted as described before (3). No previous extraction or elution steps are necessary. The prefilter is not used in this procedure.

1. Place the gel slice containing the DNA of interest into a tube. Add approximately 10 vol of low-salt buffer 1.
2. Heat the sample with occasional shaking at 65°C for approximately 30 min. Ensure that the agarose is completely melted.
3. Place the DNA sample and low-salt buffer in a 42°C water bath (Note 8 in section 4).
4. Follow the protocol as described for Method 1 in section 3.1, but use prewarmed (42°C) low-salt buffer in all steps. It is essential to work rapidly in order to keep the column temperature at 37–42°C, so that the agarose does not gel in

the column. The subsequent DNA elution with high-salt may be done as before at room temperature. It is important not to use the Elutip prefilter with this procedure.

4. Notes

1. DNA samples may be equilibrated to low-salt conditions by addition of low-salt concentrate (e.g., 10x buffer) or dialysis. A convenient sample size is 200 µL–5 mL. Alternatively the sample may be ethanol precipitated and redissolved in low-salt buffer.
2. The Elutip is designed for flow in one direction only. Buffer should not be drawn into the syringe through the column.
3. The column must be removed from the syringe prior to withdrawal of the syringe plunger, in order to avoid air flow through the column in the wrong direction.
4. Schleicher and Schuell manufacture filters designed for use with the Elutip. These filters remove particulate matter (from DNA samples) that might otherwise impede column flow. (Elution of DNA from gels often allows carryover of gel fragments.) Use of these filters is not always necessary.
5. When using Elutips following nick translation, the washing buffer used has twice the NaCl concentration of low-salt buffer 1, which helps to remove unincorporated nucleotides and short oligonucleotides from the bulk of labeled DNA adsorbed to the Elutip.
6. Following the elution of DNA, some labeled material always remains bound to the column. This may be shown by a Geiger counter (use ^{32}P), and has not been satisfactorily explained (*see also* ref. 1). This is also the situation where labeled oligonucleotides may be separated from unincorporated nucleotides.
7. The method recommended for recovery of labeled DNA from gels for Maxam-Gilbert sequencing uses this buffer.

8. Original reference and manufacturers' instructions suggest 37°C. The use of 42°C gives a greater safety margin with respect to the solidification of the gel in the tip caused by temperature reduction.

9. When purifying DNA from low-melting temperature agarose, it is important to keep the fluid flow over the tip sufficiently fast to prevent gelling caused by temperature drop, yet not so fast that the DNA does not bind efficiently and yields may be low. At 42°C, with a sample volume of 5 mL, a flow rate of approximately 1 drop/s is usually sufficient. At 37°C, the flow rate recommended is 2 drops/s (3). Working in a 37°C facility for a few minutes can help to prevent gelling of agarose.

References

1. Maxam, A. M. and Gilbert, W. (1980) Sequencing end-labeled DNA with base-specific chemical cleavages. *Meth. Enzymol.* **65**, 499–560.
2. Zarbl, H., Sukumar, S., Arthur, A.V., Martin-Zanca, D., and Barbacid, M. (1985) Direct mutagenesis of Ha-ras-1 oncogenes by *N*-nitroso-*N*-methylurea during initiation of mammary carcinogenesis in rats. *Nature* **315**, 382–385.
3. Schmitt, J. J. and Cohen, B. N. (1983) Quantitative isolation of DNA restriction fragments from low-melting agarose by Elutip-D affinity chromatography. *Anal. Biochem.* **133**, 462–464.

Chapter 7

Plasmid Preparation on Sephacryl S1000

Paul D. Van Helden and Eileen G. Hoal

1. Introduction

Production of pure plasmid DNA is a prerequisite for most laboratories engaged in recombinant DNA research. There are a number of procedures available for plasmid purification, all of which have common features; namely, (1) growth of plasmid-containing bacteria, (2) lysis of bacteria, (3) low-speed centrifugation to remove the majority of bacterial DNA and protein, (4) an ultracentrifugation step in cesium chloride and ethidium bromide to separate plasmid DNA from RNA and chromosomal DNA, and (5) recovery of plasmid DNA band (extraction of ethidium bromide and cesium chloride). Steps (4) and (5) of this procedure are expensive, time-consuming, and require expensive machinery [ultracentrifuge, rotor(s)]. In addition, incomplete extraction of ethidium bromide can result in plasmid DNA that is not manipulable by many enzymes commonly used in DNA research.

The method described here is an alternative to the ultracentrifugation step commonly used and does not require the use of ethidium bromide. The procedure is a modification of that described by Bywater et al. *(1)*, is inexpensive and simple, delivers plasmid DNA that is an excellent enzyme substrate and of high purity, and is faster than the traditional ultracentrifugation method, even when a vertical rotor is used.

2. Materials

1. Sephacryl S1000 Superfine (Pharmacia Fine Chemicals).
2. Column for gel filtration chromatography, plus flow adaptors (dimensions 100 x 1.25 cm).
3. Fraction collector plus UV monitor or UV spectrophotometer.
4. Bacterial pellet (containing plasmid).
5. Suspension buffer: 25mM Tris-HCl, pH 8.0; 50 mM EDTA, pH 8.0; 1% (w/v) glucose.
6. Alkaline SDS: 0.2M NaOH, 1% SDS, prepared prior to use from 1M NaOH and 10% SDS stock.
7. Neutralizer: 3M potassium acetate adjusted to pH 4.8 with glacial acetic acid.
8. TE buffer: 10 mM Tris-HCl, pH 7.5, 1 mM EDTA.
9. Ribonuclease A made DNAse-free by heat treatment (80°C for 10 min in TE buffer).
10. Proteinase K (solution in water, 10 mg/mL).
11. Running buffer: 35 mM Tris-HCl, pH 8.2, 30 mM NaH_2PO_4, 1mM EDTA, 0.75M NaCl.
12. 5M NaCl.
13. Phenol (water-saturated).
14. Chloroform/isoamyl alcohol (25/1).

3. Method

1. Prepare the bacterial pellet as described in Chapters 25 and 26 in Vol. 2 of this series, or refs. *1,2*.
2. Suspend the cell pellet in X mL of suspension buffer (X may be 20 mL for 1–2 g of pellet from 1 L of culture).

3. Add 2 X mL of alkaline SDS and incubate for 10 min on ice (*see* Note 1 in section 4).
4. Add 1.5 X mL of cold neutralizer. Invert the tube gently to mix, then incubate on ice for 10 min (*see* Note 2, section 4).
5. Centrifuge for 30 min at 35,000g (17,000 rpm Sorval SS34 if available).
6. Carefully decant the supernatant and filter to remove particulate material, which is inevitably present (use cheesecloth or similar material).
7. Add 1 vol of isopropanol, mix, and allow to stand for approximately 20 min at 4°C (*see* Note 3 in section 4). Centrifuge at 20,000g for 20 min, and discard the supernatant.
8. Dissolve the pellet in 10 mL of TE buffer, add 1.93 g of solid ammonium acetate, dissolve, and stand for 20 min on ice. Centrifuge for 20 min at 20,000g, and save the supernatant (*see* Note 4 in section 4).
9. Add 2 vol of ethanol (–20°C) or 1 vol of isopropanol to the supernatant, then stand for 20 min on ice and centrifuge at 20,000g for 20 min.
10. Dry the pellet briefly under vacuum and redissolve in 3 mL of TE buffer, to which 1 mL of 1M Tris, pH 7.5, is also added.
11. Add RNAse A (200 µg) and incubate for 30 min at 37°C.
12. Add SDS to 0.5% and proteinase K (200 µg) and incubate for a further 30 min at 37°C.
13. Add 5M NaCl to adjust the final concentration to 1M NaCl and then phenol extract and chloroform extract.
14. Pack the column with Sephacryl S1000 and equilibrate with running buffer (Note 5 in section 4).
15. Apply the sample to the column and elute at 4°C. A 60 cm head works well for a 100 cm column. Collect 2 mL fractions and read A_{260} of each (may be done automatically with a UV monitor) (*see* Fig. 1).
16. Transfer, if necessary, relevant fractions to tubes suitable for centrifugation and add 1 vol of isopropanol. Mix, stand

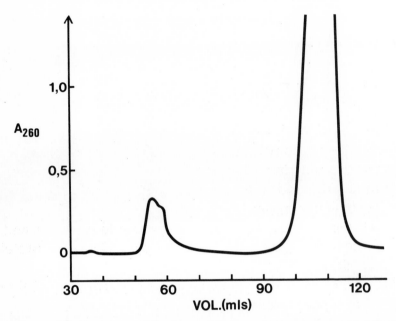

Fig. 1. Gel exclusion chromatography of crude plasmid pBR322 preparation on Sephacryl S1000 Superfine as described in section 3. A trace of chromosomal DNA may be observed at approximately 38 mL, plasmid DNA at 52–60 mL, and RNA between 100 and 120 mL. The fractions containing plasmid DNA may be pooled and recovered or recovered separately. The profile of the plasmid fraction differs in each run and reflects the composition of individual preparations, i.e., varying proportions of supercoiled, linear, or nicked-circular DNA. Individual fractions recovered separately allow one to obtain some plasmid consisting almost entirely of supercoiled-form DNA. If the alkaline lysis preparative method described is carefully adhered to, very little chromosomal DNA will be evident on chromatography and none will be seen in the plasmid fraction. Shorter columns may be used (*see* ref. 1), but the risk of chromosomal DNA and RNA contamination of plasmid DNA is increased.

at room temperature for 20 min, then centrifuge at 16,000*g* for 15 min at room temperature (Note 6 in section 4).

17. Drain the pellets well.
18. Plasmid DNA thus obtained may be dissolved in solution of choice or may be reprecipitated with ethanol to remove any traces of running buffer salts (Note 7 in section 4).

4. Notes

1. This alkaline lysis method, described in *ref*. 2, is a scaled-up modified version of the Birnboim and Doly procedure (3), which yields an excellent starting plasmid preparation with a very high content of supercoiled form and minimal nicked or linear forms. It has been used by the authors for the preparation of plasmids from 2 to 8 kb pairs in size. The use of lysozyme to lyse cells is not necessary in this procedure. After addition of alkaline SDS, the solution should become viscous and clear.

2. At step 4 in section 3 the viscosity should decrease, and a large white aggregate of chromosomal DNA and protein should be evident.

3. The method detailed in refs. 1 and 2 requires the addition of 2 vol of ethanol to precipitate DNA. It is convenient to use 1 vol of isopropanol since the total volume to be centrifuged is kept to a minimum.

4. Step 8 in section 3 is not essential, but it does remove a large amount of unwanted material (pellet discarded) by precipitation, and therefore yields a cleaner sample for column chromatography.

5. The original method for column packing (1) describes the use of a short (50 cm) but wider column. A longer column gives better separation and therefore cleaner plasmid preparations, hence the use of a 100-cm column in this description.

6. At step 16 in section 3 do not cool these solutions or allow to stand for too long; otherwise formation of phosphate salt crystals may occur.

7. As prepared, the plasmid DNA is ready for manipulation and has not come into contact with enzyme inhibitors such as ethidium bromide or others such as those found in DNA eluted from agarose gels.

8. The time required for this procedure is approximately 6–12 h from time of application of sample to the column. This

is substantially faster than the ultracentrifugation method, since even the use of a vertical rotor requires approximately 16 h, followed by many more hours of extractions and dialysis. If a number of different plasmids are being prepared simultaneously, however, it may still be faster to use the ultracentrifuge, since many tubes (i.e., plasmids) can be run together, whereas each column must be run separately, unless more than one column can be arranged for simultaneous running. Since expensive reagents are not necessary in the column technique (e.g., CsCl), and the gel may be reused, this technique is also economical.

References

1. Bywater, M., Bywater, R., and Hellman, L. (1983) A novel chromatographic procedure for purification of bacterial plasmids. *Anal. Biochem.* **132**, 219–224.
2. Thompson, J. A., Blakesley, R. W., Doran, K., Hough, C. J., and Wells, R. D. (1983) Purification of nucleic acids by RPC-5 analog chromatography: Peristaltic and gravity-flow applications. *Meth. Enzymol.* **100**, 368–388.
3. Birnboim, H. C. and Doly, J. (1979) A rapid alkaline extraction procedure for screening recombinant plasmid DNA. *Nucl. Acids Res.* **7**, 1513–1523.

Chapter 8

Electrophoresis of DNA in Nondenaturing Polyacrylamide Gels

Robert McGookin

1. Introduction

For the majority of purposes, such as restriction mapping and sizing of cloned fragments, electrophoresis of DNA in horizontal agarose gels is perfectly adequate. As the size of the fragments of interest decreases to below 1000 base pairs (bp), however, the resolution deteriorates significantly, and an alternative is required. Polyacrylamide gels from 5% acrylamide upward provide a convenient system for analysis of fragments down to 10 bp. This type of gel is also very useful for checking for efficient blunt-end ligation of linkers if the linkers are labeled with ^{32}P.

The gel system described here is nondenaturing. A denaturing polyacrylamide gel electrophoresis system is described by Maxam and Gilbert (1) and in detail in Chapter

2. Solutions

1. 20x Gel buffer: 1M Tris, 1M boric acid, 200 mM ethylene-
    diamine tetracetic acid (EDTA), pH 8.3. Once the solids
    have dissolved, adjust the pH to 8.3 with 5M NaOH, and
    make up to volume. Autoclave (15 psi for 15 min) and
    store at room temperature. *Note* that this is   *not* the same
    as 10x TBE as described by Peacock and Dingman (2).
2. 30% Acrylamide stock: a solution of 30% (w/v) acryl-
    amide, 1% (w/v)N,N'-methylene bisacryalamide is pre-
    pared, filtered, and stored in the dark at 4°C. Use elec-
    trophoresis-grade or other good-quality acrylamide and
    bisacrylamide.
3. 10x Sample buffer: 0.5M Tris, 0.5M boric acid, 100 mM
    EDTA, pH 8.3, 50% (v/v) glycerol, 0.25% (w/v) bromo-
    phenol blue, 0.25% (w/v) xylene cyanol. Prepare from
    stock gel buffer, filter, and store at 4°C.
4. 10% (w/v) ammonium persulfate (AMPS): usually pre-
    pared fresh, but will keep at 4°C for at least a week.
5. Stain: 10 mg/mL ethidium bromide. Care must be exer-
    cised since ethidium bromide is a potent mutagen. The
    solution should be stored in the dark at room temperature
    and is diluted to 1 µg/mL before use.

3. Method

Since acrylamide is a neurotoxin, gloves must be worn
when handling this material. Also, grease marks from fingers
should be avoided on the clean gel plates, so it is recommended
that gloves be worn throughout the following procedure.

1. Clean the gel plates thoroughly and degrease by wiping
    with methanol or ethanol. Assemble the gel casting appa-
    ratus as per the manufacturer's instructions. The volumes
    given below are for a gel 18 x 14 x 0.15 cm (50 mL).

Table 1
Gel Mix for 6, 8, and 10% Polyacrylamide Gels[a]

Acrylamide conc.	6%	8%	10%
20x Gel buffer	2.5 mL	2.5 mL	2.5 mL
Acrylamide stock	10.0 mL	13.3 mL	16.7 mL
H_2O (distilled)	37.1 mL	33.8 mL	30.5 mL
10% AMPS	350 μL	350 μL	350 μL
TEMED[b]	50 μL	50 μL	50 μL

[a] Mix in the order shown and pour immediately.
[b] TEMED = *N,N,N´,N´*-tetramethylethylenediamine.

2. Prepare the gel mix as described in Table 1. Pour the gel (no degassing is required) and leave for about 60 min to polymerize. This can be clearly seen by the formation of "halos" around the edges of the gel and particularly between the teeth of the comb where the gel mix is exposed to the atmosphere. Oxygen in the atmosphere inhibits polymerization at these points.

3. Prepare the required volume of 1x gel buffer for the electrode tanks and prepare the samples while the gel is setting. The samples should be in as small a volume as possible and can contain about 1 μg/band without appearing overloaded. Better resolution is obtained with about 0.1 μg/band. Add one ninth of the sample's volume of 10x sample buffer. Suitable markers should be included, such as Alu 1-digested pBR322 or Hae 3-digested phage φX174 (*see* Fig. 1).

4. Once the gel has polymerized, remove the well-forming comb and clean out any unset acrylamide with a drawn-out Pasteur pipet attached to an aspirator. "Tails" of polyacrylamide should be removed if possible. Assemble the apparatus ready for electrophoresis and rinse the wells with electrode buffer.

5. Pre-run the gel at 200 V for a minimum of 30 min. Rinse the wells again with electrode buffer before loading the sam-

1 2 3 4 5 6 7 8

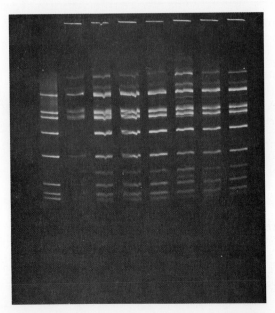

Fig. 1. Polyacrylamide gel of Alu 1-digested pBR322. A 6% polyacrylamide gel of Alu 1-digests of pBR322 and a variety of minipreps
of pBR322 was run, stained with 1 µg/mL of ethidium and photographed by incident UV light. Track 1: Alu 1-digested pBR322
markers. The highest molecular weight band is 910 bp, the lowest,
100 bp. Tracks 2–8: Alu 1-digested minipreps of amp[s] derivatives of
pBR322. Many contain partial digests only.

ples. The samples are loaded with a syringe or micropipet.
Continue electrophoresis at 200 V for about 2.5 h (until the
bromophenol blue is two-thirds through the gel). The gel
will get quite warm after a quick run at this voltage. Lower
voltages can be used and the run time extended accordingly.
6. Disassemble the apparatus and place the gel in 1 µg/mL of
ethidium bromide for about 30 min. The gel is now ready
for viewing and photography under UV illumination. An
example of the resolution to be expected is shown in
Fig. 1.

4. Notes

1. This has proved to be a robust technique. The main point to emphasize is the cleaning out of the wells after polymerization and before electrophoresis. If this is omitted, the resolution is severely affected.
2. If the salt content of any of the samples is significantly different, either extra salt should be added to make all samples the same or, better, the samples should be precipitated and dissolved in sterile water before the addition of sample buffer.

References

1. Maxam, A.M. and Gilbert, W. (1977) A new method for sequencing DNA. *Proc. Natl. Acad. Sci. USA* **73,** 560–564.
2. Peacock, A.C. and Dingman, C.W. (1968) Molecular weight estimation and separation of ribonucleic acid by electrophoresis in agarose-acrylamide composite gels. *Biochemistry* **7,** 668–671.

Chapter 9

Use of ^{35}S Nucleotides for DNA Sequencing

Paul D. Van Helden

1. Introduction

DNA sequencing is a technique that has become central to many laboratories engaged in molecular biology. The two basic sequencing methods commonly used are known as the Maxam-Gilbert method of chemical cleavage (1) and the dideoxy chain termination method developed originally by Sanger (2). One of the most critical steps in either of these two methods is the radiolabeling of the DNA such that the sequencing gel "ladder" of reaction products may be obtained after autoradiography of the sequencing gel.

The radioisotope, phosphorous-32, is in common use in recombinant DNA technology, including DNA sequencing. It is possible, however, to replace ^{32}P by ^{35}S-labeled nucleotides in the dideoxy method with results that yield more sequencing information per reaction or per gel (3). dNTP (α-^{35}S) is a nucleotide analog in which a nonbridge oxygen atom is replaced by sulfur on the α phosphate. This substitution allows bacterial polymerase to recognize these analogs as substrates and incorporate them into growing chains of newly synthesized DNA, an essential requirement for this technique. The incorporation of ^{35}S nucleotide analogs is somewhat slower than that of normal nucleotides, but this may be adequately compensated for by adjustment of incubation times.

Autoradiographs of sequencing gels obtained from ^{32}P-labeled DNA are characterized by intense, but slightly broad, diffuse bands, which are the result of the high energy and relatively deep penetration of material by the high-energy β particles of ^{32}P. This results in "scatter" and band diffusion and limits the ability of researchers to read sequences in the upper regions of gels where the close proximity of bands causes overlap. In contrast, ^{35}S with a lower energy emission [0.167 MeV (max) compared to 1.709 MeV (max) of ^{32}P] produces little scatter (low penetration). Sequencing gel bands are therefore sharply defined and more data may be obtained from upper regions of the gel. As a further consequence of the lower energy of ^{35}S, the laboratory worker does not receive the same incident radiation dose, and the same precautions as observed for ^{14}C are sufficient (perspex screens used in ^{32}P work can be dispensed with), which allows for easier laboratory handling of material.

Sulfur-35 is also a more stable isotope, with a half-life of 87.4 d in contrast to the 14.3 d of ^{32}P. The labeled, but unincorporated, dNTP can therefore be stored longer (because of the longer half-life and the lower energy that results in less breakdown of the dNTP), and the ^{35}S reaction products (DNA) may also be stored at –20°C for at least a week without the appearance of significant artifacts, in contrast to ^{32}P-DNA, which needs to be run on a sequencing gel with minimum delay. Storage longer than overnight is not recommended for ^{32}P-DNA, since the decay of ^{32}P breaks the sugar phosphate backbone, and this will give rise to artefacts and diffuse bands in gels.

The advantages of ^{35}S, namely, sharper bands, more information per gel, ease of handling, and stability, far outweigh any disadvantage in the use of this isotope. The only disadvantage of ^{35}S for sequencing gels relates to the low energy of this isotope; longer exposure times than are necessary for ^{32}P are required (approximately 1.5–2-fold longer). This can be compensated for by increasing the substrate DNA concentration in the reaction, by loading more of each reaction on the gel, or by the use of more sensitive film (*see* section 4). The low-energy β particles are also easily absorbed by the gel or other materials (even thin Saran wrap), which means that the gel must be dried and uncovered prior to autoradiography, a step that is not necessary for ^{32}P gels. This latter point is not necessarily a disadvantage, since dried gels are easy to handle, do not crack, and can therefore be re-exposed if required with no danger of the gel breaking up, may be autoradiographed at room temperature, and give sharper bands.

It will be assumed that the reader is familiar with DNA cloning techniques and dideoxy-sequencing methodology and that the DNA to be sequenced is available in single-stranded form, most conveniently in the phage M13. These procedures are described in Vol. 2 of this series. This chapter will not, therefore, detail either cloning or sequencing reactions or gels, but will emphasize only the difference between ^{32}P and ^{35}S dideoxy (M13) sequencing.

2. Materials

1. Suitable primer DNA.
2. Suitable dNTP/ddNTP mixtures (*see* Note 1 in section 4).
3. Klenow fragment (DNA polymerase) from *Escherichia coli*.

The above three may be made up as required from individual suppliers of components or purchased in kit form from commercial suppliers.

4. Cloned DNA fragment in single-stranded form (cloning *see* refs. 4,5).
5. Apparatus for running sequencing gels and drying gels (*see* refs. 1–3 and Chapters 52 and 53 in Vol. 2 of this series).

3. Method

1. Prepare the DNA to be sequenced as required (*see* Chapter 53 in Vol. 2 of this series and refs. *4,5*).
2. Anneal the primer and template and add all components for the sequencing reactions, but replace (α-^{32}P)dNTP by an equal amount of (α-^{35}S)-dNTPα-S of the same specific activity (*see* Notes 1 and 2 in section 4).
3. Allow the synthesis of complementary DNA strand to take place for 20, instead of 15, min (Note 3 in section 4).
4. Add the chase mixture and incubate for a further 15 min.
5. Stop the reaction by adding the formamide/dye mixture (Note 4 in section 4).
6. Run the sequencing gel and then separate the glass plates with the gel attached to one plate (Note 2 in section 4) (Figs. 1 and 2).
7. Fix the gel (attached to glass plate) in a large horizontal tank in 10% acetic acid/10% methanol solution for 20 min at room temperature (Note 5 in section 4).
8. Remove the gel on the glass plate from the tank and place a large sheet of filter paper (e.g., Whatman No. 1) onto the gel surface (Note 6 in section 4). The gel will adhere to the paper and may now be peeled carefully off the glass. Cover the surface of the gel with plastic wrap (e.g., Saran wrap), and dry the gel on a gel dryer.
9. Autoradiograph gel using Kodak XAR-5 or Fuji X-ray film in a sealed cassette at room temperature for the required time, develop film, and read sequence (Note 7 in section 4) (Fig. 1).

4. Notes

1. Because of the lower rate of incorporation of ^{35}S analogs, it may be advisable to decrease dideoxynucleotide analog concentration in the like mixture, i.e., if ^{35}S-dCTP is used, then decrease the concentration of ddCTP to one half that recommended for ^{32}P reactions in the C reaction mix.

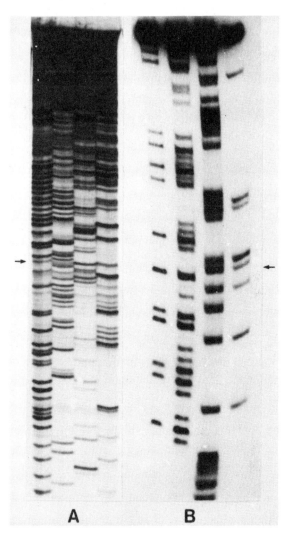

Fig. 1. Portions of autoradiographs from ^{35}S M13 (A) and ^{32}P Maxam-Gilbert (B) sequencing gels. The insert in M13 was approximately 870 bases, whereas the M-G sequencing was done on a 72-base element. The arrows represent 165 (A) and 30 bases (B) read from the lower region of the autoradiographs from separate 15 x 40 cm sequencing gels. The clarity of the ^{35}S bands relative to the ^{32}P bands can be seen. This figure also illustrates that it is possible to read data from upper regions of a ^{35}S gel, which would not be possible on a ^{32}P gel. The use of a sharks-tooth comb is illustrated in (A) and recommended for ^{35}S sequencing.

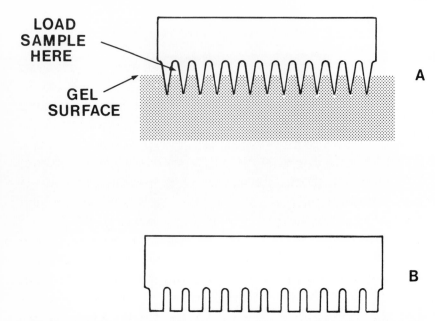

Fig. 2. Combs for DNA sequencing gels. (A) Sharks-tooth comb. (B) Conventional comb. Note that the gel is allowed to set with both combs in position, but the sharks-tooth comb is not removed prior to sample loading. Samples are loaded in the buffer-filled space between the teeth with a very fine needle. A hand-pulled glass capillary is very useful for this purpose. The differences between the results obtained are adequately illustrated in Fig. 1. The spacing between individual lanes obtained from autoradiography is small after using the sharks-tooth comb, which allows accurate alignment for reading sequences. In contrast, the space between the individual wells of a conventional comb sets the lanes well apart and creates uncertainty about alignment, particularly in upper regions of a gel. Use of a sharks-tooth comb is recommended for ^{35}S and dideoxy sequencing.

2. Primer and template concentration may be increased by approximately 35% over those used with ^{32}P to increase the intensity of bands obtained on autoradiography. Alternatively, approximately 40% more sample per lane may be applied to the sequencing gel.
3. For the synthesis of the complementary DNA strand, the time of incubation has been increased by a third to allow

for the lower rate of incorporation of ^{35}S nucleotide analogs.

4. If the samples are to be electrophoresed immediately after synthesis, then the dye mix may be added. If one should wish to store ^{35}S-DNA, however, it is best done by freezing the reaction mixes directly at this stage, followed by addition of the dye mix immediately prior to electrophoresis. ^{35}S-DNA may be stored at –20°C for at least a week without appreciable breakdown of DNA or appearance of artifacts on sequencing gels, in contrast to ^{32}P samples, which should not be stored for longer than overnight.

5. Fixing the gel in acetic acid/methanol is essential, since drying the gel would otherwise allow the urea to crystallize out. The washing may be done in a vertical tank, but in this position the gel may easily slide off the glass backing plate and be lost or break. The washing is therefore done with a greater degree of safety in a horizontal tank, in which the solution may be gently agitated to allow the urea to be washed away from the gel surface. Even if the gel should free itself from the glass plate, it may be recovered with care.

6. When preparing the gel for drying, the filter paper is best layered onto the gel surface by allowing contact first in the middle of the gel and then simultaneously lowering the two ends so that the point of contact is broadened toward the extremities of the gel. This is important, since no air bubbles should be trapped between the paper and the gel. The filter paper should be carefully patted onto the gel, with a damp cloth if necessary, to ensure a good contact. Lift one end of the filter paper and, if contact is good, the gel will easily peel off the glass backing plate.

7. Routine exposure time for autoradiography is 16–24 h, but because of the long half-life of ^{35}S and stability of the dry gel, the procedure may be repeated if longer or shorter exposure times are required. Because of the relatively low energy of ^{35}S, it is essential that the dry gel make direct

contact with the X-ray film; it is therefore necessary to remove any plastic wrap that has been placed on the gel during the drying process.

References

1.   Maxam, A.M. and Gilbert, W. (1980) Sequencing end-labeled DNA with base-specific chemical cleavages. *Meth. Enzymol.* **65**, 499–560.
2.   Sanger, F., Nicklen, S., and Coulson, A.R. (1977) DNA sequencing with chain terminating inhibitors. *Proc. Natl. Acad. Sci. USA* **74**, 5463–5467.
3.   Biggin, M.D., Gibsdon, T.J., and Hong, G.F. (1983) Buffer gradient gels and ^{35}S label as an aid to rapid DNA sequence determination. *Proc. Natl. Acad. Sci. USA* **80**, 3963–3965.
4.   Schreir, P.H. and Cortese, R. (1979) A fast and simple method for sequencing DNA cloned in the single-stranded bacteriophage M13. *J. Mol. Biol.* **129**, 169–172.
5.   Messing, J. and Vieira, J. (1982) A new pair of M13 vectors for selecting either DNA strand of double-digest restriction fragments. *Gene* **19**, 269–276.

Chapter 10

Direct Dideoxy DNA Sequencing

*Alexander Graham, J. Steven,
and D. McKechnie*

1. Introduction

Advances in DNA sequencing techniques have made it possible to routinely sequence long DNA molecules (*1,2*). Recently double-stranded DNA has been directly sequenced with chain terminators (*3*), and this eliminates the tedious task of subcloning the fragments into single-stranded M13 vectors. The procedure requires a plasmid DNA template that is denatured to separate the DNA strands and a primer complementary to one of the strands of the plasmid vector adjacent to the cloning site that has been used. Thus by using primers complementary to each strand, the sequence of the insert can be read from both ends. In the presence of the four deoxynucleoside triphosphates, one of which is radioactively labeled, and the Klenow fragment of *Escherichia coli* DNA polymerase I (or reverse transcriptase, as described in this chapter), primed synthesis occurs across the region of interest. In the presence of competing dideoxynucleoside triphosphates (ddNTPs), specific termination occurs at each of the four different nucleotides, and if this is carried out separately with each of the ddNTPs, then the four sets of reaction products can be analyzed on DNA sequencing gels.

Direct sequencing is rapid since it eliminates the need to subclone in phage M13 vectors, and indeed there are cases in which some DNA sequences cannot stably be replicated in M13 vectors, precluding M13 sequencing. Direct sequencing can also be performed on rapid small-scale plasmid preparations, which makes DNA sequence analysis of multiple samples easy to perform and has also recently been used to sequence DNA fragments cloned into λ-phage vectors (4).

This chapter describes the steps involved in direct DNA sequencing of plasmid DNA using reverse transcriptase; namely, preparation of DNA templates, template-primer hybridization, sequencing reactions, DNA sequencing gels, and analysis of results.

Previous chapters in this series, e.g., Chapters 51, 52, and 53 in Vol. 2, have dealt with various aspects of DNA sequencing methodology.

2. Materials

2.1. Sequencing Primers

The synthetic primers used here are *PstI* site primer / counterclockwise (5'-AACGACGAGCGTG-AC-3') and *PstI* site primer/clockwise (5'-GCTAGAGTAAGTAGTT-3') for sequencing pBR322 plasmids. These were either purchased from N. E. Biolabs or were synthesized on an Applied Biosystems 380A DNA synthesizer and purified by standard methods.

Clockwise and counterclockwise pBR322 primers for the EcoRI/Hind III and *Bam HI* sites are also commercially available from several companies. A primer for sequencing in λ-gt11 at the EcoRI site is available from Clontech Labs Inc.

We have also synthesized sequencing primers for many other plasmids (e.g., pUC plasmids and pKK233 plasmids).

2.2. Sequencing Reactions

2.2.1. Buffers and Stock Solutions

1. 10 x RT Buffer: 500 mM NaCl, 340 mM Tris-HCl, pH 8.3, 60 mM MgCl$_2$, 50 mM dithiothreitol.
2. dN Mix: 0.5 mM dGTP, 0.5 mM dCTP, 0.5 mM dTTP.
3. ddN Solutions: dideoxynucleoside triphosphate stocks are as follows: 0.01 mM ddATP (ddA), 0.125 mM ddGTP (ddG), 0.25 mM ddCTP (ddC), and 0.5 mM ddTTP (ddT).
4. Chase solution: 1.25 mM dATP, 1.25 mM dCTP, 1.25 mM dGTP, 1.25 mM dTTP in 1 x RT buffer.
5. AMV enz: dilute AMV reverse transcriptase in 1 x RT buffer to 5 U/µL. This stock is prepared fresh. We have used AMV reverse transcriptase from several commercial sources (BioRad, BRL, Boehringer, or Life Sciences) with no variation in sequencing performance.
6. Chase/AMV enz: 2 µL of AMVenz + 12 µL of chase solution. We have found that omission of M-MLV enz from this has no effect on the sequence obtained.
7. M-MLV enz: dilute M-MLV reverse transcriptase (BRL) in 1 x RT buffer to a concentration of 180 U/µL. Prepare fresh.
8. Chase/M-ML Venz: 2 UL M-ML Venz + 12 µL chase solution. We have found that omission of M-ML Venz from this has no effect on the sequence obtained.

2.3. Sequencing Gels

1. Formamide dye mix: formamide is deionized by gently stirring 1 L of formamide with 50 g of Amberlite MB-1 (BDH) for 30 min and then filtered to remove resin. Add 0.03 g of xylene cyanol • FF, 0.03 g bromophenol blue, and 0.75 g of Na$_2$EDTA • 2H$_2$O (20 mM). Store at room temperature.
2. Acrylamide stock solution: 38 g acrylamide, 2 g bis (i.e., N,N'-methylene bisacrylamide) made up to 100 mL with

water. 5 g of Amberlite MB-1 is added and stirred gently for 30 min. The solution is filtered to remove resin, then fine particles are removed by passing through a 0.45 μm Millipore filter.

3.   10 x TBE buffer: 108 g of Tris base, 55 g of boric acid, 9.3 g of Na$_2$EDTA · 2H$_2$O, should give a stock solution of pH 8.3.

3. Method

3.1. Preparation of DNA Templates

1.   Supercoiled plasmid DNA is prepared by the traditional cesium chloride gradient purification procedure (ref. 5 and Chapter 25 in Vol. 2 of this series).
2.   For preparation of plasmid DNAs from small (1.5-mL) overnight bacterial cultures (miniscreen DNA), the procedure of Birnboim and Doly (ref. 6 and Chapter 27 in Vol. 2 of this series) is used with the following modification. After treatment with DNAse-free RNAse, the plasmid preparations are passed down two spun columns of Sephadex G50 (7), and the resulting plasmids that are collected in the flow-through are ethanol precipitated, dried, and resuspended in H$_2$O. Enough DNA is obtained for 1–2 sequencing reactions.

3.2. Template Primer Hybridization

Two methods for template-primer hybridization can be used.

3.2.1. Linearization Method

1.   The plasmid DNA with an insert to be sequenced is digested with a restriction endonuclease that will linearize the plasmid, but not excise or cut the DNA insert (3).
2.   0.3 pmol of plasmid is mixed with 15 pmol of oligonucleotide primer in a volume of 8.4 μL of water, and 0.8 μL of 10 x RT buffer is added.

3. The mixture is transferred to a siliconized glass capillary, heat-sealed, and placed in a boiling water bath for 5 min. It is then immediately cooled (frozen) in a dry ice/ethanol bath. This prevents reannealing of the pBR322 strands and promotes primer-template hybridization on thawing.

3.2.2. Denaturation Method

1. Supercoiled plasmid DNA containing the insert to be sequenced is denatured with alkali (8) as follows: 1–2 µg of supercoiled plasmid DNA is dissolved in 20 µL of 0.2M NaOH, 0.2 mM EDTA.
2. After 5 min at room temperature, the solution is neutralized by the addition of 2 µL of 2M ammonium acetate-acetic acid, pH 4.5.
3. DNA is precipitated with 2 vol of ethanol, chilled in dry ice/ethanol for 15 min (or overnight at –20°C), and then collected by centrifugation.
4. The DNA is rinsed in 70% ethanol, dried, and resuspended in 8.4 µL of water containing 15 pmol of oligonucleotide primer and 0.8 µL of 10 x RT buffer. This is incubated at 37°C for 15 min to anneal primer to template.

3.3. Sequencing Reactions

1. Dry 25 µCi (α-^{32}P) dATP (410 Ci/mmol, Amersham) in an Eppendorf tube. Add 5 µL of dNmix and 0.8 µL of 10 x RT buffer and vortex. Add the hybridized template-primer solution; this is called HYB-LABEL mix.
2. Set up four tubes at room temperature with the following using AMV enz or M-MLV enz. The enz is added to the side of the tube and all four tubes centrifuged briefly in order to start the reactions simultaneously. The tubes are transferred to a 42°C water bath for AMV-enz (or 37°C for M-MLV-enz) for 15 min.
3. Add to the side of each tube 3 µL of chase/enz mix containing AMV or M-MLV enz and incubate for an additional 15 min at the same temperatures as before.

Tube	HYB-LABEL, µL	ddN, µL	AMVenz(or M-ML Venz), µL
A	3	1, ddA	1
C	3	1, ddC	1
G	3	1, ddG	1
T	3	1, ddT	1

4. The reactions are terminated by adding 1 µL of 0.25M EDTA, pH 7.0, and are dried for 30 min in a Speedvac concentrator (Savant). There will be a small volume of glycerol present after drying (from the enzyme storage buffer).

5. The dried samples are resuspended in 10 µL of formamide/dye mix and are heated to 65°C for 10 min just prior to applying 2 µL per lane of a sequencing gel.

3.4. DNA Sequencing Gels

1. "Wedge"-shaped slab gels are prepared as follows: Glass plates (34 x 40 cm) are separated by spacer strips made from 0.4-mm thick Mylar plastic, and two small pieces of 0.4-mm thick spacers are inserted at the bottom edge of each side and another piece 0.4-mm thick inserted at one-third of the way from the bottom. The gel plates are sealed along the two sides and the bottom with tape wedge-shaped spacers are now available from BRL.

2. For a 100-mL gel, the following are mixed: 42 g of urea (BDH, Aristar™), 15 mL of acrylamide stock solution, 5 mL of 10 x TBE buffer, and H_2O to 100 mL. This is heated to 37°C to dissolve the urea.

3. 0.6 mL of freshly prepared 10% (w/v) ammonium persulfate is added, and to 5 mL of this mixture add 10 µL of TEMED (N,N,N',N'-tetramethylene diamine) and pour into the assembly to act as a "plug."

4. To the remainder of the mixture add 100 µL of TEMED and pour between the glass plates at an angle of approximately 45°; the plates are lowered gradually until horizontal, and

the flat side of a "shark-stooth" comb (BRL) is inserted to keep the gel surface straight.

5. After polymerization (at least 1 h, or can be left overnight), the "shark-stooth" comb with the teeth 6 mm (or 3 mm) apart is used to form the wells by placing it on the gel surface such that slight indentations are made without piercing the gel surface. This results in 48 lanes for the 6-mm comb or 96 wells for the 3-mm comb. The use of "shark-stooth" combs (*see* Chapter 9) brings the lanes very close to one another, allowing more accurate reading of sequences. The variable gel thickness causes the samples to gradually slow down as they reach the thicker bottom area, and this results in a more even spacing of adjacent DNA fragments throughout the gel.

6. The gel is equilibrated by pre-electrophoresis at operating voltage for 1 h. Flush each sample lane with fresh buffer prior to applying samples. Electrophoresis is at about 1.5 kV, and until the bromophenol blue is about 1–2 in from the bottom (this corresponds to an oligonucleotide of about 20 bp) and so insert sequences should not be missed. The length of the run is determined by what part of the sequence one wants to read. If as much sequence as possible is to be determined, then part of each sample can be electrophoresed as above, and the remainder electrophoresed for twice as long. Glass plates up to 1 m long are used routinely, and by using longer gels more sequence can be determined per run.

7. Problems with sequencing gels are usually not encountered. Smearing of bands can result, however, if the gel is run too hot, if too large a sample volume is applied, or if the acrylamide solution is not deionized. It is important to flush out the wells immediately prior to applying the DNA samples since this will prevent streaking caused by urea crystals present in the slots.

8. At the end of the electrophoresis, the plates are carefully separated, and the gel is prepared for autoradiography. It

is necessary to prevent diffusion of the DNA bands either by performing autoradiography at –70°C or by "fixing."

   a. Low-temperature autoradiograph (^{32}P): The wet gel is transferred to a sheet of used X-ray film, then wrapped in Saran wrap, and auto-radiographed at –70°C. Intensifying screens may be used, but resolution will be poorer.

   b. Fixing and drying (^{32}P and ^{35}S): Place the gel attached to a glass plate in 10% acetic acid for 20 min, drain the gel, and blot off excess liquid. The wet gel can be wrapped in Saran wrap and autoradiographed at room temperature.

9. Alternatively the gel can be fixed for 20 min in 10% acetic acid, 10% methanol (to remove excess urea, which would otherwise crystallize out). Remove the plate with gel and place a slightly larger sheet of filter paper (Whatman No. 1) onto the gel, invert, and transfer the gel onto the filter paper. Dry the gel on a gel drier and autoradiograph at room temperature (do not cover in Saran wrap). The sharpness and, therefore, the resolution of the bands increases when the gels are dried prior to autoradiography.

10. We use Kodak X-omat XAR5 film routinely, although we have found Hyperfilm-Bmax film (Amersham) almost as efficient, but with a much clearer background.

4. Analysis of Results

An example of a sequencing gel is shown in Fig. 1. The sequence of nucleotides is read from a gel by starting with the smallest fragment at the bottom of the autoradiograph and recording on which track this band appears. Continue reading up the gel from track to track. This results in sequence in a 5'-to-3' direction from the primer. It is important when reading the sequence to note spaces as well as bands, ensuring that faint bands are discarded if the spacing is wrong or are read as real if the spacing is correct. For each film read the sequence at least twice, compare, and then form a concensus. Any uncertainties or mistakes in the sequence will complicate the data analysis.

A   B

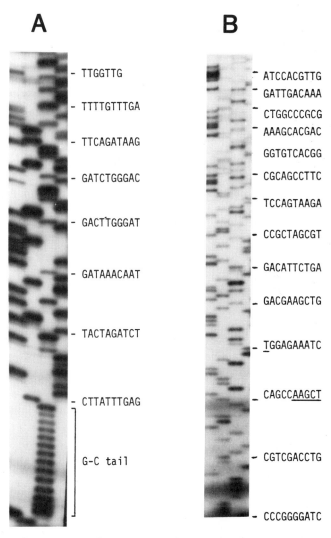

Column A:
- TTGGTTG
- TTTTGTTTGA
- TTCAGATAAG
- GATCTGGGAC
- GACTTGGGAT
- GATAAACAAT
- TACTAGATCT
- CTTATTTGAG

G-C tail

Column B:
- ATCCACGTTG
- GATTGACAAA
- CTGGCCCGCG
- AAAGCACGAC
- GGTGTCACGG
- CGCAGCCTTC
- TCCAGTAAGA
- CCGCTAGCGT
- GACATTCTGA
- GACGAAGCTG
- TGGAGAAATC
- CAGCCAAGCT
- CGTCGACCTG
- CCCGGGGATC

Fig. 1. A portion of the autoradiograph of a sequencing gel is shown, together with the sequence read from the autoradiograph. (A) Sequence analysis of cDNA insert cloned in pBR322 at the *PstI* site using homopolymer tails. The sequence can clearly be read after the G-C tail region into the insert. (B) Sequence analysis of a DNA cloned into the *Hind III* site of pUC9. The sequence at the bottom (5′ end) corresponds to pUC9 sequence. The *Hind III* recognition sequence is underlined, and the remaining sequence corresponds to the DNA insert. This DNA preparation was from a small-scale plasmid preparation.

Also search for restriction enzyme sites both at the end of the insert and at known sites in the sequence. By comparing restriction enzyme patterns of different fragments, those with similar patterns can be overlapped, and those that match can be joined. Computers are routinely used to read and analyze sequencing data. Analyses can include assembly of individual gel readings into a total sequence and then analysis of the assembled sequence for coding regions, restriction enzyme site, promoters, and so on. Data input can be done by various methods, e.g., vocal, direct entry, or semiautomate (_see_ Chapter 11).

Several computer data bases are now available, and these consist of listing of all published sequences. These can be accessed on-line, or a disk can be obtained to load into your computer. These data bases are generally updated every 6 mo.

5. Notes

1. We routinely clone ds cDNA into pBR322 using homopolymer tailing. Attempts to sequence these clones using Klenow fragment and specific primers adjacent to the _PstI_ site, into which the ds cDNA had been cloned, resulted in regions of intense radioactivity, which indicates failure of the enzyme to readily synthesize beyond that region. These problems were not alleviated by the use of more enzyme (2 U instead of 0.2 U), by enzymes from different commercial sources, or by the use of a higher reaction temperature (45°C).

A method for sequencing through such regions of ds cDNA subcloned into phage M13 vectors, has been described (9). This method uses AMV reverse transcriptase, instead of Klenow fragment, and the results show that AMV RT can read through regions that the Klenow fragment finds difficult. We have used reverse transcriptase to directly sequence such ds cDNAs in pBR322.

Using identical amounts of DNA and primer, the same sequencing reactions performed with either AMV or M-

MLV reverse transcriptases show very few "pile-up" points, and the sequence can easily be read past the homopolymer tail region (Fig. 1A). Under these conditions we have been able to sequence through dG • dC homopolymer tails containing as many as 45 dG • dC bases with no difficulty.

It is possible to use Klenow polymerase in place of reverse transcriptase for direct dideoxy DNA sequencing (*see* ref. 3 for protocol). We would suggest, however, that for difficult regions of sequence (e.g., homopolymer tails, long tracts of dG • dC) it is essential to use reverse transcriptase.

2. Both methods of template-primer hybridization described here give identical results using this sequencing protocol, although the alkaline denaturation method is easier to perform in our hands.

3. The sequence reactions can be stored at –20°C overnight (before addition of formamide dye). It is better if the samples can be applied to the sequencing gel immediately after the sequencing reactions since this reduces background and artifactual bands.

4. Buffer gradient gels (*10*) have several advantages over conventional gels. The mobility of a DNA fragment on a normal gel has a logarithmic relationship to molecular weight; thus bands at the bottom are widely spaced, whereas those at the top are so tightly packed as to be unreadable. In the case of a buffer gradient gel, the banding pattern is more evenly spaced, and so the sequence can be read further into this region. Wedge-shaped gels like buffer gradient gels result in more sequence being read from the gel. We find, however, that the wedge-shaped gels are easier to prepare and are as good as gradient gels.

5. Deoxyadenosine 5'-(α-^{35}S thio) triphosphate has been used as the radioactive label in place of (α-^{32}P) dATP (*10*). This analog of dATP is a substrate for DNA polymerase (and reverse transcriptase), and the short path length of the β particles omitted by ^{35}S results in very sharp band defini-

tion, which increases the length of DNA sequence data that can be read from a gel (*see* Chapter 9).

6. Altering the dideoxy concentrations can help to alleviate problems caused by early termination in the sequence. For sequencing short fragments of DNA (up to 50 nucleotides), higher ddNTP ratios can be used (up to three times greater than normal). For long runs (up to 400 nucleotides), lower ddNTP ratios can be used (1–2 times lower than normal), which will result in the necessary lower frequency of termination and thus generate longer fragments. The optimal ddNTP concentrations have to be determined by a process of trial and error.

7. Recently a method for sequencing large DNA fragments cloned in M13 has been reported (*11*). We have used this method for sequencing pBR322 cDNA clones using pBR322/*PstI* site primers and reverse transcriptase as described in this chapter. From the newly obtained sequence it is possible to synthesize primers homologous to the 3' ends of these new sequences and then to use these primers to sequence further into the clone. This procedure can be repeated until the insert is sequenced. This procedure has become relatively easy because of the rapid small-scale automatic synthesis of oligonucleotides, and is made simpler by the fact that these primers do not have to be purified after synthesis (*11*).

8. This method should be applicable to direct sequencing of double-stranded DNA cloned into any double-stranded vector (e. g., plasmids, λ-phage vectors) provided suitable primers are available.

References

1. Maxam, A. M. and Gilbert, W. (1977) A new method for sequencing DNA. *Proc. Natl. Acad. Sci. USA* **74**, 560–564.
2. Sanger, F., Nicklen, S., and Coulson, A. R. (1977) DNA sequencing with chain terminating inhibitors. *Proc. Natl. Acad. Sci. USA* **74**, 5463–5467.

3. Wallace, R. B., Johnson, M. J., Suggs, S. V., Miyoshi, K., Bhatt, R., and Itakura, J. (1981) A set of synthetic oligodeoxyribonucleotide primers for DNA sequencing in the plasmid vector pBR322. *Gene* **16**, 21–26.
4. Zagursky, R. J., Baumeister, K., Lomax, N., and Berman, M. L. (1985) Rapid and easy sequencing of large linear double-stranded DNA and supercoiled plasmid DNA. *Gene Anal. Tech.* **2**, 89–94.
5. Clewell, D. B. and Helsinki, D. R. (1969) Supercoiled circular DNA-protein complex in *E. coli*: Purification and induced conversion to an open circular DNA form. *Proc. Natl. Acad. Sci. USA* **62**, 1159–1166.
6. Birnboim, C. and Doly, J. (1979) Rapid alkaline extraction procedure for screening recombinant plasmid DNA. *Nucleic Acids Res.* **7**, 1513–1523.
7. Maniatis, T., Fritsch, E. F., and Sambrook, J. (1982) *Molecular Cloning. A Laboratory Manual* Cold Spring Harbor, New York.
8. Chen, E. Y. and Seeburg, P. H. (1985) Supercoil sequencing: A fast and simple method for sequencing plasmid DNA. *DNA* **4**, 165–170.
9. Karanthanasis, S. (1982) M13 DNA sequencing using reverse transcriptase. (*Bethesda Research Labs*) **4**, 6–7.
10. Biggin, M. D., Gibson, T. J., and Hong, G. F. (1983) Buffer gradient gels and ^{35}S label as an aid to rapid DNA sequence determination. *Proc. Natl. Acad. Sci. USA* **80**, 3963–3965.
11. Sanchez-Pescador, R. and Urdea, M. S. (1984) Use of unpurified synthetic deoxynucleotide primers for rapid dideoxynucleotide chain termination sequencing. *DNA* **3**, 339–343.

Chapter 11

Computer Applications
to Molecular Biology

DNA Sequences

Robert Harr and Petter Gustafsson

1. Introduction

Computers have become a necessary tool in all laboratories involved in DNA sequence work. There has been a rapid increase in the amount of DNA sequenced, and the most recent issue of the *European Molecular Biology Laboratory (EMBL) Nucleotide Sequence Data Library* (release 10, 1987) contains almost 10 million base pairs in almost 9000 entries compared with the first release from 1982 containing half a million base pairs in 500 entries. The requirement for computers is obvious not only from the total amount of DNA sequenced, but also when taking into account the number of bases needed in order to organize a gene. The size of a standard gene with no intervening sequences is 500–1000 base pairs, whereas interrupted genes may be 10 times longer.

The purpose of this article is to give a computer manual for the analysis of DNA sequences. The GENEUS system (*1*) has been chosen as the model system, and a detailed manual will be given for the handling of this particular software system. The

GENEUS software system can analyze both DNA and protein sequences, but we will restrict this chapter to DNA sequence routines. Despite the fact that the manual will describe one particular computer system, the analysis procedures and the result presentations described are general. Differences in command structure and language between individual computer systems exist, however. The manual will cover most areas needed in the handling of DNA sequences, including how to enter a DNA sequence into the computer, how to merge DNA sequences in a DNA sequence project, how to find sequences in the database, how to analyze entered DNA sequences for information needed to continue and refine subsequent cloning experiments, and, most important, how to search for interesting biological information.

The DNA sequences chosen as experimental material in this manual are *psbA* genes coding for the D1 polypeptide in photosystem II from the blue-green bacteria *Anacystis nidulans* (a family of three genes), *Anabaena* (a family of two genes), and the green algae *Chlamydomonas reinhardii* (2–4) and the *psbD* gene coding for the D2 polypeptide in photosystem II from the green algae *C. reinhardii* (5) and part of (position 501–2000) the genomic sequence for the mouse kappa immunoglobulin J and C regions (6). For more information on computer applications to molecular biology, *see* "Further Reading" in this chapter.

2. Computer Equipment and Technical Description

The presented manual will be restricted to the GENEUS software system, which has been described elsewhere (1). The GENEUS system is constructed for the Digital Equipment Corporation's (DEC) VAX computers equipped with the VMS operation system. GENEUS is written in VAX-11 PASCAL and FORTRAN and is dependent upon the command line interpreter (CLI). The database integrated with the GENEUS software system is the EMBL nucleotide sequence data library described above to which an information retrieval system has been added (*1*). The GENEUS system uses standard VT100/ 200-compatible terminals for nongraphic applications. The graphic capabilities use the Graphic Kernel Standard (GKS), which includes drivers for most graphical devices, including DEC VT 240/241-, Tektronix 4010/4014-, and Tektronix 41XX-compatible terminals. A specially designed digitizer board system, combined with an IBM PC-compatible computer, for the direct reading of DNA sequence gels into the computer, can be added to the system. Anyone possessing a DEC VAX/VMS computer with a GKS library can relink the graphical programs within GENEUS, thus adapting GENEUS to specific devices at the installation. For additional information about other software systems used for the study of both DNA and proteins, *see* the Further Reading section in this chapter.

3. Command Language

A command is used to instruct the computer what the user wants the computer to perform. In the GENEUS system, the computer *first* calls upon the analysis and, *second*, the DNA sequence. Several types of command format exist. Some computer systems have chosen a menu-driven communication system, whereas others use questions that are answered sequentially. The GENEUS system uses a qualifier-based command format, which enables the user to communicate a whole set of commands separated by qualifiers. The qualifier-based command format is combined with questions when entering certain analysis programs. The qualifier used is a slash: (./.). If a DNA sequence is to be listed, e.g., the DNA sequence of plasmid pBR322, type the following command:

LIST PBR322..ARTP <cr> (Note: Underlined text in capital letters shows instructions typed by the user on the terminal or computer keyboard. <cr> indicates a carriage return.)

where LIST is the instruction to the computer to list a DNA sequence, and PBR322 and ARTP are the names of the sequence and master, respectively, under which the DNA sequence of plasmid pBR322 is stored in the computer. No additional specifications are given, thereby instructing the computer to list the whole DNA sequence in a standard single-stranded format. Further instructions can also be added directly to the computer by typing:

<u>LIST PBR322..ARTP/POS=1:1000/STR=2</u> <cr>

LIST is again the analysis instruction and PBR322 the name of the DNA sequence. Now we have added additional instructions for the computer to perform. The slash (/) is the qualifier by which you separate additional instructions. POS indicates that only a part of the sequence, base 1–1000, should be listed. STR=2 indicates the result should be printed as a double-stranded DNA sequence.

4. DNA Sequence and Knowledge Databases

GENEUS stores sequences in databases. The GENEUS system uses three levels of management to group the DNA sequence. First are individual databases for each user and one shared database for the EMBL nucleotide sequence data library (the EMBL nucleotide sequence data library is write-protected). The second level is the master name of a sequence. The third level is the DNA sequence name. The full name of a sequence must consist of three parts. The database in which the

Database	Master name	Sequence name
EMBL	ARTP	PBR322
DEMODB	PSBA	ANACYSTIS1
		ANACYSTIS2
		ANACYSTIS3
		ANABAENA620
		ANABAENA625
		CHLAMYDOMONAS

sequence is placed, the name of the sequence itself, and the name of the master, i.e., the group name of the sequence.

The syntax for the full name of a sequence is:

DATABASE:SEQUENCE NAME..MASTER NAME

exemplified by:

EMBL:PBR322..ART or
DEMODB:PSBATAQI..DNASEQPROJ

If the DNA sequence name is unique, however, the master name can be omitted from the name. In the EMBL nucleotide data library, the sequence name is always unique. The master name can, however, always be used to simplify the search and management of the databases.

In section 6 in this chapter, describing the editor, in which a DNA sequence is entered into the database by the help of the editor, DEMODB is the database, PSBATAQI is the DNA sequence name, and DNASEQPROJ is the master name (Fig. 1).

Several commands simplify the management of the database. Type: DBMAST<cr>* to list the master names in a database or DBMAST/SQ=PBR322<cr> to find the master name (=ARTP) of the DNA sequence name PBR322. Type DBSEQS <cr> to list all sequence names in a database or DBSEQS/MA=ARTP<cr> to list the DNA sequence names found under the master name ARTP.

Besides the DNA sequence, other important information for every sequence is supplied with the EMBL library. This information can be called a knowledge base. In the knowledge base you find a description of the sequence (DE), keywords (KW), organism (OS and OC), title and authors of the reference (RT and RA), and feature tabularized data (FT) containing biological information stored in the sequence. The GENEUS system contains an information retrieval system for the EMBL library (1).

Type INFOLIST PBR322<cr> to get a presentation of the whole information sheet within the EMBL library accompanying the pBR322 sequence.

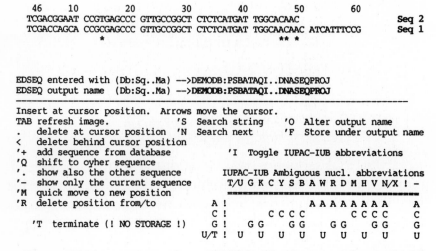

Fig. 1. DNA sequence editor, EDSEQ. The figure is identical with the appearance of the editor on the terminal screen. Characters used to execute editor commands are shown. The DNA sequence used is DEMODB:PSBATAQI..DNASEQPROJ. The sequence has been entered twice, Seq1 and Seq2. The asterisk indicates differences between the two entered sequences (*see* section 6 for further details).

Type DBQUEST<cr> to activate the information retrieval system.

The information retrieval system is now activated and is presented on the terminal screen as is shown in Fig. 2. Type a number from 1 to 7<cr> to activate/deactivate searches in different information areas. In the example in Fig. 2, search of type number 1 is chosen and the search string REMM02 is typed. After the search has been completed, type L<cr> to list the result. The result from the information search is presented on the terminal screen (*see* Fig. 2, bottom two lines).

5. Starting the Computer and the DNA Sequence Programs

Switch on the terminal or computer. Log-in or initiate the programs.

```
      See documentation from EMBL for more information.

      Search and list lines (-> = active)
      ==========================================
      1)  ID->- Identification.           L)  List
      2)  DE  - Description.              F)  File list
      3)  KW  - Keywords.                E)  Exit
      4)  RA  - Reference authors.
      5)  RT  - Reference title.
      6)  RL  - Reference location.
      7)  RX  - Reference all.

MESSAGE >
FOUND   >   1 sequence(s)              PROGRESS>    7630

MENU  > L
STRING> REMM02
      ------------------------------------------------------
>>> EMBL:REMM02..VIRR                                  <<<
ID   REMM02     standard; RNA; 1371 BP.
```

      Press return to get back to menu.

Fig. 2. Information retrieval system for the EMBL nucleotide sequence data library (*1*). A search of type 1, ID–Identification and the search string REMM02 was chosen. The result of the search is shown at the bottom of the figure. See Chapter 4 for further details.

Type: GOGENEUS<cr>* .

Type SHELP<cr> to get help information on the terminal screen. The next step in the GENEUS system is to set the right database. If the EMBL nucleotide sequence data library is to be used, type the following command:

Type: DATABASE EMBL<cr>

If a piece of DNA has been sequenced and is to be entered into the computer, or if some other sequence is to be entered, choose your own database. As an example in this manual the name DEMODB has been chosen for your own database.

Type: DATABASE DEMODB<cr>.

6. Editor

An editor is used to enter new DNA sequences or edit sequences already entered and stored in the database. As an example in this manual, a piece of DNA containing one of the three *psbA* genes from the blue-green algae *A. nidulans* (2) has been shot-gun cloned using the three restriction enzymes *Taq*I, *Sau*3a, and <u>Msp</u>I with the recognition sequences TCGA, GATC, and CCGG, respectively. Four subclones are DNA sequenced. The sequences are entered with the help of the EDSEQ editor into the GENEUS system for a subsequent use in the DNA sequence project system.

Type: <u>DATABASE DEMODB</u><cr>.

Type: <u>EDSEQ PSBATAQI..DNASEQPROJ</u><cr>, where EDSEQ is the command to start the editor, PSBATAQI is the name of one of the sequences from the cloning described above, and DNASEQPROJ is the master name to which the sequence name belongs (*see* section 4, in this chapter).

The editor is now activated and is presented on the terminal screen, as shown in Fig. 1. Enter the DNA sequence. Use the different functions that are presented on the screen if applicable. One safe way to ensure that a sequence is entered without mistakes is to enter the sequence twice. In the GENEUS system this can simply be done by typing '*Q* and' - , which allows the sequence to be entered twice; both sequences will appear on the terminal screen. The computer warns with a star and a beep whenever the two sequences do not match. In Fig. 1, this is clearly shown at position 14, where there is a mismatch between the two entered sequences and at positions 56–59, where there is a one-base insertion/deletion.

Type: '<u>F</u> to store the sequence
Type: '<u>T</u> to leave the editor

Enter the three other sequences in the same way.

7. DNA Sequence Project System

If DNA sequences read from autoradiogram are to be spliced together, the two programs SP1 and MMBAT, which are included in the GENEUS system, are used.

The DNA sequences from the shot-gun cloned *psbA* gene described in section 6 in this chapter are entered in the same way as any other sequences into the computer. This can be done with the sequence editor (EDSEQ, section 6, in this chapter), but also with a text editor or any other device that puts the sequence into a normal sequential file on the disk. In GENEUS a specially designed digitizer board system can also be used.

The first program (SP1) will compare every sequence (gel reading) with all the other gel readings and look for overlaps of *x* bases between the sequences (*x* is normally set to 15). The final result of this cross-matching is given as an "arrow-list."

```
type SP1<cr>

Database : <cr>

Mastername : DNASEQPROJ<cr>

Give merge_limit for gel splice together (15) :<cr>


     Merge_limit = 15

     WAIT , I am computing.

A Semigraphic list in file XCONT.GRA, available now.

type XCONT.GRA<cr> and the following result will be presented:


     SeqName    !-.-.-.-.-!-.-.-.-.100.-.-.-.-!-.-.-.-.200.-.-.

     PSBAM2     <--

     PSBASAU1   --->

     PSBATAQ1     -->

     PSBMSP1     ->
```

Scheme 1

The arrows and positions are an overview on how the sequences match each other. The list does not, however, give a detailed description of how the bases in one sequence match the other sequence. By running step two, MMBAT, the result will be a detailed base-to-base matching list, as presented below.

Type <u>MMBAT MM</u><cr> to run step two as a batch job
Job MM (queue SYS$BATCH, entry 860) started on
SYS$BATCH

This statement means that the job of constructing a detailed base-to-base list will be done in the background of the computer. The job is completed when you get the following message.

Job MM (queue SYS$BATCH, entry 860) completed

Type XCONT001.001<cr> and the result will be that given in Scheme 2.

8. Listing a DNA sequence

The LIST command, which is one of the simplest commands in DNA sequence computer systems, produces a list of a DNA sequence stored in the database. Usually different types of sequence presentations can be produced, e.g., single- or double-stranded.

Type <u>LIST ANACYSTIS..PSBA</u><cr> and the result will be that given in Scheme 3.

9. Producing Reversed Complement

In most cases the computer programs handle only single-stranded DNA sequences oriented in the 5' to 3' direction in their analysis. It often happens, however, that a sequence anal-

```
PSBAM2      R   129   CCGCATCTACGTGGGTTGGTTCGGCGTGCTGATGATCCCCACTCTGCTGA

PSBASAU1        1                         GATCCCCACTCTGCTGA

"CONSENSUS"     1     CCGCATCTACGTGGGTTGGTTCGGCGTGCTGATGATCCCCACTCTGCTGA


PSBAM2      R   79    CCGCCACCATCTGCTTCATCGTTGCGTTCATTGCAGCCCCTCCCGTCGAC

PSBASAU1        18    CCGCCACCATCTGCTTCATCGTTGCGTTCATTGCAGCCCCTCCCGTCGAC

"CONSENSUS"     51    CCGCCACCATCTGCTTCATCGTTGCGTTCATTGCAGCCCCTCCCGTCGAC


PSBAM2      R   29    ATCGACGGCATCCGTGAGCCCGTTGCCGG

PSBASAU1        68    ATCGACGGCATCCGTGAGCCCGTTGCCGGCTCTCTCATGT –GGCAACAA

PSBATAQ1        1      TCGACGGCATCCGTGAGCCCGTTGCCGGCTCTCTCATGTAtGGCAACAA

PSBMSP1         1                         CCGGCTCTCTCATGTAtGGCAACAA

"CONSENSUS"     101   ATCGACGGCATCCGTGAGCCCGTTGCCGGCTCTCTCATGTA-GGCAACAA


PSBASAU1        117   CATCATTT CGGCG TGTTGTTCtTCagaAGCAtCGgCtattttaTcgAT

PSBATAQ1        50    CATCATTTCCGGCGCTGTTGTTCcT tccAGCAaCGcCatcggccTgcAT

PSBMSP1         26    CATCATTTCCGG

"CONSENSUS"     151   CATCATTTCCGGCGCTGTTGTTC-TC----AGCA–CG-C-------T--AT


PSBASAU1        165   TT

PSBATAQ1        99    TTCTATCCGATTTGGGAAGCCGCTAGCCTCGA

"CONSENSUS"     201   TTCTATCCGATTTGGGAAGCCGCTAGCCTCGA
```

Scheme 2

ysis must be performed on both strands. This is true for, e.g., translation, homology searches, and pattern searches, but not for searches of, e.g., restriction endonuclease sites and hairpin-loops in which the same type of information is stored on both strands.

Type: <u>REVCOMP ANACYSTIS ANACYSTISRC</u> <cr> and the computer will make the reversed complement of the DNA sequence stored under the sequence name ANACYSTIS and

```
ID  List of sequence --> DEMODB:ANACYSTIS..PSBA

XX

XX          10        20        30        40        50        60
SQ  ATGACCAGCA TTCTTCGCGA GCAACGCCGC GATAACGTTT GGGATCGGTT TTGTGAGTGG

.

.

.

XX         1030      1040      1050      1060      1070      1080
SQ  CTCGACTTGG CAGCAGGCGA AGCGACCCCG GTCGCTTTGA CTGCGCCTTC AATTCACGGT
```

type <u>LIST ANACYSTIS..PSBA/POS=1-20/STR=2</u><cr> and the following
double-stranded sequence from position 1 to 20 will be presented:

```
ID  List of sequence --> DEMODB:ANACYSTIS..PSBA

XX

XX          10        20
SQ  ATGACCAGCA TTCTTCGCGA
SQ  TACTGGTCGT AAGAAGCGCT
```

Scheme 3

store the reversed complement under the sequence name
ANACYSTISRC. The master names will remain unaffected,
and ANACYSTISRC will be placed in the same master as
ANACYSTIS.

10. Searching for Restriction Endonuclease Sites

Restriction endonucleases are vital tools in all cloning
experiments. The DNA sequence contains the absolute information for the position, type, and number of restriction endonuclease sites within a given stretch of DNA and can thus be

used profitably in the planning of future cloning experiments. A search for restriction endonuclease sites in a DNA sequence is also effectively combined with a restriction endonuclease map determined by conventional means to verify if the cloned piece of DNA has been corrrectly sequenced.

Most restriction endonucleases have recognition sequences of either four, five, or six bases. It is often advantageous to search for a group of restriction enzymes with a specific number of bases in their recognition site, e.g., four-cutters, simultaneously.

Type <u>RSEARCH ANACYSTIS..PSBA/POS=200-400</u><cr>, specifying that you want to perform a search for restriction endonuclease sites in the sequence ANACYSTIS from position 200–400. Having entered the program you will further specify the analysis by answering questions asked by the program (Fig. 3). In the example given, you choose to search for six-cutters in the first menu. In the next questions you may select specific restriction endonucleases to search for, you may list the chosen enzymes and their recognition sequence, and you specify if your sequence is linear or circular. You may then choose different types of result presentation. You choose alternatives 1 and 6. You may also choose a graphic map presentation on a graphic terminal (not shown). As the last step, the result of the analysis is presented on the terminal screen (Fig. 3, lower part). From the result you can see that the sequence contained one site for each of the four six-cutters *Hae*III, *Nae*I, *Nhe*I, and *Pvu*II. The result is also presented as a semigraphical list that can be used on a normal nongraphic terminal.

11. Translate a DNA Sequence into Amino Acids

The transfer of information from the DNA sequence to amino acids into polypeptides is of course a vital part of the biological processes in every living cell. Given the fact that codon triplets code for the same amino acids in (almost) all organisms, translation is the most universal and most easily

```
RSEARCH DEMODB:ANACYSTIS..PSBA/Pos=200:400<cr>

Choose symbol for enzymefiles

SYMBOL          CONTENT
   A    -    "All enzymes"
   4    -    Four cutters
   5    -    Five cutters
   6    -    Six cutters

Choice : 6<cr>
Do you want to choose names of enzymes (Y/N) ? : N<cr>
Do you want to list search enzymes (Y/N) ?      : N<cr>
Circular or linear sequence (C/L) ?             : L<cr>

Result presentation
1 - List of restriction enzyme sites sorted by enzymes.
2 - List of restriction enzyme sites sorted in order.
3 - List of restriction enzyme fragments sorted in order.
4 - List of restriction enzyme fragments sorted by size.
5 - List of restriction enzymes with sequence.
6 - Semigraphic list

Choice : 1<cr>
Choice : 6<cr>
Choice :<cr>

RESTRICTION ENZYME SITES SORTED BY ENZYMES.

ENZYME      STRING          POSITION(S)

HAEII..M5   GGCGCT             238
NAEI...B3   GCCGGC             202
NHEI...B1   GCTAGC             298
PVUII..M3   CAGCTG             369

SEMI-GRAPHIC DISPLAY OF RESTRICTION ENZYMES

              200      233      267      300      333      367      400
        SITES  !--------!--------!--------!--------!--------!--------!
HAEII..M5    1                1
NAEI...B3    1    1
NHEI...B1    1                            1
PVUII..M3    1                                                  1
```

Fig. 3. Restriction endonuclease search. One of the *psbA* sequences from *A. nidulans* from position 200–400 was chosen. Instructions are executed by answering questions. The result of the search is shown in the bottom of the figure (*see* section 10 for further details).

interpreted of all searches for biological information to be performed on a computer at the present time. The commands to translate a DNA sequence into amino acids must be able to translate in all three reading frames at the same time and also to translate unique parts of the sequence in a limited number of frames. Note that if you are searching for reading frames that could code for a protein, you must also search the reversed complement. To simplify the search for a coding frame that constitutes a gene; you can use the GENESCAN program,

```
MetThrSerIleLeuArgGlu        ...    ProSerIleHisGly***

  ***ProAlaPhePheAlaSer       ...    LeuGlnPheThrVal

 AspGlnHisSerSerArgAla        ...    PheAsnSerArgLeu

ATGACCAGCATTCTTCGCGAGCA       ...    CCTTCAATTCACGGTTAA

1        10        20               1170        1180
```

Scheme 4

which searches the DNA sequence for stretches of information that has a codon usage that matches other found and sequenced genes (*see* section 14).

Type TRANSLATE ANACYSTIS..PSBA<cr> and GENEUS will translate the DNA sequence in all three reading frames, starting from position number 1 and ending at the last position, 1183.

Type TRANSLATE ANACYSTIS..PSBA/POS=1:-21/ GEN1=1:21<cr> and the program will translate the DNA sequence ANACYSTIS from position 1 to 21 (GEN1) and show the DNA sequence from position 1 to 21 (POS). (*See* Scheme 4.)

Type AUTOTRANS PBR322<cr> and the program AUTOTRANS will use the information stored in the knowledge base on lines FT for the DNA sequence of pBR322 in the EMBL nucleotide sequence data library and translate the corresponding genes (tet and TEM-1 beta-lactamase) into amino acids. (*See* Scheme 5.)

12. Searching for Secondary Structures (Hairpins)

In many biological systems, secondary structure information stored within the DNA sequence seems to play a role in

```
MetThrSerIleLeuArgGlu

ATGACCAGCATTCTTCGCGAG

1        10        20
```

Scheme 5

determining, e.g., start and stop of transcription and regulatory circuits as attenuation. Also, complex structures such as the tRNA and rRNA molecules are formed on the basis of intra-strand folding because of stable hairpins. A hairpin structure is sometimes also called a dyad symmetry, as opposed to in-verted repeats that can not form a hairpin. The only simple way so far to evaluate the importance of a secondary structure is to calculate the free energy, deltaG (kcal), of a single hairpin. The basic equations for those calculations were worked out by Tinoco et al. (7). These equations still form the basis for most computer programs dealing with hairpin structure stabilities and biological importance. The equations calculate the free energy, deltaG, in kilocalories. The obtained deltaG value can be used to describe the biological importance of the discovered hairpin structure. The earliest equation of Tinoco et al. had been developed to be able also to calculate a delta S value, but those equations are rarely used (8). More elaborate equations exist to calculate the most probable structure of more complex nucleic acid sequences such as rRNA (9).

A hairpin has the features shown in Scheme 6.

It is difficult to evaluate the biological importance of a hairpin struture only from the deltaG value. Data have been accumulating during the last years, however, that simplify the evaluation. Secondary structures occur in transcriptional con-trol signals, as transcriptional stop and attenuation in prokar-yotic cells and in, e.g., regions for initiation of DNA replication

```
                  TCC

            A    C     LOOP

            G–C

            G–C

            G–C    STEM

            G–C

        GATACGGAT GATCGAATT
```

Scheme 6

in both pro- and eukaryotic cells. Most data concerning hairpin structures have been collected from transcriptional stop signals and attenuators in *Escherichia coli*. Transcriptional stop structures, regardless of whether they are rho-dependent or independent, often have a deltaG value of lower than –15 kcal; very often lower than –25 kcal. Attenuators often contain hairpins with deltaG values of as high as –5 to –6 kcal, but one finds overlapping secondary structures regularly. Specific mutagenesis of secondary structures decreasing its stability with –5.0 kcal can sometimes decrease its biological activity dramatically. The results from a search for hairpin structures must often be complemented with other types of searches for other biological information.

Type HAIRPIN ANACYSTIS1..PSBA/GVAL=–22.0<cr>, indicating that the computer will search the sequence ANACYSTIS1 for hairpin structures with a deltaG value of lower than –22.0 kcal. In the sequence used, the gene starts at position 101 (ATG) and stops at position 1181 (TAA). The result will be as shown in Scheme 7.

In the sequence given, there exist around 15 hairpin structures with a deltaG value of below –15 kcal, and, as is evident above, five structures with a deltaG value below –22 kcal. An analysis of the positions for the structures above indicates that the last hairpin is located after the stop codon for the protein-coding region and presumably represents a transcriptional stop signal. The other structures are located within the coding region, and none is found at the non-translated 5'-end. The given example clearly shows that hairpin structures can be found in parts of the sequence where they are not expected. Hairpins with deltaG values of lower than –15 kcal, however, such as the ones found in the presented sequence, are only rarely found within protein-coding regions.

13. Codon Usage

The codon usage defines the codon triplets used for every single amino acid in a coding region within a DNA sequence.

```
        STEM                      LOOP        deltaG—value

                    175
TTGAGT GGG TAACCAGCA  CCG AC  AACCGCAT        -24.4 kcal
ACTCACCCCTAGTAGTCGTTGCGGCTTG  GGTGCATC
     *   * * *      **   *
                    196


                    182
    CAGCA   CCGA  CAACCGCA  TCT             -27.2 kcal
    GTCGTGCGGCTTGGTTGG GT  GCA
      **      **      *
                    189


                    1146
    GGCAGCAGGCGAAGCGAC  CC                   -26.6 kcal
    CCG CGTCAGTTTCGCTG  GC
      *     * *  1151


                    1179
GACTGCGCCTTCAATTCACGG  TTAATTGC             -22.6 kcal
CTGAC  GACATTAAG GCC   GTTAAGGC
     ***  **      *
                    1187


                    1222
    ATCAAAAAGCGCC  TCGA                      -26.4 kcal
    TAGTTTTTCGCGG  GGTA
                    1231
```

Scheme 7

The codon usage is often used to study the occurrence and use of alternative codons for appropriate amino acids. Because the codon usage differs between different organisms, but also depending upon the expression level of individual genes, the codon usage can be used to obtain information on the gene itself. The codon usage is also used in the GENESCAN program (section 14) to find gene coding regions within an unknown DNA sequence.

Type CODSUM ANACYSTIS..PSBA<cr>, indicating that you want to compute the codon usage of one of the *psbA* genes from *A. nidulans*. The result will be that given in Scheme 8.

As within all coding sequences, there is a large difference in the frequency of the different codons used. In *Escherichia coli* and yeast, it has been found that there is a difference in codon usage for certain amino acids between strongly/moderately and weakly expressed genes (*10,11*). The two classes of genes can often be discriminated between by looking at the frequency of the codons labeled S and W in the table above. In *E. coli*, the codons labeled S and W are predominantly used in strongly/moderately and weakly expressed genes, respectively. Codons labeled with * are rarely used in strongly expressed genes in *E. coli*. In the example above in which a highly expressed gene, *psbA* from the blue-green bacteria *A. nidulans*, was used, it can easily be seen that the same codons are avoided as for highly expressed genes in *E. coli*. Codon tables for the distribution of individual codons in different organisms have also been calculated (*12*).

14. Searching for Coding Regions

In gene technology before the determination of the DNA sequence, very little was commonly known about the location and presence of genes that might be present on the cloned piece of DNA. It has become more and more common to sequence a cloned piece of DNA very early in a project because the DNA sequence contains all information needed for further labora-

Phe	TTT	5 W	Ser	TCT	1	Tyr	TAT	2 W	Cys	TGT	2
Phe	TTC	22 S	Ser	TCC	3	Tyr	TAC	12 S	Cys	TGC	2
Leu	TTA	1	Ser	TCA	0	*	TAA	0	*	TGA	0
Leu	TTG	7	Ser	TCG	11	*	TAG	0	Trp	TGG	10
Leu	CTT	1	Pro	CCT	4 S	His	CAT	1	Arg	CGT	5 S
Leu	CTC	4	Pro	CCC	5 W	His	CAC	8	Arg	CGC	7 W
Leu	CTA	0 *	Pro	CCA	1	Gln	CAA	9	Arg	CGA	0 *
Leu	CTG	15	Pro	CCG	3	Gln	CAG	1	Arg	CGG	1 *
Ile	ATT	5 W	Thr	ACT	2	Asn	AAT	1 W	Ser	AGT	1
Ile	ATC	18 S	Thr	ACC	12	Asn	AAC	18 S	Ser	AGC	11
Ile	ATA	0 *	Thr	ACA	0	Lys	AAA	2	Arg	AGA	0 *
Met	ATG	14	Thr	ACG	1	Lys	AAG	0	Arg	AGG	0 *
Val	GTT	7	Ala	GCT	6 S	Asp	GAT	4	Gly	GGT	18 S
Val	GTC	3	Ala	GCC	7 W	Asp	GAC	5	Gly	GGC	14 W
Val	GTA	1	Ala	GCA	9	Glu	GAA	3	Gly	GGA	0 *
Val	GTG	14	Ala	GCG	2	Glu	GAG	12	Gly	GGG	0 *

Scheme 8

tory work and, it is hoped, all biological information. It is clear that statistically safe methods to locate protein-coding regions should be very useful. Several equations have been developed to solve this task (13–15). Coding frames in all organisms have been found to contain a higher frequency of purine–any base–pyrimidine triplets than noncoding frames, and the so-called RNY rule searches for a high preference of triplets with purine–any base–pyrimidine (13). It is also possible to use a codon table (as described above, section 13, to locate regions that have a high frequency of triplets matching the codon table used

(14,15). This method is usually safer than the RNY method, but is dependent upon the use of a correct codon table. In GENEUS, the method developed by Staden and McLachlan *(14)* is used.

Type <u>GENESCAN ANACYSTIS2..PSBA</u><cr>, indicating that you want to search one of the *psbA* genes from *A. nidulans* for coding regions. In the sequence example, 100 bases upstream and downstream from the ATG and TAA, respectively, were added to the coding region. The result is shown in Fig. 4.

```
GENEUS> GENESCAN DEMODB:ANACYSTIS..PSBA

The codon table from EMBL normgene (Figure 6b, p562 in reference
below) is used as the statistic table to compute codon usage

ODD WINDOW LENGTH IN CODONS (RETURN-KEY = 25 ) : 25
OUTPUT DIST. BETWEEN CODS   (RETURN-KEY = 1  ) : 30

VALUES LOG10(P/(1-P)) =    2.03   -2.40   -2.28 FOR WINDOW 25
```

POS	FRAME 1		FRAME 2		FRAME 3	
88	0.000009		0.000153		0.999838*	
178	0.000915	G	0.993052*	OA	0.006032	G
268	0.000001	O	0.999876*	OG	0.000123	
358	0.000037		0.999539*	OA	0.000424	OG
448	0.001046	G	0.993783*	OG	0.005171	G
538	0.002788	OA	0.971409*	OA	0.025803	OG
628	0.000060	O	0.999881*	OG	0.000059	G
718	0.005187	O	0.994076*	OA	0.000737	G
808	0.000010		0.999962*	OG	0.000028	OG
898	0.013070	O	0.984758*	OG	0.002172	
988	0.002508	OA	0.997229*	OG	0.000264	G
1078	0.000035		0.998432*	OG	0.001534	A
1168	0.008890		0.989847*	OA	0.001264	A
1258	0.697663*		0.240949		0.061389	A

```
A=ATG in frame, G=GTG in frame, O=open reading frame
*=highest statistical value for codon usage

REF.   R. Staden. (1984) Nucl. Acids Res., Vol 12, Number 1
       Part 2, p. 551-567.
```

Fig. 4. Search for protein coding regions. One of the *psbA* sequences from *A. nidulans* was chosen. The coding region was located by searching the sequence for a high frequency of triplets matching the codon usage in the EMBL "norm gene." <u>A</u> describes an ATG in frame, <u>G</u> a GTG in frame, <u>O</u> an open reading frame, and <u>*</u> the highest frame (*see* section 14 for further details).

The codon table from the EMBL nucleotide sequence data library for a "norm gene" was used. By answering questions, the length of the window used to calculate the statistical appearance of codons and the output length are chosen. In Fig. 4, A describes an ATG in frame, G a GTG in frame, O an open reading frame, and * the highest statistical value for codon usage in that window and frame. It can be clearly seen that the codon usage just upstream and downstream from the coding region is statistically closer to the EMBL "norm gene" in different reading frames than the one where the psbA gene is located, and that the psbA gene is easily located for looking at the ATG, the codon usage, the open reading frame, and the stop codon.

15. Searching for Sequence Patterns

Most searches for information in a DNA sequence are based on the principle of hit–nonhit. Certain types of DNA sequence information are a function of a predictable statistical pattern, however, in which information is a function of a predictable statistical pattern in which the four bases—A, G, C, and T— can occur at each position within the pattern, but with different frequencies. Search algorithms that are able to find statistical patterns have been developed (16). Such a pattern is, e.g., an E. coli promoter. An E. coli promoter is defined as two patterns, the –35 and –10 regions, spaced 15–20 bases apart. The –35 and –10 patterns have a DNA sequence of TTGACA and TATAAT, respectively. There is a difference, however, in the occurrence of the four bases at the different positions. The last T in TATAAT is almost invariant, whereas the first A can be replaced with almost any base. Other patterns existing are ribosome binding sites, CRP-cAMP binding sites, and splice points. Every pattern requires its own statistical table. The more patterns that have been sequenced and analyzed, the safer the analysis of an unknown sequence.

Type PAT PSBA2<cr>, indicating that a search for patterns in the sequence PSBA2 (PSBA2 is the same as ANACYSTIS2, where 100 bases were added upstream and downstream from the coding region) is going to be performed. Within the program, it is decided what region of the sequence to use, the pattern to search for, and the level of significance as threshold level for matches to be displayed. A correct choice of the threshold level is very important in the search of the patterns. A significance value of 1 indicates that the pattern found is identical with the consensus pattern used in the search. A long pattern and an even distribution among the bases in the pattern decreases the threshold value. In a search for *E. coli*-like promoters, a significance value of 0.0001 can be used to find weak promoters. Strong promoters usually have a significance value of higher than 0.05. When searching, however, for ribosome binding sites, GGAGG, a value of lower than 0.001 is difficult to use. The same is true for splice points (16).

A search for STANDARD PROMOTERS and a threshold level of 0.004 was chosen. The result is shown in Fig. 5. Interesting patterns at positions 17–45 and 1157–1185 were found. The gene starts at position 101 and stops at position 1180. S1 mapping has localized the transcriptional start point to position 51

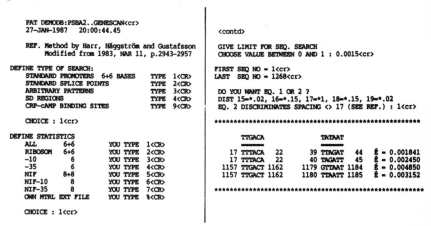

Fig. 5. Search for sequence patterns. One of the *psbA* genes from *A. nidulans* was chosen. Patterns similar to an *E. coli* promoter were searched for. A statistical value of 0.0015 was used in the search (*see* section 15 for further details).

```
        GGAGG

        =====

    53  CGAGG  57     Siginificance value = 0.063

   493  GGAGC  497         "          "  = 0.429

   823  AGAGG  827         "          "  = 0.125

   829  AGAGA  833         "          "  = 0.071
```

Scheme 9

(2). The found patterns at positions 17–45 are most likely the promoter. The significance value is lower than usually found for *E. coli* promoters, however. The next step is to search for ribosome binding site, the so-called Shine-Delgarno sequence, with a threshold value of 0.05, is performed. The result is given in Scheme 9.

The pattern found at positions 53–57 is most probably the ribosome binding site used for this particular gene.

16. Sequence Comparisons

Stretches of DNA that have the same or similar DNA sequence can be hypothesized to contain the same or similar biological functions. The same is true for the reverse; evolution of genetic information predicts that DNA sequences that are evolutionary related should also be comparable in their DNA sequences. Different types of DNA evolve at very different rates, however. Repetitive DNA as well as, e.g., mitochondrial DNA, evolves much faster than coding regions in a eukaryotic nucleus. Functions of a high rigidity, such as the photosynthetic apparatus, evolve at a slow speed.

Therefore, a search for sequence similarities constitutes a very important step in the evaluation of the biological information stored within little or even completely unknown DNA sequences. This will be even more true in the future, when more and more sequences are known from a larger number of organisms. One recent notable example has been the finding that there are homologies between the reaction center polypeptides of photosynthetic purple bacteria and the *psbA* and *psbD* genes, coding for the D1 and D2 polypeptides in reaction center II of higher photosynthetic organisms (5). The finding (also based on X-ray data of the reaction center polypeptides from Rhodopseudomonas spp.) has given way to the hypothesis that the D1 and D2 polypeptides are the reaction center polypeptides of photosystem II (17).

A search for repeated or inverted DNA sequences is also simplified by a computer search for homologous sequences. The intricate sequence structure of the immunoglobulin genes from germ line mouse, composed of repeated sets of genetic information, was in part solved by computer studies.

Several ways to search for homologous regions by the help of a computer have been developed. Two methods will be described: the dot matrix and the local homology search (*18,19*). Both are combined with a graphic display of the result.

16.1. Dot Matrix Method

The dot matrix system is a graphic display of regions of DNA within a longer sequence that are totally homologous for a continuous stretch of bases, the number of which is chosen by the user (*18*). The system is fast, simple, and easy to run on a personal computer (PC). It has the disadvantage, however, of producing much background noise if the DNA sequences show strong homologies. Much of the information search is left to the user's visual analysis of the graphic picture produced.

The strongly homologous regions between the *psbA* genes, coding for the reactions center polypeptides D1 from the very related cyanobacteria A. *nidulans* and *Anabaena*, are easily found with the dot matrix method.

Type DOTMATRIX ANACYSTIS..PSBA ANABAENA..PSBA<cr>. Having entered the program, the threshold level to six identical consecutive bases is set, i.e., the program will place a dot in a two-dimensional diagram each time there is a stretch of six identical bases. The result is shown in Fig. 6A. It is clearly shown that the program detects the similarity between the two sequences easily. A detection of homologous regions between the DNA sequences of the *psbA* and *psbD* genes is difficult with the dot matrix method, however, despite the fact that similarities have been detected between regions of the respective polypeptides, the D1 and D2 proteins (5).

Type <u>DOT MATRIX ANACYSTIS..PSBA CHLAMYDO-MONAS..PSBD</u><cr>. Use a threshold level of six. The result is shown in Fig. 6B. No similarities are found. Lower threshold levels increase the background noise to a level at which the result becomes almost impossible to evaluate.

16.2. Local Homology Search—The Harr Plot

Local homology comparisons attempt to determine if there are sections in a particular sequence that are similar to sections in another sequence or if there are sections in a sequence that are similar to other sections in the same sequence. Here mathematical algorithms are used to determine where local homologies exist. The equations take mismatches into account, but also account for small insertions and deletions. The user may define the limits for the number of point mutations, deletions, and replacements that are to be tolerated in the homology search. The speed of the process is highly dependent on how many consecutive bases trigger an attempt for an extension of a possible homology and the limits for the further extension. As is evident, a local homology search that is performed on the basis of mathematical calculations of homologous regions taking into account point mutations, deletions, and replacements produces a result that is better than that obtained with a simple dot matrix program, but requires more computer power to be performed in reasonable time. The local homology programs are preferentially combined with a graphic result presentation. Often the list of homologous sequences found between sequences or within a sequence can

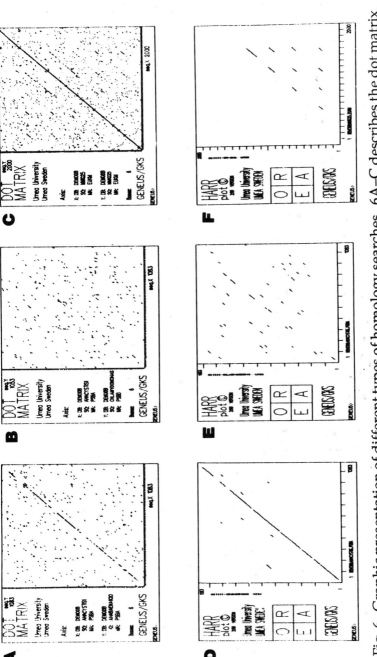

Fig. 6. Graphic presentation of different types of homology searches. 6A–C describes the dot matrix method (18); 6D–F describes the Harr plot (19); in 6A and D the psbA sequences from A. nidulans (X axis) and Anabaena (Y axis) were used. In 6B and E, the psbA sequence from A. nidulans (X axis) and the psbD sequence from C. reinhardii (Y axis) were used. In 6C and F, the immunoglobulin kappa chain gene from mouse (X and Y axes) was used to search for direct repeats (see Chapter 16 for further details).

be several pages long. The result is easily contained within one single graphic picture, however.

In the GENEUS system, the search for local homologies is divided into two parts: the actual calculation of the homologous regions and the presentations of the result, graphic (the HARR plot) (*19*) or nongraphic.

Type <u>COMPARE ANACYSTIS..PSBA ANABAE-NA.. PSBA/QVAL=5.0 AAA.DAT</u><cr>, indicating that a search for homologies between the two *psbA* sequences from Anacystis and Anabaena is going to be performed. A significance value of 5.0 (*Q* value) is chosen to discriminate found homologies from the background. The result will be placed in a file named AAA.DAT.

The QVAL defines the level of significance chosen for homologous regions. The *Q* values have been defined by Goad and Kanehisa (*20*) and are statistical evaluations of found homologies (*see* Table 1).

Other definitions set by the use of qualifiers are: POSA and POSB, indicating sequence positions in sequences A and B, respectively; CA, CB, IA, and IB, indicating reversed complement (C) and inversion (I) for the two sequences, respectively; LOOPS, indicating the maximum number of bases in one particular loop-out (insertion–deletion) within a homology. The value is usually set to three. START indicating the number of consecutive, identical bases that are required in the beginning of a homology. The value is usually set to three.

Type <u>HARRPLOT AAA.DAT</u><cr>, specifying that you want to display the result from the homology search in the form of a HARR plot.

The result of the local homology search between the two *psbA* sequences from *Anacystis* and *Anabaena* is shown in Fig. 6D. It is clearly shown that the background noise is lower and the extension of homologous regions is longer than with the dot matrix method.

Type <u>COMPARE ANACYSTIS..PSBA CHLAMYDO- MONAS..PSBD/QVAL=5.0 BBB.DAT</u><cr>, indicating that the

Table 1
Correlation Between the Statistical Q Value and Characteristics
of a DNA Sequence Homology

Q value	Length of homology	Matching bases number, %	Loop-outs	
			Number	Size, bases
75	190	167(88)	0	0
60	163	139(85)	0	0
50	177	140(79)	0	0
30	78	67(86)	0	0
	145	111(77)	2	1
			1	2
			2	3
20	40	37(93)	0	0
	50	42(84)	1	1
	86	67(78)	2	1
	120	100(83)	3	1
10	37	29(78)	0	0
	46	30(65)	1	1
			1	2
7	19	16(84)	2	1
6	10	10(100)	0	0
	17	14(82)	1	1
5	9	9(100)	0	0
	15	12(80)	1	1
	21	16(76)	2	2
	29	19(66)	1	1

two distinctly related reaction center polypeptide genes *psbA* and *psbD* from the blue-green bacteria *A. nidulans* and the green algae *C. reinhardii*, respectively, are going to be searched for homologous regions. A *Q* value of 5.0 is chosen as significance level.

Type <u>HARRPLOT BBB.DAT</u><cr>. The result is shown in Fig. 6E. It is clearly seen that there are homologous regions

between the two sequences; regions that are almost impossible
to detect without very efficient algorithms for local homology
searches combined with a powerful graphic display of the
result.

It is often interesting to search for possible repeated sequences within a piece of DNA. Such information can be the
clue to both the function and evolution of a DNA sequence.
Type <u>COMPARE MMIG25..EUKM MMIG25..-EUKM/</u>
<u>P O S A = 5 0 1 - 2 5 0 0 / P O S B = 5 0 1 - 2 5 0 0 / Q V A L = 1 0 . 0</u>
<u>CCC.DAT</u><cr>, indicating that the COMPARE program
should find the repeated regions within the kappa chain
immunoglobulin gene from mouse. The purpose of the search
is to find the so-called J-repeats. To decrease the background
noise, a Q value of 10.0 is chosen.

Type <u>HARRPLOT CCC.DAT</u><cr>. The result is shown in
Fig. 6F. The five J-repeats are easily seen. Compare Fig. 6F with
the result presented in Fig. 6C, in which the dot matrix method
was used with the same sequence.

16.3. Multisequence Alignment

GENEUS can perform multiple sequence alignment
among different members in a gene family or among different
genes that are evolutionary related. Scheme 10 displays the

```
    pos        1040      1050      1060      1070      1080

ANABAENA625  ..GCtGGtGAAGtt gCtCCtGTtGCgtTaACcGCtCCTgCtATcaACGGtTAA

ANABAENA620  ..GCtGGtGAAGtt gCtCCtGTtGCaaTaAgcGCtCCTgCtATcaACGGtTAA

CHLAMYDOMO   ..tCaactaActctAgCtCaaacaaCtaa

ANACYSTIS2   ..GCaGGcGAAGcg aCcCCgGTtGCtcTgACtGCaCCTgCaATcaACGGcTAA

ANACYSTIS3   ..GCaGGcGAAGcg aCcCCgGTtGCtcTgACtGCaCCTgCaATcaACGGcTAA

ANACYSTIS1   ..GCaGGcGAAGcg aCcCCgGTcGCtttTgACtGCgCCTtCaATtcACGGtTAA


"CONSENSUS"  ..GC-GG-GAAG--A-C-CC-GT-GC--T-AC-GC-CCT-C-AT--ACGG-TAA
```

Scheme 10

result of an alignment produced when the MULTI program has aligned six sequences in the *psbA* gene family from blue-green bacteria and green algae.

The presented result can be used to study the evolutionary relations and distances between the DNA sequences.

17. Additional Processing of Produced Results

The results produced from most analysis procedures described above are stored on disk files after the program has been terminated. A further processing of the results is thus possible using word processing programs and other computer editors. This option is advantageous to use when, e.g., additional information should be added to results that are to be used for publications or visual presentations. The results from individual programs are stored in a WORK directory and labeled with the same name as the analysis program.

Initiate the editor. Type <u><WORK>RESEARCH.DAT;1</u> <cr> to make the result from the first (;1) RESEARCH analysis available to the editor. Edit the result. Store the edited result using the normal editor functions.

18. Switching the Computer Off

Terminate the programs, log out, and switch off the computer after the sequence analysis has been completed.

Type: <u>LOGOUT</u><cr>

Further Reading

Computer Applications in Research on Nucleic Acids Vol. 1, 2, and 3 (1982, 1984, and 1986) IRL Press, Oxford and Washington DC, identical with *Nucleic Acids Research* Vol. 10:1 (1982), 12:1 (1984), and 14:1 (1986).

Harr, R. and Gustafsson, P. (1987) Computer Applications to Studying DNA, in *Techniques in Molecular Biology* Vol. 2 (Walker, J.M. and Gaastra, W., eds.) Croom Helm, Kent, England.

Appendix of commands:
Individual Computer Instructions

<u>PROGRAMS</u>   <u>QUALIFIERS</u>

<u>1. Editor</u>

EDSEQ

EDINFO

<u>2. Database</u>

DATABASE

DBSEQS   /DB=Database   /MA=Master   /OUT=File_specification   /DATE

DBMAST   /DB=Database   /SQ=Seqname   /OUT=File_specification

DBDEL   /NOCONFIRM

DBCOPY   /DEL_OLD   /NOCONFIRM

DBINSERT

DBOUT

DBQUEST

<u>3. DNA sequence project system</u>

SP1

MMBAT

<u>4. DNA sequence analysis programs</u>

LISTSEQ   /STR=<1 or 2>   /Pos=First:Last   /SPA=<1 to 15>
           /OUT=File_specification

Scheme 11

```
INFOLIST   /KEYS=Linetype   /STRING=Character_string
           /OUTPUT=File_specification

ALIGN      /PosA=First:Last   /PosB=First:Last

AUTOTRANS /Pos=First:Last   /SINGLE

TRANSLATE /Pos=First:Last    /GEN1 up to /GEN8=First:Last

TRNA       /Pos=First:Last

RSEARCH    /Pos=First:Last

..TEXLINE

..TEXCIRC

RMAP

REVCOMP

COMPARE    /ALG=<1 to 12>   /QVAL=Quality value (decimal)
           /LOOPS=<0 to 6> /START=<1 to 6>        /IA  /IB  /CA  /CB
           /PosA=First:Last   /PosB=First:Last

..HARRPLOT

..RETYPE

DOTMATRIX

HAIRPIN    /Pos=First:Last   /GVAL=Free energy of hairpin
           /BULGE=<0 to 6>

GENESCAN

CODSUM

PAT

MULTI

SCANDB
```

Scheme 11 (*continued*)

Appendix

PROGRAMS	QUALIFIERS
EDSEQ	
EDINFO	
DATABASE	
DBSEQS	/DB=Database /MA=Master /OUT=File_specification /DATE
DBMAST	/DB=Database /SQ=Seqname /OUT=File_specification
DBDEL	/NOCONFIRM
DBCOPY	/DEL_OLD /NOCONFIRM
DBINSERT	
DBOUT	
DBQUEST	
LISTSEQ	/STR=<1 or 2> /Pos=Firstpos:Lastpos /SPA=<1 to 15> /OUT=File_specification
INFOLIST	/KEYS=Linetype /STRING=Character_string /OUTPUT=File_specification
ALIGN	/PosA=Firstpos:Lastpos /PosB=Firstpos:Lastpos
AUTOTRANS	/Pos=Firstpos:Lastpos /SINGLE
TRANSLATE	/Pos=Firstpos:Lastpos /GEN1 up to /GEN8=Firstpos:Lastpos
TRNA	/Pos=Firstpos:Lastpos
RSEARCH	/Pos=Firstpos:Lastpos
..TEXLINE	
..TEXCIRC	
RMAP	
REVCOMP	
COMPARE	/ALG=<1 to 12> /QVAL=Quality value (decimal) /LOOPS=<0 to 6> /START=<1 to 6>
	/IA /IB /CA /CB /PosA=Firstpos:Lastpos /PosB=Firstpos:Lastpos
..HARRPLOT	
..RETYPE	

Scheme 12

```
DOTMATRIX

HAIRPIN    /Pos=Firstpos:Lastpos   /GVAL=Free energy of hairpin

/BULGE=<0 to 6>

GENESCAN

CODSUM

PAT

MULTI

SCANDB




SP1

MMBAT
```

Scheme 12 *(continued)*

References

1. Harr, R., Fallman, P., Haggstrom, M., Wahlstrom, L., and Gustafsson, P. (1986) GENEUS, A computer system for DNA and protein sequence analysis containing an information retrieval system for the EMBL data library *Nucleic Acids Res.* **14**, 273–284.
2. Golden, S. S., Brusslan, J., and Haselkorn R. (1987) Expression of a family of *psbA* genes encoding a photosystem II polypeptide in the cyanobacterium *Anacystis nidulans* R2, *EMBO J.* **5**, 2789–2798.
3. Curtis, S. E. and Haselkorn, R. (1984) Isolation, sequence and expression of two members of the 32kd thylakoid membrane protein gene family from the cyanobacterium *Anabaena* 7120. *Plant Mol. Biol.* **3**, 249–258.
4. Erickson, J. M., Rahire, M., and Rochaix, J.-D. (1984) *Chlamydomonas reinhardii* gene for the 32.000 mol. wt. protein of photosystem II contains four large introns and located entirely within the chloroplast inverted repeat. *EMBO J.* **3**, 2753–2762.
5. Rochaix, J.-D., Dron, M., Rahire, M., and Malnoe, P. (1984) Sequence homology between the 32K dalton and the D2 chloroplast membrane polypeptides of *Chlamydomonas reinhardii*. *Plant Mol. Biol.* **3**, 363–370.

6. Max, E. E., Maizel, J. V., and Leder, P. (1981) The nucleotide sequence of a 5.5-kilobase DNA segment containing the mouse kappa immunoglobulin J and C region genes. *J. Biol. Chem.* **256**, 5116–5120.

7. Tinoco, I., Borer, P.N., Dengler, B., Levine, M. D., Uhlenbret, O. C., Crothers, D. M., and Grotta, J. (1973) Improved estimation of secondary structure in ribonucleic acids. *Nature New Biol.* **146**, 40–41.

8. Borer, P. N., Dengler, B., and Tinoco, I. (1974) Stability of ribonucleic acid double-stranded helices. *J. Mol. Biol.* **86**, 843–853.

9. Zuker, M. and Stiegler, P. (1981) Optimal computer folding of large RNA sequences using thermodynamics and auxiliary information. *Nucl. Acids Res.* **9**, 133–148.

10. Grosjean, H. and Fiers, W. (1982) Preferential codon usage in prokaryotic genes: The optimal codon-anticodon interaction energy and the selective codon usage in efficiently expressed genes. *Gene* **18**, 199–209.

11. Sharp, P. M., Tuohy, T. M. F., and Mosurski, K. R. (1986) Codon usage in yeast: Cluster analysis clearly differentiates highly and lowly expressed genes. *Nucleic Acids Res.* **14**, 5125–5143.

12. Grosjean, H. (1980) Codon Usage in Several Organisms, in *Transfer RNA: Biological Aspects* (Soll, D., Abelson, J. N., and Schimmel, P. R., eds.) Cold Spring Harbor, New York.

13. Sheperd, J. C. W. (1981) Method to determine the reading frame of a protein from the purine/pyrimidine genome sequence and its possible evolutionary justification. *Proc. Natl. Acad. Sci. USA* **78**, 1596–1600.

14. Staden, R. and McLachlan, A. D. (1982) Codon preference and its use in identifying protein coding regions in long DNA sequences. *Nucleic Acids Res.* **10**, 141–156.

15. Fickett, J. W. (1982) Recognition of protein coding regions in DNA sequences. *Nucleic Acids Res.* **10**, 5303–5318.

16. Harr, R., Haggstrom, M., and Gustafsson, P. (1983) Search algorithm for pattern match analysis of nucleic acid sequences. *Nucleic Acids Res.* **11**, 2943–2957.

17. Deisenhofer, J., Epp, O., Miki, K., Huber, R., and Michel, H. (1985) Structure of the protein subunits in the photosynthetic reaction center of *Rhodopseudomonas viridis* at 3Å resolution. *Nature* **318**, 618–624.

18. Maizel, J. V. and Lenk, R. P. (1981) Enhanced graphic matrix analysis of nucleic acid and protein sequences. *Proc. Natl. Acad. Sci. USA* **78**, 7665–7669.

19. Harr, R., Hagblom, P., and Gustafsson, P. (1982) Two-dimensional graphic analysis of DNA sequence homologies. *Nucleic Acids Res* **10**, 365–374.

20. Goad, W. B. and Kanehisa, M. I. (1982) Pattern recognition in nucleic acid sequences. I. A general method for finding local homologies and symmetries. *Nucleic Acids Res.* **10**, 247–263.

Chapter 12

Detection of Sequence-Specific Protein-DNA Interactions by the DNA-Footprinting Technique

Mark A. Plumb and Graham H. Goodwin

1. Introduction

The initiation of transcription of eukaryotic genes by RNA polymerases is controlled by complex interactions between nonhistone proteins and specific regulatory DNA sequences (promoters and enhancers) (1). In order to characterize and purify such transacting protein factors, a sensitive and accurate assay for sequence-specific DNA-binding proteins is the DNA-footprinting technique, which can be used to analyze the interaction of a complex mixture of proteins with a gene regulatory sequence(s) that is known to be important for the expression of that gene.

In this technique (Fig. 1), a purified DNA restriction fragment that contains the sequence of interest is end-labeled with ^{32}P at a single 3' or 5' end. Protein is bound to the DNA, and the nucleoprotein complex is then digested mildly with deoxy-

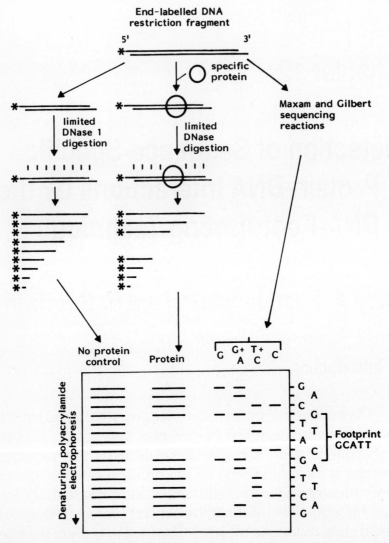

Fig. 1. Schematic representation of the footprinting technique.

ribonuclease I (DNAse I). DNAse I gives random single-stranded DNA nicks, but cleavage is inhibited if protein is bound to a sequence. The DNA is then purified and denatured, and the single-stranded end-labeled DNA fragments are resolved on a denaturing (sequencing) polyacrylamide gel and autoradiographed. The "no protein" control will give a ladder (Fig. 1) that represents cleavage at approximately every base, and the samples incubated with protein will give the same, except where protein was specifically bound. The absence of cleavage products or a gap in the base ladder (Fig. 1) is called a "footprint," and if chemical modification sequencing reactions [as described by Maxam and Gilbert (2)] are electrophoresed in parallel, the sequence of the footprint can be accurately determined. Although we will only describe footprinting with DNAse I, it should be noted that the chemical DNA cleavage reagent methidium propyl-EDTA-FeII can also be used (3). Exonuclease III protection and protection of deoxyguanosine methylation by dimethyl sulfate have also been used to detect specific binding of proteins to DNA (e.g.,refs. 4,5).

A critical element in the design of a footprinting experiment is the isolation of the end-labeled DNA restriction fragment. The DNA must be radiolabeled at only one end, and since DNA sequencing gel electrophoresis can only resolve a maximum of 200–250 bases, the footprint must be less than 250 bases from the ^{32}P-labeled end. Furthermore, the DNA must be radiolabeled at a reasonably high specific activity (greater than 1 x 10^6 cpm/μg), since in a typical reaction less than 20 ng (10^4 cpm) of a <1-kbp DNA restriction fragment will be cleaved, denatured, and electrophoresed, and result in each band in the base ladder having 5–10 cpm. This gives a reasonable overnight signal by autoradiography.

Cloned plasmid DNA can be end-labeled at restriction sites as follows (Fig. 2): (1) 5′ overhang restriction sites (e.g., *Eco*RI, *Hind* III and *Bam*H1 can be dephosphorylated by bacterial alkaline phosphatase (BAP) or calf intestinal phosphatase (CIP), and the resulting 5′-hydroxy group phosphorylated with (γ-^{32}P)-ATP and T4 polynucleotide kinase (2,6–8); (2) 3′ overhang restriction sites (e.g., *Pst*I, *Sst*I, and *Sph*I) can be labeled with (α-^{32}P)- dideoxy ATP (ddATP) and terminal deoxynucleotidyl transferase, which will phosphorylate 3′ hydroxyl groups (6,9,10); (3) Blunt-end restriction enzyme sites (e.g., *Pvu*II,*Sma*I, and *Hinc*II) can be radiolabeled by either technique, but incorporation is much less efficient; (4) On occasion it may be neces-

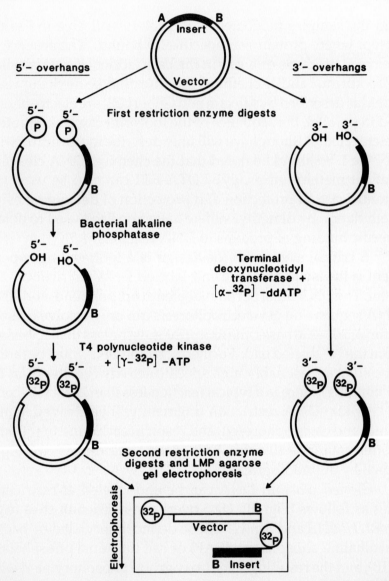

Fig. 2. Schematic representation of the two ^{32}P-end-labeling techniques.

sary to 5' end label a 3' overhang restriction site. This can be achieved by "filling in" the 3' overhang with T4 DNA polymerase or with the large fragment of the *Escherichia coli* DNA polymerase I (Klenow fragment) in the presence of excess deoxynucleotides (*6*). The resulting blunt end may then be treated with BAP or CIP and then end-labeled with (γ-^{32}P)-ATP and T4 polynucleotide kinase (*see* also ref. *8*).

Since both ends of the DNA restriction fragment site are labeled in these reactions, a secondary restriction digestion is necessary to give two fragments each end-labeled at one end. It is desirable to end-label a restriction fragment such that after secondary cleavage the two (or more) restriction fragments can be well separated by agarose gel electrophoresis. Subcloning the desired sequence into the M13 cloning sites of the pUC bacterial vectors (pUC 8-13, *see* ref. 11) is convenient, since this usually yields a recombinant that has one or more unique restriction sites on either side of the insert. It is then possible to end-label at one restriction site, do the secondary cleavage at another, and, if the insert is less than 1 kbp, separate the end-labeled vector DNA (2.7 kbp) from the end-labeled insert DNA (<1 kbp) by 1.5% low melting point agarose gel electrophoresis (*see* Fig. 2), and isolate the DNA fragment(s).

The protein samples we have used in footprint reactions are low-salt (0.14–0.3*M* NaCl) nuclear extracts. Ammonium sulfate precipitation (0.35 g/mL) removes from this extract the nonspecific DNA-binding high mobility group (HMG) proteins, which are soluble (R. Nicolas, unpublished results). The insoluble material is redissolved, and the DNA-binding proteins are further purified by DNA-cellulose affinity chromatography. A DNA-binding protein fraction (C4 protein fraction), which is enriched in sequence-specific DNA-binding nonhistone proteins, is eluted from the DNA-cellulose at 100–250 m*M* (NH$_4$)$_2$SO$_4$ (*see* refs. *12–14*).

2. Materials

2.1. Stock Solutions Stored at Room Temperature or at 4°C

1. DNA agarose gel electrophoresis sample buffer: 80% (v/v) glycerol, 10 m*M* Tris-HCl (pH 8.0), 1 m*M* EDTA, 0.1% (w/v) bromophenol blue.

2. 10x DNA agarose gel electrophoresis buffer: 0.4M Tris-acetate (pH 7.8), 0.05M sodium acetate, 0.01M EDTA.
3. 10x TBE buffer: 0.9M Tris, 0.9M boric acid, 25 mM EDTA.
4. Low-salt buffer (LSB) (store at room temperature after autoclaving): 0.2M NaCl, 20 mM Tris-HCl (pH 7.4), 1 mM EDTA.
5. High-salt buffer (HSB) (store at room temperature after autoclaving): 1.0M NaCl, 20 mM Tris-HCl (pH 7.4), 1 mM EDTA.
6. Binding buffer: 100 mM NaCl, 50 mM Tris-HCl (pH 8.0), 3 mM MgCl$_2$ (store at 4°C).
7. Phenol (*see* section 5): equilibrated with 10 mM Tris-HCl (pH 8.0), 1 mM EDTA. Store in dark (brown) bottle for up to 3 wk. Aliquots can be stored indefinitely at –20°C.
8. Chloroform/iso-amyl alcohol (*see* section 5): [CHCl$_3$/IAA, 24/1 (v/v)].
9. Footprint STOP buffer: to make 1.5 mL of STOP buffer, add (make fresh): 1.235 mL of 100 mM Tris-HCl (pH 8.0), 10 mM EDTA, 150 μL of 10% SDS, 40 μL of 4M NaCl, 10 μL of tRNA (10 mg/mL) (store at –20°C), 60 μL of proteinase K (10 mg/mL) (store at –20°C).
10. Deionized formamide (*see* section 5): stir for at least 3 h with AG 501-X8 mixed-bed resin (Bio-Rad) or "Amberlite" monobed resin MB-1 (BDH) at 30 g resin/500 mL formamide. The solution is filtered through a glass fiber "B" filter (Whatman). The shelf life of formamide can be extended by storing at 4°C (several months) or –20°C (indefinitely) and in a dark bottle.
11. Other reagents required: 4M NaCl (store at room temperature after autoclaving); 5M LiCl (store at room temperature after autoclaving); 10% (w/v) SDS; 4M potassium acetate (store at room temperature after autoclaving); 1M Tris-HCl (pH 9.2); 100 mM Tris-HCl (pH 8.0), 10 mM EDTA (store at room temperature after autoclaving); 100 mM Tris-HCl (pH 8.0), 1 mM EDTA (store at room temperature after autoclaving); 0.2M EDTA (pH 7.5); and 10%

(w/v) ammonium persulfate (either make fresh or store at 4°C).

2.2. Stock Solutions Stored at -20°C

1. 10x Kinase buffer: 0.5M Tris-HCl (pH 9.2), 0.1M MgCl$_2$.
2. 10x Terminal deoxynucleotidyl transferase buffer (10x TdT buffer): 1M potassium cacodylate (*see* section 5) (pH 7.2), 20 mM CoCl$_2$, 2 mM dithiothreitol.
3. Storage buffer: 50 mM NaCl, 20 mM Hepes (pH 7.9), 5 mM MgCl$_2$, 0.1 mM EDTA, 1 mM dithiothreitol, 20% (v/v) glycerol.
4. Sequencing gel sample buffer: 95% (v/v) formamide, 0.025% (w/v) xylene cyanol FF, 0.025% (w/v) bromophenol blue.
5. Dithiothreitol (Sigma) 50 mM in H$_2$O.
6. Spermidine (Aldrich): 20 mM in H$_2$O.
7. Yeast tRNA (BRL): 10 mg/mL in H$_2$O.
8. ATP: 0.1M in H$_2$O and neutralized with 1M NaOH.
9. Proteinase K (Boehringer): 10 mg/mL in H$_2$O.
10. DNAseI (Worthington): 2 mg/mL in storage buffer.
11. *E. coli* DNA (Sigma): 0.5 mg/mL in 10 mM Tris-HCl (pH 8.0), 1 mM EDTA (sonicated until nonviscous).
12. Restriction and DNA modifying enzymes: used as described by the commercial suppliers.

2.3. Reagents/Special Equipment

1. [γ-^{32}P]-ATP (>5000 Ci/mmol) and [α-^{32}P]-ddATP (2',3'-dideoxyadenosine 5'-[α-^{32}P]-triphosphate) (>5000 Ci/mol), aqueous stocks (10 mCi/mL) as triethylammonium salts (Amersham).
2. Elutip D mini-columns (Schleicher and Schuell).
3. Colloidon bags (Sartorius GmbH).
4. DNA sequencing system (Maxam-Gilbert procedure, NEN Products).
5. 3'-End-labeling kit (Amersham).

6. Sterilization filter units (Nalgene).
7. Low melting point agarose (LMP)-(BRL).
8. X-ray film: Kodak XAR5.

Abbreviations used include: TEMED: N,N,N',N'-tetramethylethylenediamine; ddATP: dideoxyadenosine triphosphate; EDTA: ethylenediamine tetraacetate; EGTA: ethylene glycol-*bis*-(β-aminoethyl ether) N,N,N',N'-tetraacetate.

3. Methods

3.1. 5' End-labeling of DNA Restriction Fragments Using Bacterial Alkaline Phosphatase and T4 Polynucleotide Kinase

1. Plasmid (50 μg) DNA is linearized by digestion with a restriction enzyme that yields 5' overhang sites. The DNA is deproteinized by organic ex-tractions (*see* section 5) and ethanol precipitated in the presence of 0.02 vol of 4M NaCl. The DNA is pelleted, redissolved in 100 μL of 0.1M potassium acetate, 10 mM TrisHCl (pH 8.0), and 1 mM EDTA, and ethanol precipitated. The DNA is pelleted and redissolved in 120 μL of 10 mM TrisHCl (pH 8.0) and 0.1 mM EDTA (and can be stored at –20°C).
2. To dephosphorylate the 5' ends, add 2.5 μL of 4M NaCl and 2 μL of BAP to the DNA solution (BAP is diluted in the Tris/EDTA solution), and incubate at 37°C for 30 min. (If the DNA has blunt ends, continue the incubation at 60°C for 30 min. This melts the DNA ends and facilitates the dephosphorylation.) Add 1 μL of 0.2M EDTA, pH 7.5, and incubate at 70°C for 5 min to inactivate the enzyme. The solution is deproteinized by organic extractions and ethanol precipitated.
3. The DNA is pelleted and redissolved in 30 μL of 10 mM Tris-HCl (pH 8.0) and 0.1 mM EDTA, placed in a pre-

washed Sartorius collodion dialysis bag, and dialyzed overnight against 500 mL of 2.5 mM TrisHCl (pH 9.2) at 4°C. Lyophylize the dialyzed solution and redissolve in 50 µL of water. This DNA is sufficient for seven kinasing reactions (700 footprinting experiments) and can be stored at −20°C.

4. The 5' ends are labeled with $^{32}PO_4$ as follows: Add 7 µL (7 µg) of the BAP-treated DNA to 3 µL of 10x kinase buffer, 3 µL of 50 mM dithiothreitol, 3 µL of 20 mM spermidine, and 11 µL of H_2O. Heat at 70°C for 5 min to denature the ends of the DNA and quick chill at −50°C. Thaw the solution on ice, add 5 µL of aqueous (γ-^{32}P)-ATP (50 µCi) and 7 U of T4 polynucleotide kinase. Incubate at 37°C for 30 min. Add 1 µL of 0.2M EDTA (pH 7.5) and incubate at 70°C for 10 min (to inactivate the enzyme). Add 100 µL of 10 mM TrisHCl (pH 8.0) and 1 mM EDTA, and deproteinize by organic extractions. Add 10 µL of 4M NaCl and 3 µL of tRNA (10 µg/µL), and ethanol precipitate. To remove residual unincorporated label and other contaminants, the pelleted DNA is redissolved in 100 µL of 0.1M potassium acetate, 10 mM TrisHCl (pH 8.0), and 1 mM EDTA, and reprecipitated with 3 vol of ethanol at −50°C.

5. The DNA is pelleted and redissolved in a final volume of 30 µL of enzyme digestion buffer, the second restriction enzyme added, and the digest incubated at 37°C for 2 h. It is not necessary to deproteinize the DNA prior to LMP agarose gel electrophoresis (see below). Add 15 µL of DNA agarose gel electrophoresis sample buffer prior to loading onto the gel.

3.2. Alternative Procedure for Dephosphorylating 5' Ends of DNA Restriction Fragments Using Calf Intestinal Phosphatase

This procedure is simpler than the previous procedure using bacterial alkaline phosphatase since the phosphatase can be added directly to the first restriction enzyme digestion mixture after the completion of the restriction cutting. Also the dialysis and lyophylization steps are not necessary.

1. 50 µg of plasmid DNA is digested with the first restriction enzyme in a total volume of 100 µL.

2. At the end of the reaction, 20 U of calf intestinal phosphatase (Boehringer) is added, and the incubation is continued at 37°C for 30 min. 10 µL of 10% SDS, 5 µL of 2*M* NaCl, and 5 µL of 0.2*M* EDTA is added, and the mixture is heated to 68°C for 5 min to inactivate the enzymes. The DNA is then extracted with phenol–chloroform and ethanol-precipitated in the usual manner. The precipitate is washed with ethanol, dried, and redissolved in 10 µL of 100 m*M* Tris-HCl and 1 m*M* EDTA. The DNA can then be labeled as described above.

3.3. 3' End-Labeling of DNA Restriction Fragments

1. 50 µg of plasmid DNA is digested to give a 3' overhang. The solution is extracted, ethanol precipitated, and reprecipitated as described above The DNA is dissolved in 50 µL of 10 m*M* Tris-HCl (pH 8.0) and 1 m*M* EDTA and can be stored at –20°C.

2. DNA containing 10 pmol of 3'-OH groups is end-labeled in a single reaction. For example, a 3200-bp plasmid with a single restriction cut contains ~2 pmol of 3'-OH groups µg of DNA. Thus, 5 µL (5 µg) of DNA are added to 5 µL of 10x TDT buffer, 30 µL of H_2O, 5 µL of aqueous (γ-^{32}P)-ddATP, and 5 µL (10 µL) of terminal deoxynucleotidyl transferase. Incubate the solution at 37°C for 60 min. Add 1 µL of 0.2*M* EDTA (pH 7.5) and 150 µL of 10 m*M* Tris-Cl (pH 8.0) and 1 m*M* EDTA. Deproteinize and ethanol precipitate as described above.

3. The DNA is digested with the second restriction enzyme as described in the previous sections. After digestion add 15 µL of DNA gel electrophoresis sample buffer.

3.4. Isolation of End-Labeled DNA from LMP Agarose Gels

1. After secondary restriction enzyme digestion, end-labeled DNA (45 μL) is loaded into three wells (3 x 15 μL) of a 1.5% LMP agarose gel (0.5 cm thick) containing 1x DNA electrophoresis buffer, and electrophoresis for 3–4 h at 120 V (~200 mA) at 4°C is carried out to separate the insert from the plasmid DNA.

2. Remove the gel carefully and place on a glass plate. The electrode buffer will be radioactive, so care to prevent spillage should be taken. There is 5–10 x 10^6 cpm in the gel itself, so all manipulations should be done behind perspex and with suitable precautions. Cover the gel with cling film, and place a second glass (or perspex) plate on top.

3. To locate the DNA bands, place an X-ray film (Kodak XAR5) between the cling film and the upper glass plate in a darkroom, and expose the film in the dark for 5 min. Develop the film and, using it as a template, cut out the desired radioactive agarose band. It is desirable to keep the size of the agarose strip to a minimum. Re-expose the gel to ensure the band has been correctly excised.

4. Place the agarose strip in a capped tube, add 1 mL of LSB buffer, and incubate at 65°C with periodic vortexing until the agarose is fully dissolved. Add 8 μL of LSB at 42°C, mix thoroughly, and incubate at 42°C for 10–15 min.

5. The DNA is then purified by chromatography as follows: Pre-equilibrate two Elutip D columns (Schleicher and Schuell) as follows: Using disposable syringes and plungers, wash the columns with 3 mL of HSB buffer, and then 5 mL of LSB buffer.

   Take half of the DNA solution (5 mL) in a syringe and pass it through a column. Pressure is required, so care must be taken to avoid the column being forced off the syringe. This step should be done reasonably quickly to avoid cooling the agarose, which could then block the column. Immediately wash the column with 5 mL of LSB at 42°C. The DNA is eluted with 0.6 mL of HSB.

Repeat with the remaining agarose and the second
column. The two HSB eluates are pooled (1.2 mL). Add 2
μL of tRNA and ethanol precipitate (3 vol) at –50°C. The
DNA is pelleted and redissolved in 50–100 μL of 10 mM
Tris-HCl (pH 8.0) and 1 mM EDTA.

6. Assuming a 50% recovery of a 600 bp insert in a 2700 bp
vector, the 5–7 μg of phosphatased DNA starting material
should yield 0.5–1 μg of purified end-labeled DNA insert.
The stock solution is at ~10–20 ng/μL, and a specific activity of more than 1 x 10^6 cpm/μg is expected (10^4 cpm/μL).
The sensitivity of the footprint technique depends on the
relative concentrations of the DNA fragment and specific
protein in the binding incubation (*see* section 3.5.). It is,
therefore, desirable to minimize the amount of end-labeled DNA added. Depending on the specific activity obtained, it may be possible to reduce the concentration of
the end-labeled DNA stock solution to below 10 ng/μL as
long as ~10^4 cpm are used in each footprint reaction.

3.5.. Footprint Reactions

1. Protein (0–30 μg) in storage buffer is incubated with 1 μL
(<20 ng) of end-labeled DNA in a final volume of 100 μL of
storage buffer (or 30 μL of binding buffer) on ice for 90 min,
in the presence or absence of competitor *E. coli* DNA.

2. The protein–DNA complex is then digested briefly with
DNAse I. DNAse I digestion conditions will vary according to the amounts of protein and *E. coli* DNA in the sample. It is often necessary to titrate the reaction in a pilot experiment using 1–10 μL of serial dilutions of a stock DNA-
se I solution (2 μg/μL). Add the DNAse I to the sample on
ice, rapidly vortex, and incubate at room temperature for
a total of 15 or 30 s. Add 100 μL of footprint STOP buffer.

3. Protein is degraded by proteinase K during an incubation
at 37°C for 30 min, and the DNA denatured at 90°C for 2

min. Deproteinize by organic extractions. Add 15 µL of 5*M* LiCl and ethanol precipitate (3 vol) at –50°C (Li-SDS is soluble in ethanol). Pellet the DNA and wash the pellet with 1 mL of cold ethanol. Dry the pellet in a vacuum desiccator.

4. Dissolve the pellet in 4–8 µL of sequencing gel sample buffer. Heat denature at 90°C for 3 min and quick chill at –50°C. Thaw just prior to loading on the sequencing gel.

3.6. DNA Sequencing Polyacrylamide Gel Electrophoresis

The DNAse I-digested DNA is resolved on a sequencing gel (34 x 40 cm x 0.4 mm). One glass plate should be siliconized to aid pouring and subsequent removal at the end of the electrophoresis.

1. Dissolve 42 g urea, 5.7 g acrylamide, 0.3 g bisacrylamide, and 10 mL 10 x TBE in a final volume of 100 mL (with H_2O). Warm to ~20°C to aid dissolving (do not overheat). Filter the solution under vacuum through a Nalgene sterilization filter and leave under vacuum for 1–2 min to degas (this facilitates polymerization).

2. Transfer the solution to a beaker and, while stirring, add 200 µL of 10% ammonium persulfate and 70 µL of TEMED. Pour the gel with a 50-mL syringe. Polymerization should occur in 5–10 min at room temperature. Allow the gel to polymerize for at least 1 h (gels can be stored for up to 36 h at room temperature).

3. Pre-electrophorese for 1 h at 40 mA using 1x TBE buffer in the electrode chambers. This removes persulfate ions and heats the gel. The surface temperature should reach 50–60°C.

4. Load the samples (4 µL) and electrophorese for 90–180 min at 40 mA (~1500 V). Bromophenol blue migrates at ~30 b (bases) and xylene cyanol at ~130 b. To resolve 20–250 b on one gel, it is possible to do two loadings, i.e., dissolve the samples in 8 µL of sample buffer, load 4 µL, and after 90

min of electrophoresis, load the remainder in separate wells, and electrophorese for a further 90 min. The first and second loadings will resolve ~100–250 b and ~20–100 b from the end-label, respectively.

5. After electrophoresis, remove the siliconized plate, leaving the gel attached to the plate. Place a sheet of Whatman paper over the gel, press it gently down on the gel, and peel it off together with the gel attached. Cover the gel surface with "cling-film," and dry the gel under vacuum at 80°C for 60 min. Expose against preflashed Kodak XAR5 film at -70°C in a cassette containing Lighting-plus (Dupont) screens.

3.7. Practical Example

To investigate the sequence-specific interaction between a chicken erythrocyte nuclear DNA-binding protein fraction and the 5'-flanking sequence of the chicken β^A-globin gene, a cloned genomic 585-bp PvuII restriction fragment was subcloned [pB^A 650.1 (ref. *12*)] into the HincII site of the bacterial vector pUC9 (ref. *11*) by blunt-end ligation. The recombinant has 370 bp of 5'-flanking sequence, extends 215 bp into the coding region, and resulted in a unique pUC9 Hind III restriction site at the 5' end and a unique pUC9 EcoRI restriction site at the 3' end of the insert. The partial sequence is shown in Fig. 3 (*also see* refs. *12,14,15*). End-labeled insert DNA was obtained by linearizing plasmid DNA by restriction with Hind III, treating with BAP, and labeling (7 µg) with (γ-^{32}P)-ATP and polynucleotide kinase (*see* Fig.2). After secondary digestion with EcoRI, the end-labeled insert DNA fragment was separated from the end-labeled pUC9 vector DNA by 1.5% (w/v) LMP agarose gel electrophoresis, and the DNA isolated. In this experiment a total of 2.9 x 10^6 cpm were recovered and, assuming a 50% recovery, yields an estimated specific activity of 4 x 10^6 pm/µg of DNA. The DNA was dissolved in 50 µL of 10 mM Tris-HCl (pH 8.0), 1 mM EDTA, and 1 µL aliquots (~15 ng of DNA, 6 x 10^4 cpm) used in footprint reactions.

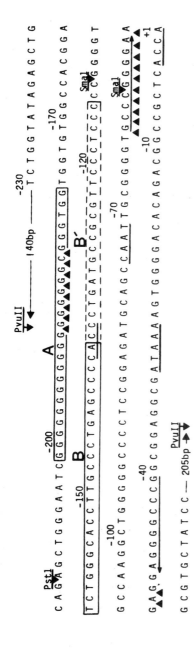

Fig. 3. Partial sequence of the 5'-flanking sequence of the cloned genomic chicken β^A-globin gene. The 575-bp PvuII DNA restriction fragment from the chicken β^A-globin gene recombinant pBIEHI (12) was subcloned into the HincII site of pUC9 (pB^A650.1, 12) to give a unique pUC9 Hind III restriction site at the 5' end (nucleotide 385) and a unique pUC9 EcoRI restriction site at the 3' end (nucleotide + 230). The partial sequence is as described (12,14,15) and "CAP" (ACCA), "TATA" (ATTA), "CAAT," and "CACCC" consensus sequences are underlined. The footprint sequences are boxed (see Figs. 4 and 5, and refs. 12 and 13). The arrowheads denote S1 nuclease cleavage sites in supercoiled plasmids (12) and (⇒ ⇐) an inverted repeat sequence.

A chicken erythrocyte nuclear DNA-binding protein fraction [C4 protein fraction (refs. *12–14*)] was isolated (~0.25 mg of protein/mL) and 0–20 µL incubated with 1 µL (~15 ng) of end-labeled DNA in a final volume of 30 µL (made up with binding buffer) for 90 min at 4°C in the presence or absence of 500 ng of competitor *E. coli* DNA. DNAse I was added (5 µL of a 0.01 mg/mL stock solution) and incubated at room temperature for 15 s, and digestion stopped with STOP buffer. The nucleic acid was purified, resolved by denaturing polyacrylamide gel electrophoresis, and analyzed by autoradiography (Fig. 4).

As shown in Fig. 4, three footprints (A, B, B') are readily identifiable according to the following criteria: (1) There is protein-dependent specific inhibition of DNAse I cleavage compared to the zero protein controls, (2) Footprints B and B' are more sensitive to the presence of competitor *E. coli* DNA than footprint A, (3) As the amount of protein added is increased, footprint A appears first, then footprint B, and then footprint B', suggesting that there are three distinct protein activities that are present in the C4 protein fraction at different concentrations or that have different affinities for their DNA binding sites. The footprinted sequences are shown in Fig. 3.

Two problems are apparent in the design of this particular experiment: First, the footprints are 180–250 bp from the 5'-end-labeled pUC9 *Hind* III site (Fig. 3), and this is beyond the accurate limits of resolution of the sequencing polyacrylamide gel. Second, as more protein is added, there is an increase in the efficiency of DNAseI digestion. The latter is attributed to the progressive decrease in the ionic strength of the footprint reactions as increasing volumes of protein in storage buffer (50 m*M* NaCl) are added to decreasing volumes of binding buffer (100 m*M* NaCl) (also *see* section 4). These problems were solved by 5'-end-labeling the *Pst*I restriction site of the same plasmid (*see* Fig. 3), which is only 11 bp 5' to footprint site A, and by doing the footprint reactions in a final volume of 100 µL made up with storage buffer. In this particular experiment (ref. *12*), the *Pst*I restriction site 3' overhang was first "filled in" by T4 DNA

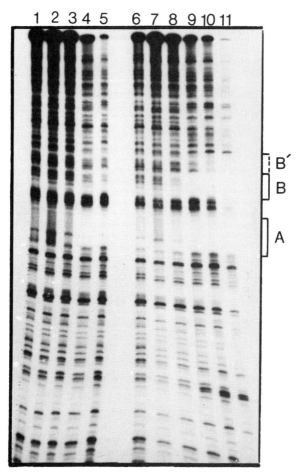

Fig. 4. Footprint analysis of the chicken erythrocyte C4 protein fraction bound to the 5' flanking sequence of the β^A-globin gene. The coding strand of the pB^A650.1 plasmid insert (Fig. 3) was 5' end-labeled at the pUC9 Hind III restriction site and purified after secondary restriction with EcoRI. Erythrocyte C4 protein (0.2 mg/mL) was incubated with 1 μL (~15 ng) of end-labeled DNA and in the presence (lanes 1–5) or absence (lanes 6–11) of competitor *E. coli* DNA (500 ng), in a final volume of 30 μL (in binding buffer) on ice for 90 min. After DNAseI digestion and nucleic acid purification, samples were electrophoresed (40 mA) for 3 h on a denaturing 6% polyacrylamide gel. Lanes 1, 2, 6, and 7, no protein controls; lanes μl 3 and 8, 0.5 μL protein; lanes 4 and 9, 4 μL protein; lanes 5 and 10, 10 μL protein; lane 11, 20 μL protein.

polymerase in the presence of excess deoxynucleotides (see refs. 6,12), the resulting blunt end treated with BAP, and the 5'-OH group phosphorylated with (γ-^{32}p)-ATP and T4 polynucleotide kinase. After secondary digestion eith EcoRI, the PstI-EcoRI restriction fragment (Fig. 3) was isolated after LMP agarose gel electrophoresis. Unless it is specifically necessary to 5'-end-label a 3' overhang restriction site, 3'-end-labeling by the terminal deoxynucleotidyl transferase method would normally be recommended.

The autoradiographs of two footprint experiments with the end-labeled PstI-EcoRI restriction fragment are shown in Figs. 5A and B, and include the four "Maxam and Gilbert" sequencing reactions (2) resolved in parallel (lanes 1–4). Although the footprints can be accurately mapped (see Figs. 3.5A,B), the DNAse I digestion conditions had to be varied for the following reasons: As shown in Fig. 5A, footprints B and B' can be readily deduced using the criteria outlined above. Footprint A is more complicated, however, since its polydeoxyguanosine sequence (Fig. 3) is comparatively resistant to DNAse I cleavage in the presence or absence of protein (Fig. 5A). To confirm footprint A, therefore, it was necessary to repeat the experiment shown in Fig. 5A, but with the addition of twice as much DNAse I to the footprinting reactions (10 µL instead of 5 µL of a 0.2 mg/mL DNAseI stock solution), as shown in Fig. 5B. In this case of excessive DNAse I digestion (lanes 6–8), it is now possible to confirm footprint A (and B and B') compared to the zero protein control (lane 5) and compared to the more dilute thymus C4 protein footprint reactions (lanes 11,12). The thymus C4 protein footprints (Fig. 5A, lane 10; Fig. 5B, lanes 9–12) indicate that the β^A-globin gene binding activities are present, but extrapolation of the C4 protein nuclear equivalents needed to obtain the footprints suggests that the activities are at least 10-fold more dilute in thymus than in erythrocyte nuclei (12).

This practical example illustrates some of the more extreme limitations of the footprinting technique. DNAseI does

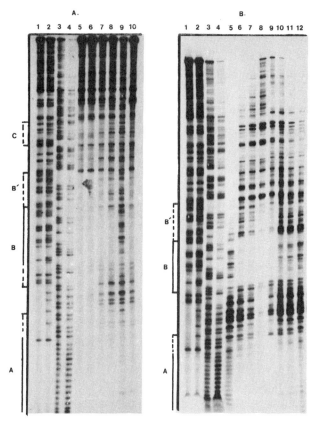

Fig. 5. Footprint analysis of the chicken erythrocyte and thymus C4 protein fractions bound to the 5'-flanking sequence of the chicken β^A-globin gene. The coding strand of the pB^A650.1 plasmid insert (Fig. 3) was 5' end-labeled at the PstI restriction site (nucleotide 212, Fig. 3) and the *PstI-EcoRI* restriction fragment isolated as described in the text and ref. *12*. The fragment was sequenced by the Maxam and Gilbert procedure (2). Lanes 1, 2, 3, 4, C, C+T, G+A, and G sequencing reactions, respectively. End-labeled DNA (1 µL, ~17 ng DNA, 1.8 x 10⁴ cpm) was incubated with erythrocyte (0.25 mg/ml) or thymus (0.08 mg/ml) C4 protein in a final volume of 100 µL (made up with storage buffer) and analyzed as described in the text and ref. *12*. The sequencing gels were electrophoresed for 90 min. (A) Lane 9, no protein control; lanes 5–8, 100, 50, 20, and 2 µL of erythrocyte C4 protein; lane 10, 50 µL of thymus C4 protein. (B) Lane 5, no protein control; lanes 6–8, 2, 5, and 40 µL of erythrocyte C4 protein; lanes 9–12, 50, 25, 12.5, and 5 µL of thymus C4 protein.

exhibit some sequence specificity of cleavage, and, although it is uncommon to find a sequence such as the poly(dG) sequence at footprint A, it is something that should always be controlled out by varying the digestion conditions and perhaps varying the end-labeling site. Using this and other data, two independent laboratories obtained the same results (within 2–3 bp) for the binding of chicken erythrocyte nuclear proteins to the 5'-flanking region of the chicken β^A-globin gene (*12,13*).

4. Notes

1. Low specific activity of end-labeled DNA restriction fragments.

   a. The plasmid DNA preparation may be contaminated with ribonucleotides and other low molecular weight material generated during bacterial lysis. Since these can inhibit (or compete for) CIP, BAP, T4 polynucleotide kinase, or terminal deoxynucleotidyl transferase, plasmid DNA should be purified by gel filtration (e.g., Bio-Gel A15M) or by CsCl-ethidium bromide gradient centrifugation.

   b. Plasmid DNA may be contaminated with bacterial genomic DNA, which will compete for enzyme and radionucleotide in the end-labeling reaction. Supercoiled plasmid DNA can be purified by CsCl density gradient centrifugation in the presence of ethidium bromide (*6,16*) or prepared by the alkali (*6*) or sarkosyl (*16*) cleared lysate procedures, which minimize *E. coli* genomic DNA contamination.

   c. Removal of the bacterial alkaline phosphatase (BAP) before the kinasing step is critical. The dialysis step is essential after the BAP reaction (dialysis is not required if calf intestinal phosphatase is used).

   d. If the first restriction cut is not unique and more than one restriction fragments are going to be end-labeled, the concentrations of radionucleotides, DNA, and kinase should be revised (*see* refs. *6,7–10* and the specifications for the enzyme used).

2. Low recovery of end-labeled DNA restriction fragment from LMP agarose gels: After electrophoresis and autoradiography, the minimum amount of LMP agarose that contains the end-labeled DNA fragment should be excised, since this will directly affect recovery. It is essential that the agarose be completely molten at 65°C, and periodic vortexing will help the process. The use of 1x TBE as the 1.5% LMP gel electrolyte is not recommended since the agarose never fully melts and column blockage often results. If difficulty is experienced, raise the temperature to 70°C. Diluting the molten agarose with 5–10 vol of LSB at 42°C and equilibrating the solution at 42°C prior to chromatography through the Elutip-D mini-columns reduces blockage of the column. After filtering the DNA solution, the 42°C LSB wash removes residual agarose that would affect subsequent footprinting reactions. It should be noted that if the mini-column becomes blocked, we have yet to devise a method of rescuing the experiment.

3. DNAse I digestion in footprinting reactions: Since it is difficult to predict optimum DNAse I digestion in different conditions, it is necessary to approach this problem on a trial-and-error basis whenever condition(s) are changed.

   The absence of DNAse I cleavage in footprinting reactions could be caused by one or more of the following:

a. Inactive DNAse I enzyme: Freezing and thawing DNAse I stock solutions can cause inactivation, so DNAse I should be stored in small aliquots at –20°C. DNAse I dilutions of less than ~1 mg/mL should be made fresh by diluting the 2 mg/mL stock solution with ice-cold storage buffer.

b. DNAse I activity requires trace amounts of Ca^{2+} ions. Although there are usually sufficient trace Ca^{2+} ions in the reagents used to make the footprinting solutions, the efficiency of DNAse I cleavage can be greatly enhanced by adding $CaCl_2$ (to 0.1 mM) to the footprinting reaction. This is especially important if EGTA is present or was used in the preparation of the DNA-binding protein fraction.

There is some evidence that EGTA/Ca^{2+} will interfere with the sample's migration during denaturing poly-acrylamide gel electrophoresis, so it is best to avoid the use of EGTA.

c. Large concentrations of nonspecific DNA-binding proteins can "coat" the end-labeled DNA and give a general protection from DNAseI cleavage. This can be overcome by either diluting the protein added or by adding nonspecific competitor *E. coli* DNA (up to 1 μg). The latter assumes that the specific DNA-binding protein has a higher affinity for the end-labeled DNA sequence than for the nonspecific competitor DNA, which is in a 10–100-fold excess (by weight of DNA). It should be noted that, since the labeled DNA fragment is prepared with carrier tRNA, this in effect can be acting as a competitor in the footprinting reaction. Its exact concentration in the footprinting reaction is not known, however (because of the possibility of nuclease degradation), and therefore, in our more recent experiments, carrier RNA is omitted from the preparation of the end-labeled DNA fragment.

d. High concentrations of salt [(NH$_4$)$_2$SO$_4$ or NaCl] will inhibit DNAse I. DNA-binding protein fractions prepared by (NH$_4$)$_2$SO$_4$ precipitation or elution from DNA-cellulose affinity chromatography at high ionic strengths should be extensively dialyzed against storage buffer (50–100 vol, overnight at 4°C).

Excessive digestion is usually solved by adding increasingly dilute DNAse I to the footprint reaction or by varying the time or temperature of DNAse I digestion. If the problem persists, there can be two causes:

a. The end-labeled DNA restriction fragment was excessively nicked or degraded during preparation and isolation (some nicking is inevitable and, interestingly, is commonly sequence-specific). This can be diagnosed by denaturing polyacrylamide gel electrophoresis of a control untreated (but denatured) aliquot of the end-labeled DNA

stock solution. The presence of excessive (>10%) radio-label migrating as low molecular weight bands indicates that insufficient care was taken to minimize nuclease activity during the preparation of the DNA fragment.
b. The nuclear DNA-binding protein fraction can contain endogenous nucleases that degrade the end-labeled DNA during the binding incubation. This is minimized by doing the DNA–protein binding incubation on ice. The addition of excess (0.1–1 μg) nonspecific competitor *E. coli* DNA can help, since it will compete for the nonspecific DNA-binding nuclease activities. Alternatively, it is possible to do the DNA–protein binding in the absence of divalent cations (i.e., in the presence of EDTA, but not EGTA) to inhibit nuclease activities, but care must be taken to add excess $MgCl_2$ (and possibly 0.1 mM $CaCl_2$) during the DNAse I digestion. Finally, further fractionation of the DNA-binding protein fraction may be necessary.

4. Denaturing polyacrylamide gel electrophoresis: The main problem encountered at this stage is fuzzy or smeared bands after electrophoresis and autoradiography. Possible causes are:
    a. Impurities in the polyacrylamide gel and sample buffer: the reagents used should be ultrapure and, although pre-electrophoresis may not be absolutely necessary, it is recommended (but should not exceed 1 h).
    b. Sample denaturation and loading: To prevent reannealing, samples are quick-chilled on dry ice after the 90°C incubation. Samples should not be thawed until immediately before loading and should be loaded as quickly as possible in a volume of less than 5 μL.
    c. Salt in the sample: Salt in the sample interferes with electrophoresis. When washing the nucleic acid pellet with ethanol prior to drying and dissolving in sample buffer, it is also desirable to wash the sides of the microfuge tube.

d. Protein in the sample: Residual protein in the sample will also interfere with electrophoresis. If necessary, the concentration of SDS in the footprint STOP buffer can be increased; the 37°C incubation in the presence of (proteinase K) STOP buffer can be increased (e.g., 30–60 min), and, the number of organic extractions can be increased (e.g., twice with 1 vol phenol, twice with 2 vol of phenol and chloroform/isoamyl alcohol, and once with 1 vol of chloroform/isoamyl alcohol prior to ethanol precipitation).

e. Excess nucleic acid: It is possible to overload the sequencing gels, and this will interfere with migration during electrophoresis. The amount of carrier tRNA added to the footprint STOP buffer should be kept to a minimum, particularly if competitor DNA (>1 µg) is added to the footprint reaction or if the protein fraction used contains large amounts of nucleic acid (>0.2 µg/µL).

5. General Practical Notes

1. Toxic materials: Phenol is toxic and can cause severe burns. Potassium cacodylate is toxic. Acrylamide and bisacrylamide are neurotoxins. Formamide is a teratogen. Inhalation of chloroform vapor should be avoided. The weighing out, dispensing, and manipulations involving these reagents should be done with suitable precautions. Wear gloves and face masks when weighing out powdered reagents. Organic extractions should be done in a fume cupboard.

2. Radioactivity: DNA end-labeling reactions and the purification of end-labeled restriction fragments involve the use and manipulation of up to 100 µCi of ^{32}P. All standard precautions should be observed for containment and to minimize exposure to ionizing radiation. Gloves, safety glasses, and perspex or glass shields should be employed whenever possible. Radioactive solutions are extracted

with organic reagents and ethanol precipitated. These steps require centrifugation, usually in a microfuge, and it is possible that leakage will occur. The microfuge should therefore be in a contained area (fume hood) and monitored routinely for contamination. The microfuge can be easily decontaminated by wiping down with wet tissues.

3. Electrophoresis: Sequencing polyacrylamide gels are electrophoresed at 40 mA and 1500 V. Safety precautions to avoid contact with the electrode buffer during electrophoresis should be taken.

4. Organic extractions: Unless stated othewise, DNA solutions are deproteinized by sequential organic extractions, twice with a mixture of 1 vol of phenol and 1 vol of chloroform/isoamyl alcohol (24:1), and once with 1 vol of chloroform/iso-amyl alcohol prior to ethanol precipitation.

5. Ethanol precipitations: Unless stated otherwise, aqueous DNA solutions are precipitated with 3 vol of ethanol at –50°C (on dry ice) for at least 10 min.

Acknowledgments

This research was supported by grants from the Medical Research Council and the Cancer Research Campaign. The authors would like to thank other members of the laboratory who have contributed to establishing the techniques described in this chapter, R. Nicolas, C. Wright, and V. Lobanenkov.

References

1. Dynan, W.S. and Tjian, R. (1985) Control of eukaryotic messenger RNA synthesis by sequence-specific DNA-binding proteins. *Nature* **316**, 774–778.
2. Maxam, A.M. and Gilbert, W. (1977) A new method for sequencing DNA. *Proc. Natl. Acad. Sci. USA* **74**, 560–564.
3. Van Dyke, M.W. and Dervan, P.B. (1983) Methidiumpropyl-EDTA-Fe(11) and DNAse1 footprinting report different small molecule binding site sizes on DNA. *Nucl. Acids Res.* **11**, 5555.

4. Wu, C. (1985) An exonuclease protection assay reveals heat-shock element and TATA box DNA-binding proteins in crude nuclear extracts. *Nature* 317, 84–87.

5. Von der Ahe, D., Renoir, J.M., Buchon, T., Baulieu, E.-E., and Beato, M. (1986) Receptors for glucocorticosteroid and progesterone recognise distinct features of a DNA regulatory element. *Proc. Natl. Acad. Sci. USA* 83, 2817–2821.

6. Maniatis, T., Fritsch, E.F., and Sambrook, J. (1982) in *Molecular Cloning: A Laboratory Manual* Cold Spring Harbor Laboratory, Cold Spring Harbor, New York.

7. Berkner, K.L. and Folk, W.R. (1977) Polynucleotide kinase exchange reaction. *J. Biol. Chem.* 252, 3176–3184.

8. Berkner, K.L. and Folk, W.R. (1979) Quantitation of the various termini generated by type II restriction endonucleases using the polynucleotide kinase exchange reaction. *J. Biol. Chem.* 254, 2561–2564.

9. Tu, C-P.D. and Cohen, S.N. (1980) 3'-End labelling of DNA with [α-^{32}P] cordycepin-5'-triphosphate. *Gene* 10, 177–183.

10. Roychoudhury, R., Tu, C-P.D., and Wu, R. (1979) Influence of nucleotide sequences adjacent to duplex DNA termini on 3' terminal labelling by terminal transferase. *Nucleic Acids Res.* 6, 1323–1333.

11. Vierira, J. and Messing, J. (1982) The pUC plasmids, and M13 mp 7-derived system for insertion mutagenesis and sequencing with synthetic universal primers. *Gene* 10, 259–268.

12. Plumb, M.A., Nicolas, R.H., Wright, C.A., and Goodwin, G.H. (1985) Multiple sequence-specific DNA binding activities are eluted from chicken nuclei at low ionic strength. *Nucleic Acids Res.* 13, 4047–4065.

13. Emerson, B.M., Lewis, C.D., and Felsenfeld, G. (1985) Interaction of specific nuclear factors with the nuclease-hypersensitive region of the chicken adult β-globin gene: Nature of the binding domain. *Cell* 41, 21–30.

14. Emerson, B.M. and Felsenfeld, G. (1984) Specific factor conferring nuclease hypersensitivity at the 5' end of the chicken adult β-globin gene. *Proc. Natl. Acad. Sci. USA* 81, 95–99.

15. Dolan, M., Dodgson, J.B., and Engel, J.D. (1983) Analysis of the adult chicken β-globin gene. *J. Biol. Chem.* 258, 3983–3990.

16. Clewell, D. and Helinski, D.R. (1970) Properties of supercoiled deoxyribonucleic acid-protein relaxation complex and strand specificity of relaxation event. *Biochemistry* 9, 4428–4440.

Chapter 13

Oligonucleotide Synthesis Using the Manual Phosphotriester Method

Daniel M. O'Callaghan
and William J. Donnelly

1. Introduction

During the past few years, synthetic DNA, used as a primer in DNA polymerization, in site-directed mutagenesis or as a probe in gene selection, has assumed a central role in recombinant DNA technology (1). This was made possible by the development of methods for efficient solid phase chemical synthesis. Deoxyribonucleotides are ideally suited to solid-phase synthesis since their relative chemical uniformity allows for application of oligodeoxyribonucleotide (oligonucleotide) purification techniques, which are largely dependent on chain length. Hence, the potential disadvantage of omitting purification after each coupling step is minimized. This contrasts with peptide synthesis in which purification of the final product is more difficult, thereby placing greater demands on coupling efficiency. In practice, coupling efficiencies of at least 95% are now attainable in oligonucleotide synthesis because of the availability of highly reactive mononucleotides and specially developed coupling catalysts. In optimized systems, this allows for synthesis of oligomers greater than 50 bases in length.

Much emphasis is now placed on automated synthesis machines, and there are several currently available (2,3). It is easy to conclude from the optimistic tone of the associated literature that synthesis is a routine, uncomplicated process requiring little more than the pressing of buttons. This is far from the truth. Synthesis requires rigorous attention to detail, to the quality of every reagent and solvent, to the exclusion of moisture, and to the operation of each step in the process. The difference between an overall coupling efficiency of 95 and 85% may be the difference between success and failure. It is a tribute to the quality of automated instruments that, with careful use, many of the potential problems are avoided. However, this is because a rigid protocol is enforced on the operator and all reagents are supplied ready-to-use. Even then, few will embark on automated synthesis without teething problems, and even in the best situation unanticipated difficulties will arise because of reagent variations or sequence-specific effects. Because of the importance of using suitable reagents and solvents, the sources of reagants and solvents that have proved successful in our hands are detailed in this protocol.

Manual synthesis confronts the operator with the potential problems in a more direct manner. Without push-button technology, attention is focused on every component, and if, in addition, solvents and reagents are prepared *in situ*, the pitfalls of inadequate purity are quickly learned. The benefit, however, is the much more comprehensive understanding that results, and even when an automated machine is eventually purchased, it is likely to present far fewer problems in a laboratory accustomed to the manual procedures. Apart from that consideration, manual synthesis is mandatory when the cost of an automated machine is prohibitive and is preferable from an economic point of view when there is only a limited or occasional requirement for oligomers (say, 20–30 per year).

Currently, two general approaches to solid-phase synthesis exist, one using phosphite triester and the other using phosphate triester chemistry. Detailed descriptions of both approaches have been published (*see* ref. *4* for reviews), and for manual use they are equally effective. The phosphite triester method is described in Chapter 14.

This chapter describes phosphotriester synthesis by the method of Gait et al. (*5*) as refined by Sproat and Bannwarth (*6*). The principal steps are illustrated in Fig. 1. The starting material is a solid support, controlled pore glass (CPG), to which is attached the first nucleoside through an ester linkage at the 3' position (Fig. 1i). This is substituted with the transient 5'-protecting group dimethoxytrityl (DMT) and with base-protecting groups isobutyryl or benzoyl. Synthesis procedes by sequential addition of nucleoside-3'-monophosphate monomers (Fig. 1ii) through formation of internucleotide phosphate links. Monomers carry the 5' and base protecting groups and the phosphate protecting group *o*-chlorophenyl. Before coupling, the 5'-DMT-protecting group on the support-bound nucleoside is removed by a brief treatment with dichloroacetic acid. Nucleoside monomer is then added, together with coupling agents I-(mesitylene sulfonyl)-3-nitro-1,2,4,-triazole (MSNT) and *N*-methylimidazole, allowing 15 min for completion of the reaction. Unreacted 5'-OH may then be capped off by acetylation with acetic anhydride. This capping step is routinely used in phosphite triester synthesis, but has only a limited role in phosphate triester synthesis, as will be discussed later. 5' Deprotection of the newly formed dinucleotide (Fig. 1iii) is then carried out to initiate the next addition cycle. Synthesis reactions are carried out in an enclosed continuous-flow system under argon, with the support-linked nucleotides contained in

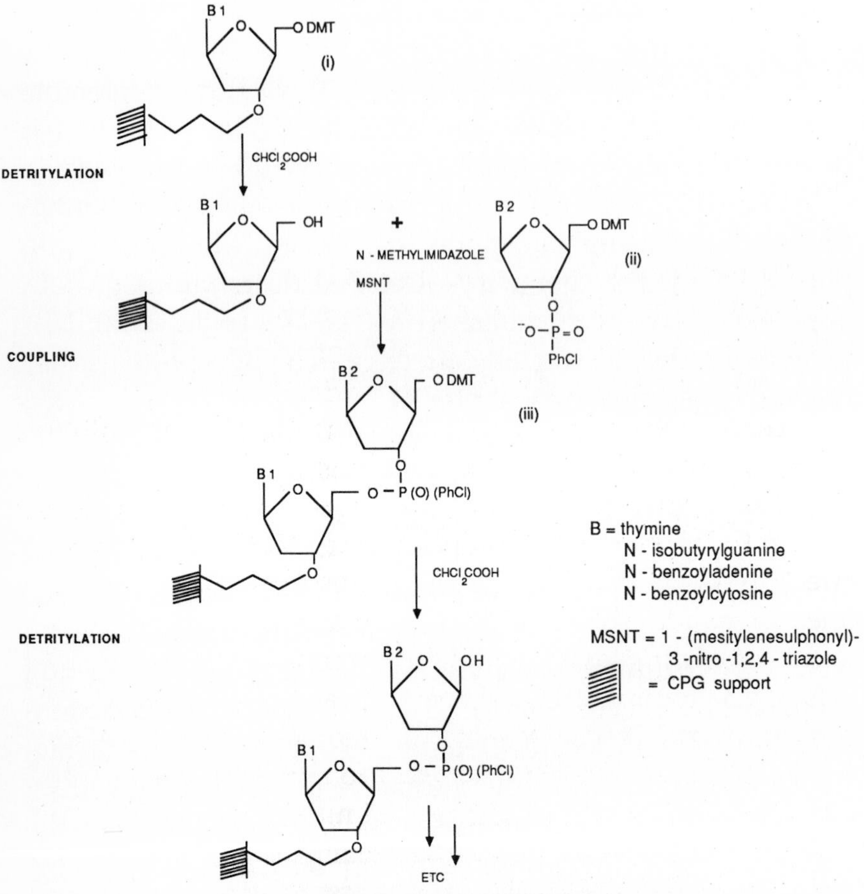

Fig. 1. Phosphotriester synthesis reactions.

a glass column. This facilitates exclusion of moisture, and allows for precise control of reaction times and quick change of solvents by simple manipulation of a selector valve. At the end of chain elongation, a series of reactions are carried out to generate the finished product. Liberation of oligomer from the support and removal of the *o*-chlorophenyl residues is achieved by treatment with *syn*-2-nitrobenzaldoxime and tetramethylguanidine. The base-protecting groups are removed with concentrated ammonia. Finally the 5'-DMT-protecting group is removed with acetic acid. Before 5' deprotection, it is advantageous to carry out a preliminary purification step over a hydrophobic resin, exploiting the binding characteristics of the DMT residue. The purity of the product at this stage depends on coupling efficiency and base composition. In the case of purine-rich sequences, the required oligomer may be contaminated with shorter sequences from which it can be separated by preparative gel electrophoresis or HPLC. Frequently, pyrimidine-rich oligomers are less contaminated and do not require further purification, particularly when used as probes. The length of oligomer that can be prepared by the above method is also dependent on base composition. Oligomers of about 20 bases should present no difficulty regardless of sequence. For pyrimidine-rich sequences it should be possible to obtain reasonable yields of up to 30 bases.

It is essential to carry out the synthesis in an enclosed system to ensure exclusion of moisture, which has a dramatic effect on coupling yields. This also allows for continuous-flow operation and facilitates the frequent solvent changes that are necessary during synthesis. The main components of a manual synthesizer (supplied by Omnifit Ltd.) are shown in Fig. 2. This consists of one to four 6.5-mm precision bore, variable-length chromatography columns coupled to a solvent delivery system. Solvents are delivered from a number of glass reagent bottles fitted with screw caps containing ON/OFF valves and are maintained under a positive pressure of dry argon via a pressure stat. Solvent selection and flow is controlled manually by a rotary valve and distributed to the column(s) through a three- or eight-way valve. Approximately 1 m of 0.3-mm id Teflon tubing fitted with a two-way valve is connected to the end of each column; this creates the necessary back pressure to enable maintenance of constant pressure in the system. Detailed assembly instructions are supplied with each DNA-synthesis kit and therefore will not be described here. Note that all tubing must be made of Teflon, and all solvent line connections should be of the high-pressure type. In our experience, type 4 connections (cf. Omnifit manual, Flangless miniature fluid fittings) are most reliable in preventing leaks. Solvent

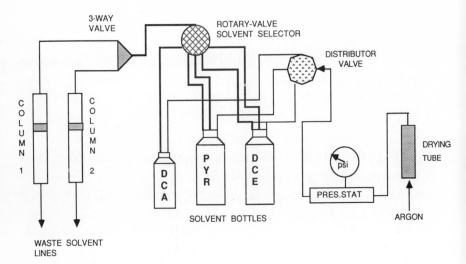

Fig. 2. Schematic diagram of the manual DNA bench synthesizer (Omnifit).

lines are connected to the rotary valve selector anticlockwise in the order pyridine, dichlorethane, 3% dichloracetic acid in dichlorethane, dichlorethane, pyridine.

2. Materials

In the past it was necessary to prepare monomers, coupling reagents, and functionalized support in the laboratory because commercial materials were not satisfactory. Currently, special synthesis grades of these materials are available from a small number of suppliers, and their availability greatly facilitates operations. It is recommended that the commercial materials be used since laboratory preparation is time-consuming. Procedures for laboratory preparation are well documented for those wishing to prepare their own (4). A detailed list of materials is given in Table 1.

Careful exclusion of moisture from all synthesis reagents is vital. When reagents are stored at –20°C, these must be allowed adequate time to equilibrate to room temperature, before opening, to prevent condensation.

Table 1
Materials Available from Commercial Suppliers

Solvents and reagents	Grade	Supplier
Acetic acid	Analar	BDH
Acetic anhydride	Analar	BDH
Acetonitrile	HPLC	BDH
Acrylamide	Electran	BDH
Amberlite mixed-bed resin	MBI	Serva
Ammonia solution (35% NH_3)	Analar	BDH
Ammonium persulfate (APS)	Electran	BDH
Barium oxide	GPR	BDH
Basic alumina	MB-5	Sigma
Boric acid	Analar	BDH
Bromophenol blue		Sigma
Calcium chloride	Analar	BDH
Dichloroacetic acid	Puriss	Fluka
Dichloroethane	Analar	BDH
Diethyl ether	Analar	BDH
4-Dimethyl amino pyridine (DMAP)		Aldrich
1,4-Dioxan	Analar	BDH
Dithiothreitol (DTT)		BDH
Ethanol	Analar	BDH
Ethylene diamine tetraacetic acid (Na$_2$ salt) (EDTA)	Analar	BDH
Formamide	Analar	BDH
Hydrochloric acid	Analar	BDH
2,6-Lutidine		Aldrich
Magnesium chloride	Analar	BDH
N, N'-Methylene bis-acrylamide	Electran	BDH
N-methylimidazole	SPS	Cruachem
Mixed-bed ion-exchange resin	MBI	Serva
Mesitylene-sulfonyl-3-nitro-1,2,4-triazole	SPS	Cruachem
Molecular sieves 5A	60-80 mesh	Phase Sep
Ninhydrin	Analar	BDH
syn-2-Nitrobenzaldoxime	SPS	Cruachem
Perchloric acid (60%)	Analar	BDH

(cont'd. on next page)

Table 1 *(cont'd.)*

Solvents and Reagents	Grade	Supplier
Phenylisocyanate (iso-cyantobenzene)	GPR	BDH
Phosphorous pentoxide	GPR	BDH
Potassium hydroxide	Analar	BDH
Pyridine	Analar	BDH
Silica gel (self-indicating)		BDH
Sodium acetate	Analar	BDH
Sodium metal	GPR	BDH
Tetrahydrofuran	Analar	BDH
N,N,N',N'-Tetramethyl-1,2-diaminoethane (TEMED)	Electran	BDH
1,1,3,3-Tetramethylguanidine (TMG)	DNA	BDH
Triethylamine	Analar	BDH
Urea	Analar	BDH

Functionalized CPG supports[a] (20–40 µmol/g)

DMTr-T-lcaa-CPG		Cruachem
DMTr-ibu-dG-lcaa-CPG		Cruachem
DMTr-bz-dA-lcaa-CPG		Cruachem
DMTr-bz-dC-lcaa-CPG		Cruachem

Protected deoxyribonucleotides

DMTr-T-p-(2 clPh) TEA salt		Cruachem
DMTr-ibu-dG-p-(2 clPh) tea salt		Cruachem
DMTr-bz-dA-p-(2 clPh)TEA salt		Cruachem
DMTr-bz-dC-p-(2 clPh) TEA salt		Cruachem

Miscellaneous

Manual DNA-bench synthesizer kit (3600 Series)		Omnifit

[a]Functionalized CPG supports with medium loading (50–100 mol/g) and high loading (150–200 µmol/g) are also available.

(cont'd. on next page)

Table 1 *(cont'd.)*

Miscellaneous	Supplier
Teflon-faced silicone septa (cat. no. 3302)	Omnifit
Teflon frits (cat. no. 6652)	Omnifit
Polynucleotide kinase	Boehringer,
enzyme (T4) (cat. no. 174645)	Mannheim
$[(\gamma)^{32}P]$-ATP (cat. no. PB10218)	Amersham
Kieselgel 60 F_{254} TLC plates	Merck
C_{18} Sep Pak cartidges	Waters

Syringes

10 µL (801RN)	Hamilton
50 µL (1705RN)	Hamilton
250 µL (1725RN)	Hamilton
500 µL (1750RN)	Hamilton
1 mL (1001RN)	Hamilton

Stoppers, sleeves, and film

Subaseal stoppers (to fit 14/23 Quickfit joints)	Gallenkamp
Teflon sleeves (to fit Quickfit joints)	Fisons
X-OMAT AR film	Kodak
X-ray developer (LX-24)	Kodak
X-ray liquid fixer (FX-40)	Kodak

Glassware

All glassware should be scrupulously clean and in most cases dried at 150°C for 2 h before use. Small items, such as Reacti-vials and pear-shaped flasks, are cooled in a vacuum desiccator and kept there until required. Standard glassware, such as that required for distillations, will not be listed here (cf., any Quickfit catalog). Note, however, that small glassware such as 5-mL pear-shaped flasks, and 25-mL round-bottom flasks should have 14/23 sockets so that they may be stopped with subaseal stoppers.

(cont'd. on next page)

Table 1 *(cont'd.)*

Miscellaneous	Supplier

Solvent storage bottles

Duran-type borosilicate glass storage bottles with Resinol screw caps and PTFE protecting seals are recommended as general solvent storage bottles (Schott Glassware Ltd.) since they can be attached directly to the DNA bench synthesizer. (Note, pouring ring should be removed first).

Special glassware	Cat. no.	Supplier
1-mL Reacti vials	13221	Pierce
3-mL Amber storage vials	13082	Pierce
Silicone/Teflon septa to fit above	12712	Pierce
Sintered glass funnels with 14/23 joints and side arm	FDK360G	Gallenkamp

2.1. Preparation of Reagents

1. Dehydration of monomers: Experience has shown that nucleotide monomers must be rigorously dried, even when obtained from specialist suppliers. This is best carried out by azeotropic evaporation with pyridine within a day of synthesis. The required amounts of each of the four monomers are weighed into clearly labeled 5-mL pear-shaped flasks and stoppered immediately with subaseal stoppers. For a synthesis on the 1 μM scale, 13.3 μmol of monomer are used for each coupling reaction. When calculating the amount of monomer required per synthesis, it is advisable to allow for two to three extra couplings (*see* Table 2 for molecular weights). Co-evaporation with pyridine is carried out on a rotary evaporator using the setup shown in Fig. 3. Careful exclusion of moisture is essential. This is achieved by adding dry pyridine through

Table 2
Deoxyribonucleotides (Monomers)

Reagent	Molecular weight
DMTr-T-p-(2ClPh) TEA salt	836.32
5'-dimethoxytrityl-thymidine 3'-(2-chlorophenyl)-phosphate triethylammonium salt	
DMTr-ibu-dG-p-(2ClPh) TEA salt	931.43
5'-Dimethoxytrityl-*N*-isobutyryl-2'-deoxyguanosine, 3'-(2-chlorophenyl)-phosphate triethylammonium salt	
DMTr-bz-dA-p-(2ClPh) TEA salt	949.44
5'-Dimethoxytrityl-*N*-benzoyl-2'-deoxyadenosine-3'-(2-chlorophenyl)-phosphate triethylammonium salt	
DMTr-bz-dC-p-(2ClPh) TEA salt	925.42
5'-dimethoxytrityl-*N*-benzoyl-2'-deoxycytidine 3'-(2-chlorophenyl)-phosphate triethylammonium salt	

the feed tube and by using dry argon for vacuum release. The sample flask is transferred to the rotary evaporator and pyridine (3 mL) is admitted under vacuum through the feed tube. The mixture is evaporated to a gum at 30°C (water bath), proceeding with caution to avoid splashing. This step is carried out three to four times and is repeated for each monomer. After evaporation the flask is wiped dry, stoppered with a subaseal stopper, and stored in a vacuum desiccator wrapped in tin foil until required. Immediately before synthesis, dry pyridine is added to

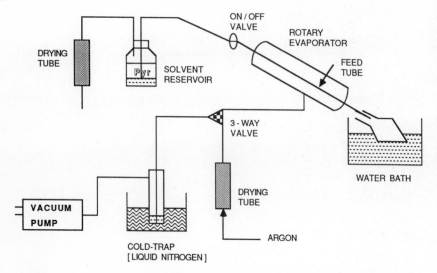

Fig. 3. Schematic diagram of the apparatus for azeotropic evaporation.

each flask to a final volume of $100x$ µL where x is the number of nucleotide additions. This should be carried out using a dry syringe and ensuring continued exclusion of atmospheric moisture.

2. MSNT: For each nucleotide addition, 20 mg of MSNT is weighed into a 1-mL Reacti-vial, which is immediately sealed with a screw cap and a Teflon/silicone septum. Unnecessary exposure of MSNT to atmospheric moisture should be avoided.

3. N-methylimidazole: N-methylimidazole specially supplied for oligonucleotide synthesis may be used directly. Otherwise any high-grade product may be used, but this must first be purified by distillation *in vacuo* under dry nitrogen. The product is stored desiccated under dry argon, and for convenience it is recommended that this be done in separate aliquots using screw-capped amber vials sealed with Teflon septa in amounts sufficient for each synthesis.

4. 4-Dimethylamino pyridine (DMAP): The analytical grade reagent is further purified by recrystallization from anhy-

drous diethyl ether. A little decolorizing charcoal is added to the warm mixture before filtration. The crystallized material is dried overnight in a vacuum desiccator over silica gel and then stored desiccated under dry argon.

5. Acetic anhydride: High-purity acetic anhydride (Puriss) available from Fluka may be used directly from the bottle. Alternatively any analytical grade reagent may be used, but this must first be purified by distillation at atmospheric pressure under anhydrous conditions. It is stored in dark screw-capped bottles with Teflon seals. Toxicity/ hazards: irritant vapor, very corrosive, flammable, avoid contact with skin and eyes (TLV 5 ppm).

2.2. Preparation of Solvents

All solvents used during coupling must be anhydrous and free from acidic or basic contaminants. Glassware and reagent bottles for solvent storage are dried at 150°C for 2–3 h before use. During solvent preparation, drying tubes of $CaCl_2$ or molecular sieve 5 Å are used on apparatus to prevent contact with moisture. Most solvents are currently available in special synthesis grade. However, these are expensive, and satisfactory purity can be achieved using the following procedures.

1. 1,2-Dichlorethane (DCE): Analytical grade 1,2-dichlorethane is passed through a column of alumina (basic type MB-5—Sigma; 100 g/L of solvent) and distilled from P_2O_5 (5 g/L) at atmospheric pressure. The anhydrous solvent is stored in screw-capped reagent bottles in the dark under dry argon at 20°C, and is stable for several weeks. Prepare 1 L/synthesis. Toxicity/hazards: Irritant harmful vapor, harmful by skin absorbtion, highly flammable (TLV 50 ppm).

2. 1,4-Dioxane: Analytical grade dioxan is passed through a column of basic alumina (200 g/L) to remove peroxides and stored in dark screw-capped bottles at 20°C. Prepare

100 mL at a time. On exposure to air and light, it rapidly forms peroxides. It is advisable to use freshly prepared solvent. Toxicity/hazards: Irritant vapor, forms explosive peroxides on exposure to light and air, highly flammable (TLV 100 ppm).

3. Pyridine (PYR): Analytical grade pyridine is refluxed with ninhydrin (5 g/L) for 3 h to remove ammonia and primary and secondary amines. It is then distilled at atmospheric pressure. The distillate is refluxed with drying agent, barium oxide, or potassium hydroxide (20 g/L) for 2–3 h and fractionally distilled at atmospheric pressure through a Vigreux column. The first 5–10% of distillate is discarded, and the main fraction is collected at 116°C and stored under dry argon in screw-capped bottles in the dark at 20°C. Prepare 1 L/synthesis. The purified solvent remains stable for several weeks. Toxicity/hazards: Sharp penetrating odor; harmful vapor, highly flammable (TLV 5 ppm).

4. 2,6-Lutidine: Prepared in the same manner as pyridine. It is stored in small, dark, screw-cappped bottles under dry argon. Prepare 100 mL at a time.

5. Phenylisocyanate (iso-cyanatobenzene): Great care must be taken when handling this *extremely* toxic substance (use fume cupboard). The analytical grade reagent is vacuum distilled (H_2O pump) using a dry nitrogen bleed. The first 5% of distillate is discarded, and the main fraction is collected at approximately 55°C at 13 mm Hg pressure. The distillate is transferred to small screw-capped amber vials with Teflon septa and stored desiccated at 4°C under argon. Prepare 50 mL at a time. It remains stable for several months. Toxicity/hazards: Irritant harmful vapor, affects eyes, skin, and respiratory system, very flammable.

6. Diethyl ether: For washing of the support after synthesis, peroxide impurities and most of the water present are removed from analytical-grade diethyl ether by passage through a column of basic alumina (200 g/L). For DMAP

crystallization, complete dryness is achieved by refluxing the alumina-treated solvent with fresh sodium metal wire and distilling at atmospheric pressure, discarding the first few percent of distillate. It is stored at 20°C in dark reagent bottles under dry argon. Freshly dried solvent should be used within 2 d. Alumina-treated solvent may be stored for several weeks. Toxicity/hazards: Dry solvent forms explosive peroxides on exposure to light or air, highly volatile liquid, harmful vapor, highly flammable (flash point is 45°C) (TLV 400 ppm).

7. Tetrahydrofuran (THF): This is purified and dried as described for ether and stored in screw-capped bottles in the dark. The pure anhydrous solvent rapidly forms peroxides and should be used within 2 d of preparation. Prepare 100 mL at a time. Toxicity/hazards: Harmful vapor, liable to form explosive peroxides on exposure to air and light, highly flammable (TLV 200 ppm).

2.3. Preparation of Buffers

1. Triethylammonium bicarbonate: Triethylammonium bicarbonate (TEAB) is prepared by bubbling CO_2 gas, filtered through glass wool, into a suspension of approximately 300 mL of redistilled triethylamine in 1 L of distilled water kept at 0°C in a salt-ice-water bath until the pH of the solution drops to 8.0. The TEAB concentration is determined by standardized titration with methyl orange indicator and adjusted to 2.0M. It is stored in dark screw-capped bottles, and is stable for 2–3 wk at 4°C and for several months at –20°C. Toxicity/hazards: Triethylamine has an irritant vapor and is highly flammable. Avoid contact with skin and eyes (TLV 25 ppm).

3. Method

All operations are described for a single-column synthesis and must be replicated for multiple parallel synthesis. To avoid confusion during synthesis, it is important that as each cycle is completed it is marked off on a guide sheet. Visual monitoring of the eluant flow rate throughout the synthesis is important to ensure that no blockage develops. In the case of multiple columns, it is important that the eluants are kept separate to facilitate this monitoring.

3.1. Assembly of Synthesizer

1. The bench synthesizer is set up as described in the manufacturers manual. When satisfied that the plumbing is leak-free, bottles of anhydrous solvents are connected to their respective lines. The head space of each bottle is flushed with argon, and each solvent is passed through the system for 10–15 s. The top endpiece of the column is removed while this is carried out.

2. The glass column is thoroughly cleaned and dried at 150°C for 1 h and cooled in a vacuum desiccator. The bottom endpiece is attached, and a new porous Teflon frit placed on top of the adjustable plunger. The screw cap of the septum injector endpiece is fitted with a new Teflon-coated septum, ensuring that the Teflon side is facing toward the solvent flow.

3. An amount of functionalized CPG equivalent to 1 µmol of bound nucleoside (normally between 20 and 30 mg depending on the nucleoside loading) is weighed into the column. The bottom endpiece is connected to the waste line, and the column attached to the system.

4. Pyridine is added dropwise onto the CPG with gentle agitation to remove air bubbles until the column is filled. The plunger is fitted ensuring exclusion of air. The bottom plunger is adjusted to leave a gap of 1 mm above the bed of the column, and is finally secured using the brass pinch clamps.

5. The flowrate of the system is set on pyridine at approximately 1 mL/min using the valve on the waste line and

Table 3
Sequence of Operations for One Addition Cycle

Operation	Time
Pyridine wash	3 min
1,2-Dichloroethane	2 min
3% Dichloroacetic acid in 1,2-dichloroethane	40–75 s[a]
1,2-Dichloroethane	2 min
Pyridine wash	3 min
Coupling (monomer addition)	15 min[b]

[a]Deprotection time for 5'-A or G is 40 s, and for 5'-C or T, 75 s.
[b]20 min for mixed-base addition.

keeping the pressure stat at 6 psi. This flow rate is re-checked on the pyridine cycle throughout synthesis. Note that this means that the flowrate of the dichloroethane cycles will exceed 1 mL/min because of viscosity differences.

3.2. Dehydration of Functionalized CPG

1. A freshly prepared solution of phenylisocyanate in anhydrous pyridine (10%) is made up in a dry 500-μL syringe and slowly injected onto the column.
2. After 15 min the phenylisocyanate is washed off by passing dry pyridine through the column for 3–4 min.

3.3. Deprotection

Table 3 shows the sequence of operations for one addition cycle.

1. The first step is deprotection of the 5'-OH of the bound nucleoside. This is achieved by passing the dichloroacetic solution through the column for 40 s for a purine residue and 75 s for a pyrimidine residue. The time is important since overexposure to acid causes cleavage of purine residues.

2. When coupling yields are being monitored, the deprotection eluant is collected for later analysis of DMT content. This is assayed as the orange-colored DMT cation (*see* section 3.12). In addition, the orange color of the eluant provides an on-the-spot visual check on the coupling efficiency.

3. The deprotection step is terminated by passing dichloroethane through the column (2 min). This is followed by a pyridine wash in preparation for the coupling step.

3.4. Coupling (Monomer Addition)

1. Immediately before coupling, 100 µL of monomer solution is injected into a vial of MSNT using a dry 250-µL syringe, leaving the syringe in place. Ensure that the MSNT is completely dissolved. After 1 min further, using a separate dry syringe, 10 µL of *N*-methylimidazole is added. The activated mixture is drawn into the 250-µL syringe. Use care to avoid air bubbles.

2. At the end of the pyridine wash, the solvent selector switch is switched to stop, the septum cap unscrewed one and a half turns, and the syringe inserted fully. The septum cap is retightened, the activated monomer is slowly injected onto the CPG (15 s), and the stopwatch is started. The septum cap is again unscrewed one and a half turns, the syringe removed, and the cap retightened as before. It is advisable to cover the column with tin foil during the coupling reaction.

3. The syringe is rinsed several times in anhydrous pyridine and inserted through a subseal stopper into a 25-mL Quickfit conical flask filled with pyridine, where it is left until the next cycle.

4. After a coupling time of 15 min, excess monomer is removed with a pyridine wash (3 min) and the next cycle commenced by repeating the deprotection step.

5. After the final coupling is complete, the CPG is washed with pyridine for 3–4 min followed by dichloroethane for

3–4 min. It is important that the final product is not 5'-deprotected at this stage.

3.5. Cleavage of Oligonucleotide from CPG and Removal of Permanent Protecting Groups

1. The septum injector endpiece is removed from the column, and the waste line is disconnected, leaving the bottom plunger in place. The column is fitted by the bottom endpiece to a small Buchner flask via a rubber bung with an 8-mm bore, and diethyl ether (5–6 mL) is passed through. The CPG is finally dried using a venturi pump for a few seconds.

2. The CPG is then pushed from the column into a 5-mL pear-shaped flask while using care not to lose any CPG. *syn*-2-Nitrobenzaldoxime (70 mg, 0.422 mmol) is added to 1 mL of dioxan:H_2O (1:1 v/v) followed by 50 µL of 1,1,3,3-tetramethylguanidine (TMG) (made fresh each time). A portion of this solution is added to the flask containing the CPG, and the mixture is incubated at 37°C for approximately 16 h or at 20°C for 24 h, preferably with gentle shaking.

   The quantity of oximate solution used depends on the number of protected phosphate groups in the sequence and amounts to approximately 25 µL/base added for a synthesis scale of 1 µM. This removes the 2-chlorophenyl-protecting groups from the internucleotide phosphates and releases the oligonucleotide from the CPG. Longer sequences (>20 bases), and especially G-rich sequences, require longer incubation periods to ensure complete deprotection and cleavage from the CPG.

3. After incubation the mixture is filtered through a sintered glass funnel fitted to a 25-mL round-bottom flask, and the CPG is washed with dioxan/H_2O (1/1 v/v) until free of yellow color.

4. The CPG at this stage can be quickly checked for the presence of any remaining DMT. A few grains of CPG are

placed on a white background, and a drop of 60% per-chloric acid in ethanol added. The immediate appearance of an orange color is indicative of oligonucleotides still bound to the CPG. If this occurs, the CPG should be rein-cubated for a further 16 h using a fresh mixture of oximate solution.

5. The filtrate is evaporated to dryness *in vacuo,* and concen-trated ammonia (5 mL) is added. The flask is fitted with a ground-glass stopper with a Teflon sleeve, sealed with in-sulating tape, and incubated at 55–60°C for 6 h. This re-moves the acyl-protecting groups. Loss of any ammonia during this incubation should be avoided since incom-plete removal of acyl groups may occur. When cool, the flask is opened carefully and excess ammonia is blown off in an air stream before evaporating to dryness *in vacuo.* At this stage the product retains the 5'-DMT group and is suit-able for purification by Sep-Pak or reverse-phase HPLC.

3.6. 5' Deprotection

This may be carried out immediately after base deprotec-tion or at some later stage after preliminary purification.

1. Distilled H_2O (1 mL) and concentrated acetic acid (4 mL) are added to the dried oligonucleotide and the solution incubated at 20°C for 30 min to effect the removal of the 5' DMT groups.
2. The mixture is then diluted with 5 mL of H_2O and extract-ed with diethyl ether (10 mL) five times in a separating funnel to remove hydrophobic contaminants.
3. The final aqueous phase is evaporated to dryness, resus-pended twice in distilled H_2O (1 mL), and reevaporated to dryness. It is stored at –20°C until required.

3.7. Analysis of Product

Analysis of the oligonucleotide product is carried out by polyacrylamide gel electrophoresis (PAGE)-autoradiography after 5' end labeling or, alternatively, by HPLC. HPLC may be performed by reverse phase on the DMT derivative or by ion exchange on the fully deprotected product and is adequately described elsewhere (4,7). Radiolabeling for electrophoresis is normally carried out by 5'-phosphorylation of the fully deprotected oligomer using the enzyme polynucleotide kinase and $[(\gamma)^{32}P]$-ATP. Figure 4 shows some typical autoradiograms. It may be necessary to denature the oligonucleotide mixture before radiolabeling, especially when strong inter- or intramolecular hybridization is possible; otherwise the autoradiogram may not give a true representation of the product mixture. Similarly, in some cases resolution by normal-phase HPLC may be affected by hybridization, making analysis difficult. A further feature of HPLC is that, unlike electrophoresis, resolution of oligomers greater than approximately 20 bases in length declines.

3.8. Purification of Oligonucleotides

Invariably, even with the best synthesis, analysis of the crude product will reveal the presence of oligonucleotides shorter than the desired product. Requirement for further purification will be determined by the level of these contaminants and the intended use of the product. For example, when the oligonucleotide is intended for use as a probe, modest levels of contaminants may be tolerated. A convenient preliminary purification step that should be routinely employed in all synthesis is chromatography of the crude product on a Sep Pak C18 cartridge (Waters Assoc.). This removes all sequences not containing a DMT group as well as various contaminants associated with the deprotection reagents. It is used before the 5'-deprotection step. Alternative techniques for preliminary purification of oligonucleotides include ethanol precipitation, microscale DEAE cellulose chromatography (8), Biogel P-2 chromatography (9), and Sephadex gel filtration (i.e., NAP™ nucleic acid purification columns from Pharmacia). Purification by ion exchange is described in Chapter 14.

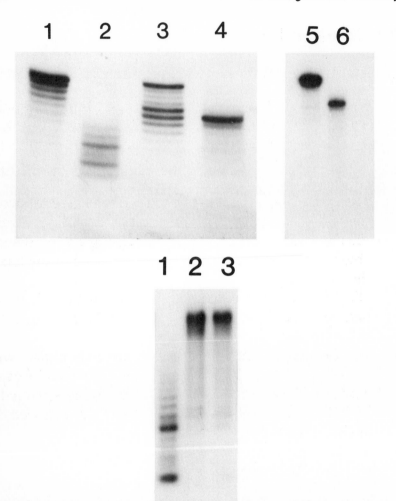

Fig. 4. Oligonucleotide syntheses by the phosphotriester method. PAGE-autoradiography of various reaction products. Samples were chosen to show the various types of reaction mixtures encountered. Unless otherwise stated, syntheses did not include a capping step. (top) Sep Pak purified products. 1, 19 mer showing a prominent n-1 component; 2, failed synthesis of 19 mer with a 3'-purine, capped after each coupling; 3, 19 mer showing prominent shorter sequences; 4, 15 mer with a 3'-pyrimidine, capped after each coupling; 5 and 6, 22 and 20 mer, both pyrimidine rich. (bottom) Sep Pak purification of crude products from synthesis of a 19 mer. 1, sequences eluted in 10% Acn/TEAB; 2, total reaction products; 3, sequences eluted in 30% Acn/TEAB.

3.9. Sep Pak Purification

See ref. *10* for Sep Pak purification details.

1. The Sep Pak cartridge is connected to a 10-mL disposable syringe and is equilibrated by washing with 10 mL of acetonitrile (Acn), HPLC grade; 5 mL of 30% Acn in 100 mM triethylammonium bicarbonate buffer (TEAB), pH 8.0; and 10 mL of 25 mM TEAB.

2. The crude 5'-protected oligonucleotide mixture (approximately 50 A_{260} units) is dissolved in 25 mM TEAB (2 mL). This is passed through the Sep Pak using a 5-mL disposable syringe, forming a yellow-brown band at the top of the cartridge.

3. Oligonucleotides lacking the DMT group and other contaminants are eluted with 10% Acn in 25 mM TEAB (15 mL).

4. DMT-containing oligonucleotides are then eluted with 30% Acn in 100 mM TEAB (5 mL) and recovered by evaporation to dryness *in vacuo*. The residue is resuspended in H$_2$O (2–3 mL) and redried. 5' Deprotection is carried out as described above before PAGE.

3.10. Further Purification

Both electrophoresis and HPLC may be applied to further purification of the final oligonucleotide product, and the purification conditions are essentially those described above for analysis. The potential advantages of electrophoresis are its cheapness, the possibility for simultaneous purification of several preparations, and the fact that resolution is not affected by the length of the oligonucleotide.

3.11. Polyacrylamide Gel Electrophoresis

Electrophoresis is carried out on a 20% polyacrylamide denaturing gel containing 7M urea. Analytical gels, i.e., for autoradiography, are 0.4 or 0.8 mm and preparative gels are 1.5 mm thick. One centimeter sample wells are suitable for both purposes. Short oligonucleotides (15 bases) are adequately resolved on gels of length, 20 cm whereas longer oligonucleotides require a 40-cm gel. Reagents should be of the special electrophoresis grade when available.

1. Urea (210 g), acrylamide (100 g), and *N,N'*-methylene bisacrylamide (2.5 g) are dissolved in distilled H_2O to a volume of approximately 400 mL. Amberlite MB1 resin (20 g) is added, and the mixture stirred for 10 min. The mixture is filtered and made up to a final volume of 450 mL with distilled H_2O.

2. To 45 mL of this solution are added 5 mL 10x TBE buffer (1 *M* Tris-borate, pH 8.3, 100 m*M* EDTA), 400 µL of 10% ammonium persulfate (freshly prepared), and 40 µL of TEMED. The gel is poured and left to set for 2 h.

3. The running buffer is 1x TBE, pH 8.3. The gel is pre-run for 30–40 min at 800 V (1.5-mm gels) or 1200 V (0.4-mm gels). Sample wells are thoroughly rinsed with buffer before sample application.

4. Before electrophoresis, crude oligonucleotide product should be given a preliminary purification using a C18 Sep Pak cartridge or one of the alternative techniques mentioned above.

5. For analytical gels, 5' end labeling of oligonucleotides with T4 polynucleotide kinase and [(γ)[32]P]-ATP is carried out by the method of Maniatis (*11*). It is advisable to heat the sample by immersion to 100°C for 2 min followed by immediate cooling before addition of enzyme or ATP to disrupt covalent association. The reaction mixture after labeling is lyophilized and resuspended in 5 µL of gel-loading solution (90% deionized formamide, 25 m*M* TBE buffer, pH 8.3, 0.6 m*M* EDTA), boiled for 3 min, cooled on ice, and immediately applied to the polyacrylamide gel. The usual strict safety precautions should be taken when handling

[(γ)^{32}P]-ATP, i.e., use of face shield and protective gloves. The gel is run until the dye front has migrated two-thirds the length of the gel.

6. For preparative gels the oligonucleotide product (5–10 A_{260} units) is lyophilized in a microtube, suspended in 10 μL of gel-loading solution (90% deionized formamide, 25 mM TBE bufffer, pH 8.3, 0.6 mM EDTA), boiled for 3 min, cooled on ice, and immediately loaded on the gel.

7. 5–10 μL of gel-loading solution containing a little Bromophenol Blue is applied to a spare sample well as a marker and migrates near the position of an eight-base oligonucleotide.

8. The gel is run until the dye has migrated two-thirds the length of the gel.

9. Analytical gels are wrapped in cling film and exposed to a Kodak X-OMAT AR X-ray film for 2–5 min and developed using the procedure recommended by Kodak.

10. Preparative gels are placed on a Kieselgel 60 F$_{254}$ TLC plate (Merck) wrapped in cling film and illuminated from overhead with a short-wave UV lamp (260 nm). DNA bands appear as dark pink shadows against a fluorescent background. The desired oligomer band is neatly sliced out with a sharp scalpel, crushed, and extracted overnight with 1M TEAB (1 mL) in a microtube. The extract is filtered through a 0.45-μm filter and purified using a Sep Pak cartridge. Salts and polyacrylamide gel impurities are removed using 25 mM TEAB (10 mL), the oligonucleotide fraction is eluted with 30% acetonitrile in 100 mM TEAB (3 mL) and recovered by evaporation *in vacuo*. The residue is resuspended in 1 mL of H$_2$O and redried. Before the last evaporation, the final yield of product is determined by measuring the A_{260}. One A_{260} unit is equivalent to 30 μg of oligonucleotide. It is stored lyophilized at –20°C until required.

3.12. Dimethoxytrityl Assay

1. The washings from each deprotection cycle are collected and made up to a standard volume (5 mL) with 3% DCA in dichloroethane. Dilutions (1:15) are made with 60% perchloric acid in ethanol, and the absorbance at 495 nm is recorded.

2. The coupling efficiency of each step is determined by:

$$\frac{A_{495}}{A_{495} \text{ of previous step}}$$

The yield should be approximately 95% for most couplings, but may be 100% (+) in the case of G-rich sequences and less than 85% for the first coupling.

4. Notes

1. Synthesis on the 1-μM scale should provide an adequate amount of oligonucleotide for all requirements in which the product is used as a primer, linker, or probe. The theoretical yield of a 20 mer from a synthesis with an average coupling efficiency of 95% is 38% or approximately 2 mg. In practice the yield may fall well short of this, particularly when there is a high concentration of shorter sequences in the final product mixture.

2. In our experience, even when average coupling efficiencies of 95% (+) are recorded, the concentration of shorter sequences is sometimes greater than expected. This is most noticeable with purine-rich sequences. A major contributory factor appears to be growth of secondary chains caused by coupling of monomer to the O^6-position on G. Branches thus formed are cleaved during the oximate deprotection step, but yield satellite oligonucleotides that appear as contaminants of the desired product. In extreme cases, the high concentration of these make subsequent purification of the desired product difficult.

3. A further potential contributor to contamination is that the yield of the first coupling is often low (<85%), presumably because of steric effects. This means that unreacted 5'-OH groups exist after the first coupling, and their reaction in subsequent couplings can generate an appreciable amount of n-1 and n-2 sequences. This and the problem referred to in the previous paragraph might be overcome by the introduction of an acetylation or capping step after each coupling. This would "cap-off" unreacted 5'-OH and G-OH groups. In our experience, however, general capping is not to be recommended when using the phosphotriester method of synthesis. Often this depresses coupling efficiencies and leads to failure of synthesis. This problem is most severe when the 3'-nucleoside is a purine. A limited role for capping exists, however, when confined to the first one or two coupling steps. In this way n-1 and n-2 contaminants are reduced, making purification of the final oligomer easier. When used, the capping step is carried out immediately after the coupling reaction once the excess monomer has been washed from the column with pyridine. 150 μL of 6.5% DMAP in THF is mixed with 150 μL of acetic anhydride/2,6-lutidine/THF (1/1/8) and injected immediately onto the column. After 2 min the capping solution is washed from the column with pyridine (3 min), and the cycle is continued as normal. It is important not to mix the acetic anhydride with the DMAP solution until immediately before injection onto the column, since the mixture soon discolors and loses reactivity on standing.

4. Mixed-sequence oligonucleotide probes, incorporating degenerate sites, may be synthesized by using the required mixture of nucleotide monomers at the appropriate stage of the synthesis (*12*). To allow for different reactivities of the monomers, it is recommended that they be mixed in the approximate molar proportions 5:6:4:4 for A:G:C:T, keeping the total molar concentration at the

normal coupling level. In addition, the coupling time is increased to 20 min.

References

1. Itakura, K., Rossi, J.J., and Wallace, R.B. (1984) Synthesis and use of synthetic oligonucleotides. *Ann. Rev. Biochem.* **53**, 323–356.
2. Van Brunt, J. (1985) The new generation of DNA synthesizers. *Biotechnology* **3**, 776–783.
3. Smith, J.A. (1983) Automated solid phase oligonucleotide synthesis. International Biotechnology Laboratory. December, 1983, p. 12.
4. Gait, M.J., ed. (1984) *Oligonucleotide Synthesis—A Practical Approach* IRL Press, Oxford.
5. Gait, M.J., Mattthes, H.W.D., Singh, M., Sproat, B.S., and Titmas, R.C. (1982) Rapid synthesis of oligodeoxyribonculeotides. V11. Solid phase synthesis of oligodeoxyribonucleotides by a continuous flow phosphotriester method on a kieselguhr-polyamide support. *Nucleic Acid Res.* **10**, 6243–6255.
6. Sproat, B.S. and Bannwarth, M. (1983) Improved synthesis of oligodeoxyribonucleotides on controlled pore glass using phosphotriester chemistry and a flow system. *Tetrahedron Lett.* **24**, 5771–5774.
7. Newton, C.R., Greene, A.R., Heathcliffe, G.R., Atkinson, T.C., Holland, E., Markham, A.F., and Edge, M.D. (1983) Ion-exchange high pressure liquid chromatography of oligodeoxyribonucleotides using formamide. *Anal. Biochem.* **129**, 22–30.
8. Matthes, H.W.D., Zenke, W.M., Grundstrom, T., Staub, A., Winzerith, M., and Chambon, P. (1984) Simultaneous rapid chemical synthesis of over one hundred oligonucleotides on a microscale. *EMBO J.* **3**, 801–805.
9. Gait, M.J. (1982) Synthesis of Oligodeoxyribonucleotides by a Continuous Flow, Phosphotriester Method on a Kieselguhr/Polyamide Support, in *Chemical and Enzymatic Synthesis of Gene Fragments* (H.G. Gasson and A. Long, eds.) Verlag Chemie, Weinheim, FRG.
10. Lo, K.M., Jones, S.S., Hackett, N.R., and Khorana, H.G. (1984) Specific amino acid substitutions in bacterioopsin: Replacement of a restriction fragment in the structural gene by synthetic DNA fragments containing altered codons. *Proc. Natl. Acad. Sci. USA* **81**, 2285–2289.
11. Maniatis, T., Fritsch, E.G., and Sambrook, J. (1982) *Molecular Cloning, A Laboratory Manual* Cold Spring Harbor Laboratory Press, Cold Spring Harbor, New York.
12. Cruachem Highlights (1984) Synthesis of mixed probes, Cruachem, Livingston, Scotland, UK, May edition.

Chapter 14

Manual Oligonucleotide Synthesis Using the Phosphoramidite Method

H. A. White

1. Introduction

Oligodeoxyribonucleotides may be synthesized on a solid support, which allows for the elongation of the chain without intermediate purifications. The exocyclic amino groups on cytosine and the purines are protected as the alkali-labile benzoyl or *iso*-butyryl amides. The 3'-OH group of the first nucleoside is attached to the solid support by a spacer chain, and the synthesis procedes by coupling the 5'-OH group of the growing chain to the 3'-phosphorus of the monomer being added, which has its 5'-OH group protected by an acid-labile group. The phosphorus must also be protected to avoid side reactions. When the synthesis is completed, the chain must be cleaved from the support and the base, phosphorus, and terminal 5'-OH-protecting groups removed. The desired oligonucleotide is then separated from the mixture of shorter chains and modified chains (1).

There are two general chemistries in use for oligonucleotide synthesis: the phosphotriester (*see* Chapter 13) and the phosphite triester. The phosphoramidite methods are versions of the phosphite triester chemistry (2). Both systems are in use for manual and automated synthesis. Many automated systems use β-cyanoethyl phosphoramidites, which simplify the deprotection of the phosphates and are very convenient for manual synthesis (3).

A synthesis starts with an appropriately base-protected [N^6-benzoyl adenine, N^2-isobutyryl guanine, N^4-benzoyl cytosine (thymine is not usually protected)] nucleoside that also has its 5'-OH group protected with the dimethoxytrityl group and is coupled through its 3'-OH to controlled-pore glass support via a long-chain alkylamine arm and succinyl group (*see* Figs. 1 and 2). The elongation cycle consists of the following steps.

1. Removal of the 5'-dimethoxytrityl group with dichloroacetic acid in dichloroethane.
2. Coupling of the next monomer by adding the appropriate nucleoside 3'-*O*-phosphoramidite with the activating agent tetrazole.
3. Oxidation of the phosphite to phosphate with iodine/water.
4. Acetylation of any unreacted 5'-OH groups ("capping") with acetic anhydride and dimethylaminopyridine as a catalyst. The elongation cycle is repeated until the oligonucleotide is completed (Fig. 3). The efficiency of the coupling cycle is critical to the success of a synthesis. If the efficiency of the coupling cycle is 95%, the overall yield after 15 couplings is 46%, and 21% after 30 couplings.

Treatment with aqueous ammonia effects removal of the oligonucleotide chain from the support, deprotection of the bases, and removal of the β-cyanoethyl groups from the phosphates. The product may be purified either by high-performance liquid chromatography or by preparative polyacrylamide gel electrophoresis.

(1)

dA monomer

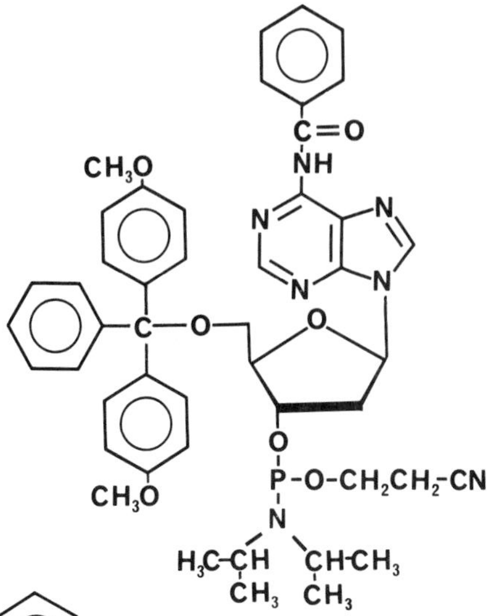

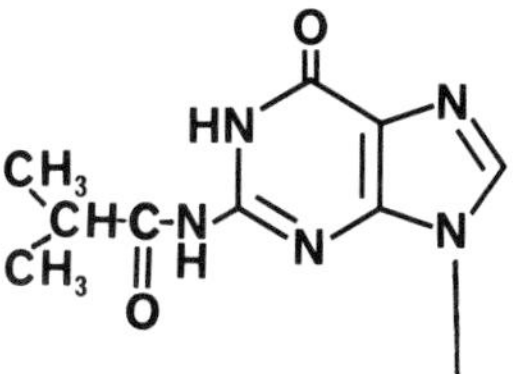

(2)

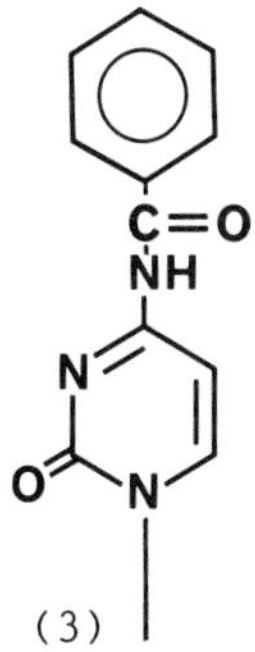

(3)

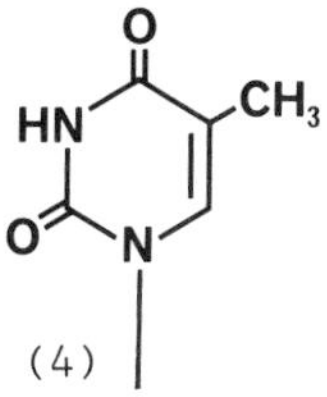

(4)

Fig. 1. (1) dA monomer [5'-*O*-(4,4'-dimethoxytriphenylemthyl)-N^6-benzoyl-2'-deoxyadenosine-3'-*O*-(2-cyanoethyl)-di-isopropyl-amido-phosphite]. (2) Protected base, dG monomer (N^2-isobutyryl). (3) Protected base, dC monomer (N^4-benzoyl-). (4) Thymidine. Not usually protected.

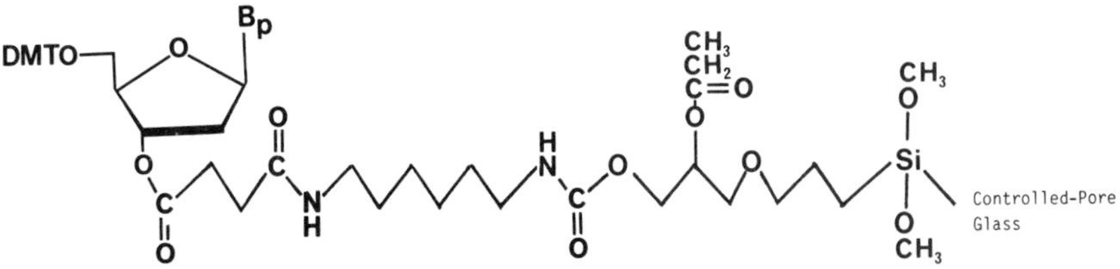

Fig. 2. Protected nucleoside linked to controlled-pore glass support with a long-chain alkylamine spacer and succinyl link.

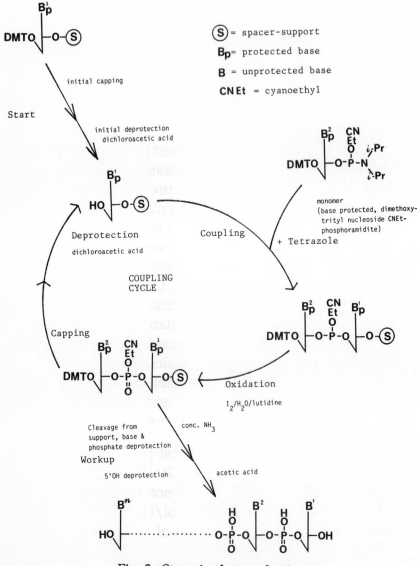

Fig. 3. Steps in the synthesis.

There are many versions of the phosphoramidite protocol. The one described below is fairly typical and is suitable for manual synthesis of oligonucleotides up to 30 or so nucleotides long on a scale of 1–5 μmol (starting). Many workers carry out the synthesis in a sintered funnel, but the flow apparatus described below is satisfactory and requires less room and fewer manipulations.

2. Materials

2.1. Reagents

All reagents must be of the highest quality available since solid-phase methods may "concentrate" the effects of trace impurities. Many protocols for the preparation and purification of the solvents, protected deoxyribonucleoside-3'-*O*-cyanoethyl phosphoramidites, and other reagents are in the literature (for example, *see* ref. 4), but most workers will find it more convenient to purchase these materials. They are available from, among others, BDH Chemicals, Ltd. (Broom Road, Poole, Dorset, BH12 4NN, England) and Cruachem, Ltd. (11 Napier Square, Livingston, EH54 5DG, Scotland).

The deoxynucleoside phosphoramidites and the derivatized supports must be stored desiccated and are usually stored at –20°C, although they are stable at room temperature. It is critical that they be kept dry and not exposed to acid vapors. They are available in septum-sealed bottles.

1. Tetrazole must be of the highest purity (sublimed) and stored in sealed amber bottles. It should be microcrystalline in appearance and should not leave a residue when dissolved in acetonitrile (0.5 mol/L).
2. Dichloroacetic acid should be of high purity and kept anhydrous.
3. Aqueous ammonia ($\cong$35% w/w) should be BDH "Aristar" grade or equivalent and must be kept tightly sealed at 0–5°C to avoid loss of ammonia.
4. Argon may be standard "industrial" grade. This is normally oxygen-free and perfectly dry. As a precaution, most workers pass it through a drying tube of silicagel (previously dried *in vacuo* over P_2O_5).

2.2. Solvents

1. Acetonitrile must be of "HPLC" grade or equivalent and must be *anhydrous* for the coupling reaction. Since anhydrous acetonitrile is time-consuming to prepare and costly to purchase, it is economical to use analytical reagent or HPLC grade acetonitrile when possible during the synthesis. Anhydrous acetonitrile can be prepared by distillation of HPLC-grade acetonitrile from calcium hydride (2) and should be stored in septum-capped bottles under dry argon. Anhydrous acetonitrile in septum-sealed bottles is available from Cruachem. Note that acetonitrile is highly flammable and is toxic by ingestion, skin absorption, and inhalation.

2. 1,2-Dichloroethane should be rendered anhydrous by first passing it through a column of basic alumina, followed by distillation from phosphorus pentoxide. It should be stored sealed under dry argon. Note that dichloroethane is highly flammable and toxic by inhalation and may cause skin irritation. A satisfactory grade is available from Cruachem.

3. Tetrahydrofuran and 2,6-lutidine must be anhydrous for use in the capping step. These may be purchased (Cruachem) or prepared by distillation from calcium hydride. Store both in amber bottles under dry argon. Both solvents are toxic. Tetrahydrofuran is highly flammable and may form explosive peroxides.

4. Other reagents (acetic anhydride, acetic acid, dimethylaminopyridine) should be of analytical reagent-grade quality.

2.3. Solutions

2.3.1. Coupling Mixture

1. Monomer: 0.2 mol/L of the 5'-dimethoxytrityl deoxyribonucleoside 3'-O-(N,N-diisopropylamino)-cyanoethyl

Table 1
Monomer Molecular Weights and Acetonitrile Volumes

Monomer, base-protected space nucleoside cyanoethyl phosphoramidites	Molecular weight	Acetonitrile needed to make 0.2 mol/L solution, mL/g
T Dimethoxytrityl-T-	744.83	6.7
C Dimethoxytrityl-BzdC-	833.92	6.0
A Dimethoxytrityl-BzdA-	857.95	5.8
G Dimethoxytrityl-iBudG-	839.93	6.0

phosphoramidites in anhydrous acetonitrile. Table 1 gives the molecular weights of the four monomers and the volumes of acetonitrile required to prepare the 0.2 mol/L solutions (5). If kept anhydrous (stored under dry argon), these monomer solutions are stable for several days.

2. Tetrazole: 0.5 mol/L tetrazole (mol wt, 70.06) in anhydrous acetonitrile. This solution should be kept under dry argon and is stable for several days.

The two solutions listed above are mixed immediately before use. Note: the solutions described in Table 1 may be purchased from Cruachem, Ltd. and are stable for several months if stored in amber glass bottles (except for the iodine solution). It is good practice to replace these solutions after they have been in use 3–4 wk as a precaution.

2.3.2. Oxidation

Iodine: 0.1 mol/L iodine (sublimed) in 4:2:1 tetrahydrofuran/2,6-lutidine/water (filter after preparation). The iodine solution may throw a precipitate on standing, so should be freshly prepared each day. It is therefore convenient to store the iodine solution in two parts: iodine in tetrahydrofuran and lutidine/water. These are mixed before use. Different workers use slightly different iodine mixtures; there seems to be little difference among them.

2.3.3. Capping

1. 6.5% (w/v) 4,N,N-dimethylaminopyridine in anhydrous tetrahydrofuran.
2. Acetic anhydride/2,6-lutidine/tetrahydrofuran (1:1:8)

These solutions are mixed immediately before use.

2.3.4. Deprotection

1. 3% (v/v) Dichloroacetic acid in 1,2-dichloroethane.

2.4. Apparatus

2.4.1. Flow Apparatus

The apparatus consists of a small chromatography column (6.5 mm bore; 50 mm long) fitted with a septum injector that can be connected to up to six reagent bottles (five are used in this method) via a six-position rotary selector valve. Figure 4 shows the layout of the flow apparatus. The support rests on a PTFE fritted disk at the bottom of the column. The column height can be varied to allow for different amounts of support. This apparatus may be obtained from Omnifit Ltd. (51 Norfolk Street, Cambridge, CB1 2LE, England) (catalog no. 3600 DNA Synthesis Kit) or a similar type of apparatus from Cruachem.

Dry argon is used to pressurize the reagent reservoirs via a pressure regulator and an eight-way distribution manifold (unused positions are blocked off or may be used to repressurize septum-sealed solvent bottles). The selector valve allows each reservoir to be connected in turn to the synthesis column.

Addition of two-way valves between the rotary valve and the column allows for use of two or more columns for multiple synthesis (*see* Chapter 13). Solvents and reagents leave the column via a length of narrow-bore PTFE tubing (which provides a reasonable pressure drop) and a cutoff valve ("exit valve").

The method described here requires five solvent bottles. If the Omnifit 3600 DNA Synthesis Kit is used, an additional 250-mL solvent bottle with a three-valve cap will be needed, as well as an additional Omnifit HP tube end fitting (2100), a 2310 gripper fitting, and a 2320 sealing plug. It is wise to stock up on PTFE tubing and Omnifit connector spares since inevitably some will be lost or damaged in the assembly. We have found that it is better to avoid the O-ring type fittings. Those supplied with apparatus should be replaced with PTFE cone-type fittings or with the flangeless high-pressure fittings.

The apparatus should be assembled as described in the instructions and shown in Fig. 4. The argon plumbing should be thoroughly flushed with argon before the solvent bottles are connected. It is convenient to connect (and label) the rotary valve with "STOP" at the twelve o'clock position (using an Omnifit 2320 sealing plug) and the solvents/reagents connected in the following order (clockwise): *reagent grade* acetonitrile; iodine/lutidine/water; dichloroacetic acid; dichloroethane; *anhydrous* acetonitrile. The solvent bottles must be thoroughly cleaned and dried before filling. Reagent grade acetonitrile, anhydrous acetonitrile, and dichloroethane should be placed in the liter bottles and the 250-mL bottles used for the dichloroacetic acid and the oxidizing mixture. If the humidity is high, it is wise to fill the anhydrous acetonitrile and dichloroethane bottles under a blanket of dry argon.

It is convenient to connect a septum-sealed bottle of anhydrous acetonitrile to the argon distribution manifold using a short syringe needle. This bottle is used as a source of acetonitrile for making up the tetrazole and monomer solutions. The risk of moist air entering the bottle is minimized if it is always pressurized with argon.

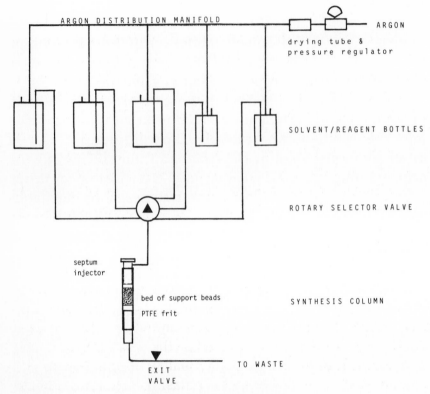

Fig. 4. Flow apparatus.

2.4.2. Syringes and Needles

Syringes with PTFE-tipped plungers—such as the Hamilton "1000 series" "gastight" syringes—are essential. Minimum requirements are for three 500-µL, one 1-mL, one 2.5-mL, and one 5-mL syringes.

All of the syringes should be equipped with needles designed for *septum penetration*. It is preferable to use needles with a *side hole* (such as the Hamilton style 5 point), since these minimize damage to the apparatus septum.

Syringes should be stored filled with anhydrous acetonitrile during a synthesis when not actually in use.

2.4.3. Mixing Vials

Vials with conical interiors and PTFE-lined septum caps are used for mixing the monomer with tetrazole for each coupling and for preparing the capping reagent mixture. Vials such as the Pierce "Reacti-Vials" are ideal (Pierce and Warriner Ltd., 44 Upper Northgate Street, Chester, CH1 4EF, England). The 0.3- or 1-mL size serves for preparing the monomer/tetrazole mixture, and the 3- or 5-mL size serves for the capping mixture. It is important that the vials be scrupulously cleaned and dried. They should be stored in a desiccator over silica gel that has previously been dried *in vacuo* over phosphorus pentoxide.

2.4.4. Deprotection Flask

Removal of the base-protecting groups, as well as cleavage from the support and deprotection of the phosphates, is accomplished by heating with concentrated ammonia solution. Considerable pressure is developed when the vessel is heated. We have found it best to use a heavy-walled 50-mL flask with a ground-glass stopper. Stopper and flask each have four hooks so the stopper can be wired in place.

3. Method

3.1. General Approach

The key to a successful synthesis lies in organization and attention to detail. The synthesis steps are:

1. Preparation of the support and apparatus.
2. Initial capping.

3.1.1. Coupling Cycle

1. Removal of the 5'-OH group ("deprotection") with dichloroacetic acid solution.

2. Washing.
3. Addition of the next nucleotide ("coupling") by injection of the monomer/tetrazole mixture.
4. Washing.
5. Oxidation of the phosphityl group.
6. Washing.
7. Capping of unreacted 5'-OH groups with acetic anhydride and catalyst mixture.
8. Washing.
9. Repeat steps 1–8.
10. Workup: deprotection of phosphates, bases, cleavage from support, and removal of 5' terminal dimethoxytrityl group.

We have found that a fairly simple recording system is effective in ensuring that the synthesis is carried out without omission or duplication of steps. The sequence is written on a blackboard. We then cross each nucleotide out *twice*—once when the monomer has been injected and again when we have noted the release of the orange-colored dimethoxytrityl cation. The method is described for a 1.5-µmol synthesis (i.e., start with 1.5 µmol of the 3' immobilized nucleoside). Preparation of an oligonucleotide in the 15–30 nucleotides range on this scale should afford between 50 µg and 2 mg of product depending upon the chain length and sequence. This will be in the range of 1–50 A_{260} units (an A_{260} unit is that amount of an oligonucleotide that gives an absorbance of 1.0 at 260 nm when dissolved in water and read in a 1-cm cuvet).

3.2. Preparation for Synthesis

1. Weigh out the appropriately derivatized controlled-pore glass support to give 1.5 µmol of the 3' end nucleoside. For controlled-pore glass beads with a long chain alkylamine/succinyl-linked nucleoside, typical loadings are in the range of 20–50 µmol/g. With a loading of 30 µmol/g, 50 mg would be required for a 1.5-µmol synthesis.

2. With the exit valve off, carefully moisten the PTFE fritted disk with reagent-grade acetonitrile (avoiding wetting the column walls).
3. Carefully pour in the support and add about 0.5 mL of acetonitrile with a Pasteur pipet. Stir the support briefly to remove air bubbles.
4. Adjust the position of the column base and PTFE fritted disk so that there will be a space of about 1–2 mm between the top of the support bed and the septum injector.
5. Next gently fill the column with acetonitrile to the top. Remove the septum cap from the flow injector and *gently* push it into the top of the column. (Take care that the displaced acetonitrile does not spurt!) When the septum injector is clamped home, replace the septum cap.
6. Open the exit valve and turn on the reagent-grade acetonitrile.
7. The general procedure for selecting a reagent or solvent is to rotate the rotary valve to the correct position, then open the argon supply valve to the bottle (each bottle has three valves: fluid, argon, and vent) and finally open the bottle fluid valve. At the end of the period the rotary selector valve should be turned first: the bottle fluid and argon valves are then turned off. We have found it best to turn off the argon to each bottle when it is not in use. The argon flow rate is low, so it is remotely possible for some vapor phase mixing to occur between bottles when fluid is withdrawn from one if this precaution is not taken.
8. Adjust the argon pressure until the flow rate through the system is 1 mL/min with acetonitrile (the flow rate with dichloroethane will be higher). The pressure should be about 6 psi.
9. Allow reagent grade acetonitrile to flow for a minimum of 1 min and then switch to anhydrous acetonitrile and wash 2 min.
10. Carry out the Capping step described below. This blocks any free amino or hydroxyl groups on the support. Wash

the support 2 min with reagent grade acetonitrile and 2 min with dichloroethane.

3.3. Performing the Synthesis

3.3.1. Injection Procedure

All reagent injections should be made as follows:

1. With the exit valve closed and the rotary valve on STOP, loosen the septum cap by one-half to one turn.
2. Insert the syringe needle fully and then tighten the cap (it should not be overtightened).
3. Open the exit valve and slowly inject the syringe contents over 15–30 s.
4. Close the exit valve, loosen the cap, withdraw the syringe, and retighten the cap.

3.3.2. Coupling Cycle

3.3.2.1. Deprotection

1. Turn on the dichloroacetic acid and allow it to flow. Insert the waste tube into a collection vessel containing about 1 mL of 10% (w/v) trichloroacetic acid. Observe the bed closely. When the orange color at the top of the bed begins to diminish, switch to dichloroethane.
2. It is important to keep the exposure to the acid to a minimum. Purines and pyrimidines differ slightly in rate of deprotection, so this step is best controlled visually. The time will be on the order of 30 s.
3. Continue collection of the effluent until the bed is free of color. The samples should be diluted with 10% trichloroacetic acid in dichloroethane. Read at 417 nm (ε = 224,000) or 504 nm (ε = 507,000) (6).
4. Wash for 1 min with dichloroethane and for 2 min with anydrous acetonitrile. The synthesis may be interrupted at this point. If the interruption is longer than a few

minutes, wash with anhydrous acetonitrile again before the coupling step.

3.3.2.2. Coupling

1. Inject 0.18 mL of 0.5 mol/L tetrazole in acetonitrile into a dry, argon-filled, septum-sealed vial (a 0.3- or 1.0-mL "Reacti-vial").
2. Inject 0.15 mL of the appropriate phosphoramidite solution (0.2 mol/L in acetonitrile) into the tetrazole solution and mix. It is essential to minimize exposure to water vapor and oxygen at this stage.
3. Rapidly withdraw the vial contents into the phosphoramidite syringe and inject it into the column. Close the exit valve and allow a total of 3 min for the coupling reaction. Rinse the syringe thoroughly with acetonitrile.
4. Wash with reagent grade acetonitrile for 2 min.

3.3.2.3. Oxidation

1. Allow the iodine/tetrahydrofuran/luditine mixture to flow for 1 min.
2. Wash with reagent grade acetonitrile for 2 min and with anhydrous acetonitrile for a further 2 min. Check that no iodine color remains. If it does, prolong the wash.

3.3.2.4. Capping

1. Add 0.25 mL of acetic anhydride/lutidine to a septum-sealed vial and inject 0.75 mL dimethylaminopyridine solution into it and mix. Withdraw the contents and inject into the column. Allow 2 min reaction time.
2. Wash for 2 min with anydrous acetonitrile and 2 min with dichloroethane.
3. This cycle is repeated until the chain is completed. The capping step may be omitted after the last oxidation. Allow a final 2-min wash with dichloroethane.

3.3.3. Workup of the Oligonucleotide

1. Extrude the support (washing with dichloroethane using a Pasteur pipet) into a small glass sinter-based funnel (a funnel 20 x 20 mm with a 6-mm diameter fine glass sinter base works well). Draw the solvent off and wash three times with small volumes of diethyl ether. Draw air through the support until it is dry.

2. Transfer the dry support to a 50-mL round-bottom flask and add 5 mL of concentrated aqueous ammonia. Grease (using a nonsilicone grease) the upper part of the ground-glass stopper and seal the flask. Use rubber bands to hold the stopper tightly in place (use a length of light wire—those sold for sealing plastic bags for use in domestic freezers are ideal—as a precaution).

3. Heat the flask to 50°C for 16 h.

4. Cool the flask and open cautiously. Allow the ammonia to vent off for about 30 min assisted by a gentle stream of air.

5. Filter the contents into a 100-mL round-bottom flask using a small glass sinter funnel (as above), and wash the reaction flask and the support with 3 x 2 mL water. Rotary evaporate to dryness using a bath set at 35–40°C. Some syntheses give products that foam badly during rotary evaporation. In such cases, transferring to a 250-mL flask helps.

6. Add 5 mL of acetic acid-water (80 mL of glacial acetic acid + 20 mL of water) to the flask to remove the terminal dimethoxytrityl group (color will not be visible). Allow to stand for 30 min at room temperature.

7. Add 5 mL of water and 10 mL of diethyl ether and shake vigorously. Remove the ether (upper) layer with a Pasteur pipet, and repeat the extraction once.

8. Rotary evaporate to dryness. Add 5 mL of water and again evaporate to dryness. Repeat this procedure until there is no smell of acetic acid. Two to three evaporations should be sufficient.

9. The oligonucleotide is now ready for purification and should be taken up in an appropriate volume of water (usually 1–2 mL) and is best stored at –20°C.

3.3.4. Purification

For many applications, preparative polyacrylamide gel electrophoresis is the method of choice for convenient purification (*see* Chapters 13 and 15). If kinase labeling is used for analytical purposes (7,8), it is best to freeze dry the product mixture two or three times. This removes volatile materials that may inhibit the kinase.

For greater chemical purity, HPLC is employed. As a first purification step, an anion-exchange method is best. This separates mainly on the basis of chain length—the product peak is the last to be eluted with a salt gradient. Reverse-phase methods are preferred as a second purification (9). Reverse-phase methods are described in Chapter 13.

3.4. Ion-Exchange HPLC Protocol

Any HPLC machine capable of gradient programing and monitoring at 260 nm will serve. An analytical-size column is adequate to purify the small amounts of oligonucleotide needed for many applications (1–10 A_{260} units).

1. Equilibrate a Whatman "Partisil 10 SAX" column (5 mm x 25 cm) at 1.5 mL/min with 0.003 mol/L KH_2PO_4, pH 6.3, in 6:4 formamide:water (adjust the pH before addition of the formamide—reagent-grade formamide is satisfactory, but the potassium dihydrogen phosphate must be of very high purity, such as BDH "Aristar" grade).
2. Inject up to 100 µL of the crude product, and run a linear gradient between 0.003 and 0.21 mol/L KH_2PO_4.
3. The elution profile should show a pattern of small peaks that precede the large product peak. A typical elution requires 30–40 min. One or two runs suffice to collect sufficent product.

4. The formamide is removed on a gel-filtration column
    (Biogel P2 or Sephadex G15) run with 20% (v/v) ethanol
    (redistilled or BDH "Aristar" grade) in water. The etha-
    nol/water is removed by rotary evaporation.

4. Notes

1. The density of argon is its chief advantage over nitrogen.
    Oxygen-free nitrogen will serve instead.
2. The volume of monomer and tetrazole solutions lost
    through the syringe dead volume is significant. To com-
    pensate for this, it is necessary to prepare an extra 30 µmol
    of phosphoramidite for every two to four withdrawals to
    be taken from a bottle.
3. Controlled-pore glass is the support of choice for most
    syntheses. Other types of silica support (such as "Frac-
    tosil" and "Vydac") are used in some methods.
4. Depurination, particularly of the benzoyl-protected ade-
    nine, occurs during the acid-deprotection step. It is there-
    fore important to keep the exposure to the acid to a min-
    imum. Purine (particularly adenine)-rich sequences are
    more difficult to synthesize successfully for this reason—
    especially if the 3' end is purine-rich. The more acid-labile
    "pixyl" (9-phenylxanthen-9-yl) group is sometimes em-
    ployed in place of the dimethoxytrityl group to reduce the
    exposure to acid conditions. Protecting groups for ade-
    nine that do not lead to depurination are being developed.
5. The orange color of the dimethoxytrityl cation is some-
    times somewhat fugitive. Detergent or solvent residues in
    the collection vessel may cause loss of color. A synthesis
    should not be abandoned on the basis of a single low
    measurement of the dimethoxytrityl release. If, however,
    color release from the *support bed* is not observed, a further
    treatment with dichloroacetic acid is worth a try, but the
    synthesis may have failed. Low yields toward the end of
    a synthesis may often be acceptable since very small
    amounts of product are usually sufficient.

6. The terminal dimethoxytrityl group may be left on the oligonucleotide if reverse-phase HPLC is to be used. This very apolar group will shift the product peak away from the others. It is removed after collection of the peak, and the oligonucleotide is rechromatographed. Other workers favor removal of the final dimethoxytrityl before the ammonia treatment.

7. "Mixed chains" are often required if an amino acid sequence is being used to generate an oligonucleotide probe. In such a case some positions in the oligonucleotide sequence may be two, three, or four different bases. Rather than make each chain separately, it is usual to use the appropriate monomer mixture (equimolar) at the coupling step for these positions. This is often effective if a family of eight or 16 chains is being made. The risk increases as the number of chains increases. There is no certainty that each monomer is coupling with equal frequency, so some sequences may be present in vanishingly small yields. Those chains rich in adenine may be lost through depurination. It is best to make the members of a group eight at a time rather than all at once.

8. Only two problems with the flow apparatus are likely to appear:

    If a large bubble forms at the top of the support bed, it will impede reagent access. The bubble can be removed as follows: on an acetonitrile wash step set the exit valve to OFF and rotary valve to STOP. Fill a 500-µL syringe about half full of acetonitrile and inject slowly to compress the bubble. Then slowly withdraw the syringe plunger; the bubble should enter the syringe. Before withdrawing the syringe, set the rotary valve to acetonitrile. Open the exit valve after removing the syringe, and repeat if required.

    Fragments of septum may block the column and cause the flow to slow or stop. The septum injector must then be removed (exit valve closed) and a needle or fine wire used to remove the offending fragment. This usually introduces an air bubble, which should be removed as above. An extra anhydrous acetonitrile wash should be used to remove any water that may have entered. It is wise to replace the septum every 10 or so nucleotides to avoid this problem.

9. Two common reasons for utter failure of a synthesis are amines in the solvents and entry of water. If yields diminish greatly with each coupling, water entry is the likeliest problem. The dead volume in the septum injector may trap some iodine solution and allow water to bleed during the coupling. The acetonitrile washes and the acetic anhydride in the capping step usually prevent this problem. Should it occur, the washes may be extended and the capping mixture injection begun before the syringe is fully inserted so the acetic anhydride washes the dead space. When water entry is not the source of low yields, the solvents should be replaced or repurified.

Acknowledgment

Cruachem Ltd. kindly provided samples of their cyanoethyl phosphoramidites for testing this protocol.

References

1. Gait, M. J. (1984) An Introduction to Modern Methods of DNA Synthesis, in *Oligonucleotide Synthesis—A Practical Approach* (Gait, M. J., ed.) IRL, Oxford and Washington.
2. Atkinson, T. and Smith, M. (1984) Solid-Phase Synthesis of Oligodeoxyribonucleotides by the Phosphite-Triester Method, in *Oligonucleotide Synthesis—A practical Approach* (Gait, M. J., ed.) IRL, Oxford.
3. Sinha, N. D., Biernat, J., McManus, J., and Koster, H. (1984)Polymer support oligonucleotide synthesis. XVIII. Use of β–cyanoethyl-*N*, *N*-dialkylamino-/*N*-morpholino phosphoramidite of deoxynucleosides for the synthesis of DNA fragments simplifying deprotection and isolation of the final product. *Nucleic Acids Res.* **12,** 4539–4557.
4. Jones, R. A. (1984) Preparation of Protected Deoxyribonucleosides, in *Oligonucleotice Synthesis—A Practical Approach* (Gait, M. J. ed.) IRL, Oxford.

5. Cruachem Ltd. (1986) Cyanoethylphosphoramidites. Cruachem Highlights, May 1986.
6. Jiricny, J. and Jones, M. B. (1984) The use of complementary 5' protecting groups, in the synthesis of mixed oligodeoxyribonucleotide sequences.
7. Gait, M. J., Matthes, H. W. D., Singh, M., Sproat, B. S., and Titmas, R. C. (1982) Synthesis of Oligodeoxyribonucleotides by a Continuous Flow, Phosphotriester Method on a Kieselguhr/Polyamide Support, in *Chemical and Enzymatic Synthesis of Gene Fragments* (Gassen, H. G. and Lang, A., eds.) Verlag Chemie, Basel.
8. Gaastra, W. and Josephsen, J. (1984) Radiolabeling of DNA Using Polynucleotide Kinase, in *Methods in Molecular Biology* Vol. 2. *Nucleic Acids* (Walker, J. M., ed.) Humana, Clifton.
9. McLaughlin, L. W. and Piel, N. (1984) Chromatographic Purification of Synthetic Oligonucleotides, in *Oligonucleotide Synthesis —A Practical Approach* (Gait, M. J., ed.) IRL, Oxford.

Chapter 15

Purification of Synthetic Oligonucleotides by Preparative Gel Electrophoresis

Robert McGookin

1. Introduction

The development of the chemistry for solid-phase oligonucleotide synthesis has enabled its widespread use in molecular biology laboratories for construction of novel linkers, site-directed mutagenesis, and many other techniques. The material obtained from a particular synthesis contains traces of premature termination products and other unwanted byproducts of the organic reactions. For most uses in genetic engineering, the material must be purified free of these contaminants. The most efficient method is undoubtedly by HPLC or FPLC, but in circumstances in which these systems are not available, the method described below provides very pure oligonucleotides in reasonable yield. The method described here is based loosely on that described by Narang et al. (1) and on information supplied by Applied Biosystems (oligonucleotide synthesis is described in Chapters 13 and 14).

The oligonucleotide mixture is run out on a denaturing polyacrylamide gel, the bands are visualized by UV shadowing and cut out, and the oligonucleotides are eluted into a suitable buffer. Impurities associated with the acrylamide are subsequently removed on an anion-exchange column. The oligonucleotide is then ready for use.

2. Materials

1. 10x TBE: 0.9M tris, 0.9M boric acid, 25 mM ethylenediamine tetracetic acid (EDTA), pH 8.3. This is standard TBE as described by Peacock and Dingman (2), but is used at a final concentration of 0.5x TBE.
2. 40% Stock acrylamide: [38% (w/v) acrylamide, 2% (w/v) N,N'-methylene bisacrylamide]. Filter the solution and store in the dark at 4°C.
3. 25% (w/v) Ammonium persulfate (AMPS). Prepare fresh for each gel.
4. Sample buffer: 98% (v/v) formamide, 10 mM EDTA. Prepare from deionized formamide and 0.5M EDTA, pH 8.0, and store in aliquots at –20°C.
5. Marker dye solution: 98% (v/v) formamide, 10 mM EDTA, 0.25% (w/v) xylene cyanol. Prepare from stocks as in step 4, and store at 4°C.
6. 0.25M Triethylamine bicarbonate (TEAB), pH 7.0. Mix the appropriate weight of triethylamine with sterile water to give about 80% of the final volume, and bubble CO_2 into the mixture until all the triethylamine has gone into solution (as judged by the disappearance of oily droplets on the surface). Adjust the pH by bubbling more CO_2 into the solution, sterilize by filtration, and store at 4°C. The pH should be checked each time before use and readjusted as necessary.
7. 2M TEAB, pH 9.0. This solution is prepared and stored as described in step 6.

3. Method

1. After recovery of the synthesized material from the column, deprotect according to the manufacturer's instructions and evaporate the sample to dryness. Dissolve the oligonucleotide mixture at about 0.5 A_{260} units/µL in sample buffer. Unpurified material can be stored in this form at –20°C.

2. Assemble an appropriate vertical gel slab. The gel should be at least 30 cm long for best resolution. Prepare the gel as described in Table 1, and leave to set for about 30 min.

3. Assemble the apparatus for electrophoresis using 0.5x TBE as the electrophoresis buffer, rinse out the wells thoroughly with electrophoresis buffer, and prerun the gel at 700 V for 30 min. Prepare the sample by heating to 90°C for 3 min and snap cooling on ice. After the pre-run period, rinse the wells again and load 2–3 A_{260} units/well, leaving the two outside wells free for 5 µL of marker dye solution. Run at 700 V until the bromophenol blue is almost at the bottom of the gel (about 4 h).

4. Disassemble the apparatus and place the gel over a Saran wrap-covered unused thin-layer chromatography plate and view under a 260 nm light source. The oligonucleotides on the gel can be clearly seen shadowed against the green fluorescent background. Cut out the least mobile (i.e., largest) band, also usually the most intensely absorbing, and place the gel pieces in a sterile Universal container. Pulverize the gel pieces with spatulate tweezers and add 5 mL of 0.25M TEAB, pH 7.0. Place the container on a rotating platform or shaking table at 4°C overnight.

5. Spin out the gel fragments at 16,000g for 10 min in a swinging bucket rotor, and carefully remove the supernatant for storage at 4°C. Add another 5 mL of 0.25M TEAB, pH 7.0, to the pellet and continue rotating for about 8 h.

6. Repeat the centrifugation step, and store the supernatant as before. Add another 5 mL of 0.25M TEAB, pH 7.0, and

Table 1
Gel Electrophoresis Solutions[a]

Stock	Per gel	Final concentration
Urea (solid)	33.6 g	7M
10x TBE	4 mL	0.5x TBE
40% Acrylamide	40 mL	20%
Water[b]	to 80 mL	
25% AMPS	120 µL	
TEMED[c]	80 µL	

[a]Volumes given are for a gel 32 x 14 x 0.15 cm.
[b]Swirl the solution in 37°C bath to dissolve the urea before adding AMPS and TEMED.
[c]TEMED = $N,N,N'N'$-tetramethylethylenediamine.

continue eluting overnight. Centrifuge once more and retain the supernatant.

7. Prepare about 1 mL bed volume of Whatman DE52 anion-exchange resin, equilibrated with 0.25M TEAB, pH 7.0, according to the manufacturer's instructions. Pour a column of about 0.5 mL in a disposable 1-mL plastic syringe barrel plugged with glass wool and wash through with 20 mL of 0.25M TEAB, pH 7.0.

8. Pass the stored gel-eluted material through the column followed by 5 mL of sterile water. Elute the oligonucleotides with 4x 0.5 mL of 2M TEAB, pH 9.0, into sterile tubes and dry down. Dissolve each sample in an appropriate volume of sterile water (usually 50 µL), and assay a small aliquot spectrophotometrically to determine which sample contains the eluted material (usually the second) and the yield.

4. Notes

1. A yield of 30–50% can be expected, most often near 30%. The final product should be checked for purity by end-

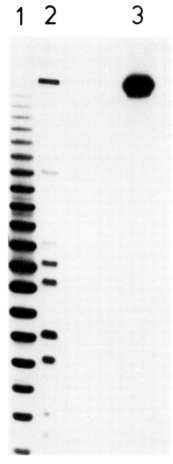

Fig. 1. Denaturing polyacrylamide gel of purification of ^{32}P-labeled 23-mer oligonucleotide. An autoradiograph of a 20% denaturing polyacrylamide gel is shown, Track 1: Oligo $(dT)_{4-22}$ markers; Track 2: 23-mer oligonucleotide before purification; Track 3: 23-mer oligonucleotide after purification by gel electrophoresis.

labeling a small quantity and running on a denaturing polyacrylamide gel. A typical result is shown in Fig. 1.
2. This is a long procedure and should only be considered when the faster methods mentioned in the introduction are not available. At the expense of yield, the process can

be speeded up by performing only one cycle of elution from the crushed gel slices.

3. Very occasionally an oligonucleotide synthesis will produce a single band of much higher molecular weight than expected on the gel. This is seen as a single, usually fairly faint band near the top of the gel, whereas the expected ladder occurs much lower down. The top band in the ladder (usually the most intense) should be purified, and this upper band should be ignored.

References

1. Narang, S.A., Hsuing, H.M., and Brousseau, R. (1979) Improved phosphodiester method for the synthesis of gene fragments. *Meth. Enzymol* **68**, 90–98.
2. Peacock, A.C. and Dingman, C.W. (1968) Molecular weight estimation and separation of ribonucleic acid by electrophoresis in agarose-acrylamide composite gels. *Biochemistry* **7**, 668–674.

Chapter 16

Isolation of High Molecular Weight DNA Suitable for the Construction of Genomic Libraries

John Steven, Douglas McKechnie, and Alexander Graham

1. Introduction

Recent advances in molecular biology have made it possible to construct complete gene libraries for any organism that uses DNA as its carrier of genetic information. A gene library should contain a large number of cloned DNA fragments that in total contain the entire donor genome. The construction of a genomic library first requires the isolation of DNA from the donor organism. To be of maximum use in the construction of genomic libraries, DNA isolated from the donor organism should fulfill the following criteria. First, the DNA must represent all sequences in the genome to be cloned. Second, it must be of high molecular weight. Third, no contaminants must taint the DNA so that its use as a substrate for restriction endonucleases and other enzymes used in genetic engineering is uninhibited.

Methods for the isolation of high molecular weight DNA from bacteria, blood, and tissue culture cells have already been described in an earlier volume in this series (1,2). These techniques are inappropriate, however, for use when DNA has to be extracted from whole animal or plant tissue. Here we describe two procedures that enable high molecular weight DNA to be isolated from such sources.

The first method, method A, allows for simultaneous isolation of both DNA and RNA from plant or fungal tissues. The tissue is either freeze dried or flash frozen in liquid nitrogen and ground to a fine powder. After resuspension in a high salt-containing buffer, cetyltrimethylammonium bromide (CTAB) is added and the salt concentration is reduced by dilution. The reduction in salt concentration results in the formation of a CTAB/nucleic acid precipitate that can be collected by centrifugation (3). After separation from RNA, the DNA is then banded on CsCl gradients, recovered, and prepared for cloning into a suitable vector. This method allows nucleic acids to be prepared from tissues that are rich in carbohydrates and would perhaps cause problems if more conventional phenol extraction procedures were used.

The second method, method B, is used mainly to isolate DNA from animal tissues. In this procedure nuclei are isolated from homogenized tissue (4) and lysed in detergent-containing buffer (5), and the DNA is then isolated from the resultant solutions after enzymatic treatment and phenol extraction to remove RNA and protein. Both these techniques should result in the preparation of high molecular weight DNA that only has to be cut into suitable size fragments for cloning into the vector of choice.

2. Materials

2.1. Method A

1. 2 x Extraction buffer: $1.5M$ NaCl, 25 mM EDTA, 100 mM Tris HCl, pH 8, 2% (w/v) cetyltrimethylammonium bromide (CTAB).
2. Precipitation buffer: 1% (w/v) CTAB, 10 mM EDTA, 40 mM Tris HCl, pH 8, 1% (v/v) β-mercaptoethanol added fresh just before use.
3. 10% CTAB solution: 10% CTAB (w/v), $0.7M$ NaCl.
4. CsCl solution A: $1M$ CsCl, 50 mM Tris HCl, pH 8, 5 mM EDTA, 50 mM NaCl.
5. CsCl solution B: $5.7M$ CsCl, 50 mM Tris HCl, pH 8, 5 mM EDTA, 50 mM NaCl.
6. TE buffer: 10 mM Tris-HCl, pH 8, 1 mM EDTA.
7. Waring blender with metal cup and lid.
8. Chloroform/isoamyl alcohol (24 vol chloroform/1 vol isoamyl alcohol).
9. An incubator or water bath at 55°C.
10. A medium-speed centrifuge, an ultracentrifuge, and a range of rotors, tubes, bottles, and adaptors for both instruments that will accommodate all centrifugation steps in the extraction protocol.
11. UV light source, dark room, and refractometer.
12. Dialysis tubing. This should be boiled in 5% (w/v) NaHCO$_3$, washed with two changes of 5 mM EDTA, autoclaved in distilled water, and stored at 4°C until required.

2.2. Method B

1. MNKT buffer: 5 mM MgCl$_2$, 10 mM NaCl, 12.5 mM KCl, 50 mM Tris-HCl, pH 7.6.
2. Homogenization buffer: $0.25M$ sucrose, 5 mM MgCl$_2$, 12.5 mM KCl, 50 mM Tris-HCl, pH 7.6, 10 mM NaCl.

3. Heavy sucrose buffer: 2.3M sucrose, 5 mM MgCl$_2$, 10 mM NaCl, 50 mM Tris-HCl, pH 8, 12.5 mM KCl.
4. ENST buffer: 10 mM EDTA, 10 mM NaCl, 0.5% (w/v) SDS, 10 mM Tris-HCl, pH 8.
5. Phenol extraction buffer: 10 mM EDTA, 10 mM NaCl, 0.5% SDS (w/v), 500 mM Tris-HCl, pH 8.
6. Dialysis buffer: 10 mM EDTA, 10 mM NaCl, 50 mM Tris-HCl, pH 8.
7. ETN buffer: 0.5 mM EDTA, 10 mM Tris-HCl, pH 8, 10 mM NaCl.
8. 10% SDS (w/v).
9. Triton X-100.
10. Phenol saturated with phenol extraction buffer.
11. RNAse A solution at a concentration of 10 mg/mL in sterile water.
12. Proteinase K made up at 25 mg/mL in sterile distilled water.
13. Dialysis tubing boiled in 5% (w/v) NaHCO$_3$, washed with 5 mM EDTA, autoclaved in distilled water, and stored at 4°C.
14. Ethanol.
15. 110 Mesh nylon bolting cloth.
16. A glass Teflon motor-driven homogenizer with a clearance of 0.15 mm or 0.006 inch.
17. A medium-speed centrifuge and an ultracentrifuge, both with a range of rotors, tubes, and bottles to fit.

3. Methods

3.1. Method A

1. Flash freeze a preweighed amount (10–50 g) of fresh plant tissue in liquid nitrogen. Transfer the tissue to a prechilled metal Waring blender cup, and grind the frozen tissue to a fine powder. Freeze-dried tissue can also be used (*see* Note 1 in section 4.1).

2. Add 2.5% (v/w) β-mercaptoethanol plus an equal volume (v/w) of 2x extraction buffer preheated to 75°C. Mix well and transfer to an incubator or water bath at 55°C.

3. Stir the mixture with a glass rod until it reaches a temperature of 55°C. Transfer to a screw-cap centrifuge tube and add an equal volume of chloroform/isoamyl alcohol. Shake gently until the organic phase and aqueous phase are mixed. Vent occasionally during mixing to avoid a pressure buildup inside the screw-cap container. Ensure that the shaking is very gentle to minimize DNA shearing and ensure that the recovered DNA has as high a molecular weight as possible.

4. Centrifuge at 12000g for 20 min.

5. Cut the tip off a sterile 10-mL disposable plastic pipet and use the pipet with the enlarged bore to transfer the aqueous phase (uppermost) to a clean sterile bottle. A 1/10 volume of 10% CTAB solution is now added, along with an equal volume of chloroform/isoamyl alcohol and the extraction repeated as in step 3 above.

6. Repeat the centrifugation as in step 4 above, and once again recover the aqueous phase and transfer into a clean tube.

7. Now add an equal volume of precipitation buffer. This reduces the salt concentration, and this reduction in salt concentration promotes the formation of a CTAB/nucleic acid complex that precipitates out of solution. With occasional gentle mixing, this precipitation is allowed to stand at room temperature for 1 h.

8. Collect the nucleic acid/CTAB precipitate by centrifugation at 6000 g for 30 min. Discard the supernatant.

9. Resuspend the pellet in 10–20 mL of CsCl solution A. When fully resuspended, layer 3.5 mL of the DNA-containing CsCl solution A carefully onto a 1.5-mL cushion of CsCl solution B in an ultracentrifuge tube. Centrifuge the tubes at 120,000g for 18 h. These volumes are suitable for the Beckman SW50.1 rotor and tubes, but these volumes of

CsCl solutions A and B can be altered to suit any comparable rotor and centrifuge. RNA will pellet through the cushion of CsCl solution B.

10. Remove most of the top layer of CsCl solution A to within 3–4 mm of interface with CsCl solution B. Collect the DNA from the interface, using a Pasteur pipet and transfer to a clean sterile tube. If more than one step gradient of the same DNA preparation has been run, then all the DNA fractions from the interfaces can be pooled at this point. The RNA that has pelleted through the cushions of CsCl solution B can now be drained and resuspended in 10 mM Tris-HCl, 1 mM EDTA, 100 mM KCl, pH 7.6. Again the RNA from several pellets can be pooled and collectively precipitated by addition of 2.5 vol of ethanol. After recovery by centrifugation, this RNA is suitable for fractionation into poly (A)$^+$ RNA and poly (A)$^-$ RNA by affinity chromatography as previously described (6,7).

11. The refractive index of the DNA solution is now adjusted to 1.399 by the addition of CsCl solution B or TE buffer as required. The DNA solution is now transferred to an ultracentrifuge tube and filled up to volume with a solution of CsCl that has a refractive index of 1.399 and a density of 1.70 g/mL. This CsCl solution is made up in TE buffer containing 100 µg/mL ethidium bromide. The ultracentrifuge tube is now capped or sealed, then centrifuged to equilibrium. At equilibrium the CsCl will have formed a gradient, and the DNA will have banded in this gradient. The length of time taken for the CsCl gradient to achieve equilibrium depends on the rotor and centrifuge being used. Normally conditions are chosen to allow for an overnight run of the ultracentrifuge.

12. After reaching equilibrium, the gradients are viewed with long wavelength UV light, and the fluorescent band (caused by ethidium bromide binding to the DNA) is removed from the first gradient. This DNA is then rebanded on a fresh second gradient of CsCl/ethidium

bromide. The formation and unloading of CsCl gradients has already been described in a previous volume of this series (*8,9*).

13. After the DNA has been recovered from the second gradient, the ethidium bromide is removed by repeated extractions with amyl alcohol or *n*-butanol (both of which should be equilibrated with CsCl solution A before use). When the organic and aqueous phases are both colorless, all the ethidium bromide has been removed.

14. The DNA solution is now diluted twofold and adjusted to 0.35*M* potassium acetate, and 2.5 vol of cold ethanol is added. DNA is allowed to precipitate at –20°C for 30 min, then collected by centrifugation at 6000*g* for 15 min. The pellet of DNA is then washed x3 with 70% ethanol and collected by centrifugation after each wash. This should remove all Cs ions from the DNA preparation. Finally the DNA is dried in vacuo, then resuspended in sterile TE buffer at a concentration of 200 µg/mL. A UV scan of DNA sample may now be performed and the resulting spectrum used to determine the yield and purity of the DNA preparations.

15. As an alternative to step 14, the DNA solution may be placed in a dialysis bag and dialyzed against 2 L of TE buffer at 4°C over a 48-h period with six to eight changes of TE buffer. This should ensure that all CsCl is removed from the DNA preparation. When dialysis is complete, DNA is removed from the dialysis bag, and a sample is scanned in a UV spectrophotometer, as above.

16. DNA solution should now be stored at 4°C until required for the next step in the cloning procedure.

3.2. Method B

1. Remove the tissue from which DNA is to be isolated from an anesthetized or killed donor organism. Add the tissue

to a preweighed beaker containing ice-cold homogenization buffer. Reweigh to determine weight of tissue.

2. Chop the tissue finely with scissors or a scapel blade.
   Drain buffer and wash the tissue with homogenization
   buffer to remove blood. Once again drain the buffer from
   the minced tissue, then add a volume of homogenization
   buffer equal to 2x its weight.

3. Transfer the tissue plus buffer to the loose-fitting glass
   Teflon motor-driven homogenizer and disrupt cells with
   8–12 strokes.

4. Filter the homogenate through 110-mesh nylon bolting
   cloth using a glass funnel and a measuring cylinder previously cooled to –20°C and collect the filtered homogenate.

5. Note the volume of the filtrate and add to it 2x this volume
   of heavy sucrose buffer. This gives a final concentration of
   1.62M sucrose.

6. Pipet 25-mL aliquots of the homogenate into 35-mL ultracentrifuge tubes and underlay with 10 mL of heavy sucrose buffer. A Beckman SW28 rotor or its equivalent is
   ideal.

7. Centrifuge at 130,000g for 1 h.

8. Nuclei will pellet through the heavy sucrose buffer. Remove by aspiration all of the top layer and about half of the
   heavy sucrose buffer layer. Decant off the remaining
   heavy sucrose buffer and wipe the inside of the centrifuge
   tube dry with clean tissues, taking care not to disturb the
   pellet of nuclei.

9. Add about 5 mL of MNKT to each tube and suspend the
   nuclei by very gentle agitation. Pour the nuclear suspension into the glass homogenizer, and add Triton X-100 to
   give a final concentration of 0.2%. The nuclei are gently
   resuspended giving one stroke, manually, of the homogenizer.

10. Centrifuge the nuclei in sterile glass centrifuge tubes at
    1000g for 10 min. Discard the supernatant and retain the
    nuclear pellet.

11. Suspend the nuclei in MNKT buffer and add Triton X-100 to 0.1%. Again give one stroke in the same glass homogenizer to suspend the nuclei, and centrifuge as in step 10.
12. Discard the supernatant and wash pellet in 5 mL of MNKT alone. Repellet nuclei by centrifugation, and discard supernatant. Resuspend the pellet of nuclei in ENST buffer to give a final concentration of DNA of less than 200 µg/mL, determined by measuring A_{260} of an aliquot. Adjust as necessary using ENST buffer. Lysis of the nuclei will be evident by the increased viscosity of the solution.
13. Add stock proteinase K to give a final concentration of 50 µg/mL and incubate at 37°C for 18 h. (Overnight incubation is usually convenient for this step.)
14. Transfer the sample to a stoppered sterile glass bottle, and add an equal volume of phenol saturated with phenol extraction buffer. Shake gently, to avoid shearing DNA, for 10 min until the organic and aqueous phases are well mixed. Transfer the phenolic emulsion to centrifuge tubes and centrifuge at 6000g for 10 min to separate the organic and aqueous phases.
15. Remove the aqueous phase and extract with an equal volume of phenol saturated with phenol extraction buffer. Re-extract the first phenolic phase with an equal volume of phenol extraction buffer. Centrifuge both extractions at 6000g as before.
16. Pool aqueous phases and transfer to a dialysis bag. Dialysis is carried out against several changes of dialysis buffer at room temp. for 3 h, then at 4°C against several more changes of dialysis buffer until the OD_{280} absorbance of the buffer outside the dialysis bag is less than 0.05 U/mL.
17. Empty samples from the dialysis bag into a sterile nuclease-free flask, noting the volume. Add RNAse A to give a final concentration of 50 µg/mL. Incubate at 37°C for 4 h.
18. Add SDS from the 10% stock solution to give a final concentration of 0.5% and proteinase K to give a 50 µg/ml final concentration. Incubate overnight at 37°C.

19. Phenol extract, as in steps 14 and 15 above.

20. Transfer the pooled aqueous phase to dialysis tubing and dialyze against at least six changes of ETN buffer at 0°C until the UV absorbance at 280 nm outside the bag is zero.

21. Take the contents of the dialysis bag and add 1/10 volume of 3*M* potassium acetate, pH 4.8, and 2.5 vol of ethanol. Chill at −70°C for 30 min after mixing. Recover the DNA precipitate by centrifugation at 6000*g* for 10 min. Wash the pellet of DNA x 2 with 70% ethanol. Collect the DNA between each wash by centrifugation as above. Dry the final pellet of DNA *in vacuo*, then resuspend at 200 µg/mL in sterile TE buffer, pH 8.

22. Check purity and yield of DNA by UV spectrophotometry. Store the DNA at 4°C until it is required for the next step in the construction of a genomic DNA library.

4. Notes

4.1. Method A

1. If lyophilized tissue is used, it should be ground up in a mill or mortar and pestle, and the resulting fine powder resuspended in 20 mL of 1 x extraction buffer containing 1% β-mercaptoethanol per gram of dry weight tissue. Then transfer suspension to water bath at 55°C and proceed from step 3 above.

2. Using lyophilized tissue may be slightly advantageous since the DNA will be dehydrated and less liable to mechanical breakage during pulverization. This may help increase the average molecular weight of the isolated DNA.

3. All steps should be carried out at room temperature unless otherwise indicated. Temperatures below 16°C may facilitate premature precipitation of the nucleic acid/CTAB complex.

4. Using this method, 20–50 µg of DNA and/or 100–300 µg of RNA may be isolated per gram of plant or fungal tissue. The yield will vary depending on the tissue or species from which the nucleic acids have been extracted. This method ensures that DNA from all subcellular organelles, as well as the nucleus, is extracted from the tissue.

5. UV absorbance is a useful guide to the purity of a DNA preparation. An approximation of DNA concentration can be calculated from UV spectrum data using the general rule that a 50 µg/mL aqueous solution of DNA has an OD A_{260} of 1. This value may vary with the G + C content of the DNA, but these variations need not be considered important in molecular biology. The ratio of A_{260}/A_{280} of pure DNA in solution should be between 1.6 and 1.8 unless the DNA has an odd G + C content. Ratios higher that 1.8 may be caused by RNA contamination, and ratios lower that 1.6 indicate presence of phenol or protein in the DNA preparation.

6. Disposable plastic or rubber gloves should be worn at all times during the extraction and recovery of DNA (in methods A and B) to minimize the risk of nuclease contamination and subsequent degradation of the nucleic acid.

7. For methods A and B, appropriate measures should be taken to render all solutions, glassware, and centrifuge tubes free of nuclease contamination. Glassware and glass pipets should be baked at 160°C for 4 h. Centrifuge tubes, pipet tips, and buffer solutions should, if possible, be autoclaved. Solutions should be prepared from Analar-grade reagents, and sterile distilled water used for their preparation. SDS-containing buffers should be autoclaved at 115°C for 10 min, and all other buffers, at 121°C for 15 min.

8. Gentle mixing of the organic and aqueous phases during the chloroform extraction step will minimize the risk of mechanical breakage of the DNA.

4.2. Method B

1. After ethanol precipitation and drying *in vacuo*, high molecular weight DNA may take a long time to go back into solution, and rehydration of the DNA can be carried out overnight at 4°C with as little mixing as possible. Do not vortex mix DNA to try to speed up the resuspension step, since this will only result in strand breakages that will reduce the molecular weight of the isolated DNA. The ethanol precipitation step may be substituted with extensive dialysis against TE buffer (10) at step 20 above. During the extraction procedure, phenolic and aqueous phases should be mixed very gently so that the DNA molecules are not subjected to mechanical breakage.
2. The UV scan of DNA will detect any impurities that may be present (details as in method A).
3. Other procedures may have to be used if nonnuclear DNA is required for cloning, since this method selects nuclei for the preparation of DNA, and other DNA-containing subcellular organelles may be lost at the initial step gradient on the heavy sucrose cushion.

4.3. Notes on Methods A and B

1. Both procedures for the preparation of high molecular weight DNA should give material that, when analyzed by electrophoresis (11) in a 0.4% agarose gel, should migrate more slowly than bacteriophage lambda DNA (50 kbp).
2. Before cloning into a suitable vector, the high molecular weight DNA needs to be cut into smaller fragments using restriction endonucleases or mechanical breakage. The size of fragments required for cloning depends on the vector to be used. Plasmid vectors are generally suitable for fragments of less than 10 kbp long and have been used successfully to construct gene libraries (12,13). Larger pieces of DNA (up to 25 kbp) can be cloned into suitable bacteriophage lambda vectors (14,15). Cloning DNA into

lambda vectors has already been dealt with in a previous volume of this series (*16*). (*See also* Chapters 17 and 18 of this volume.

3. The number of recombinant clones that are required to represent the entire genome of the donor organism depends on the genome size and the size of the fragments that are cloned.

References

1. Mathew, C. G. P. (1984) The Isolation of High Molecular Weight Eukaryotic DNA, in *Methods in Molecular Biology* Vol. 2 (Walker, J. M., ed.) Humana, New Jersey.
2. Dale, J. W. and Greenaway, P. J. (1984) Preparation of Chromosomal DNA from *E. coli*, in *Methods in Molecular Biology* Vol. 2 (Walker, J. M., ed.) Humana, New Jersey.
3. Murray, H. G. and Thompson, W. F. (1980) Rapid isolation of high molecular weight plant DNA. *Nucleic Acids Res.*, 4321–4325.
4. Blobel, G. and Potter, V. R. (1966) Nuclei from rat liver. Isolation method that combines purity with high yield. *Science* 154, 1662–1665.
5. Gross-Bellard, M., Oudet, P., and Chambon, P. (1973) Isolation of high molecular weight DNA from mammalian cells. *Eur. J. Biochem.* 36, 32–38.
6. Slater, R. J. (1984) The Purification of Poly (A)-Containing RNA by Affinity Chromatography, in *Methods in Molecular Biology* Vol. 2 (Walker, J. M., ed.) Humana, New Jersey.
7. Aviv, H. and Leader, P. (1972) Purification of biologically active globin messenger RNA by chromatography on oligothymidylic acid cellulose. *Proc. Natl. Acad. Sci. USA* 69, 1408–1412.
8. Cooney, C. A. and Mathews, H. R. (1984) The Isolation of Satellite DNA by Density Gradient Centrifugation, in *Methods in Molecular Biology* Vol. 2 (Walker, J. M., ed.) Humana, New Jersey.
9. Boffey, S. A. (1984) Plasmid DNA Isolation by the Cleared Lysate Method, in *Methods in Molecular Biology* Vol. 2 (Walker, J. M., ed.) Humana, New Jersey.
10. Maniatis, T., Fritsch, E. F., and Sambrook, J. (1982) Construction of Genomic Libraries, in *Molecular Cloning, A Laboratory Manual* Cold Spring Harbor Laboratory, New York.
11. Maniatis, T., Fritsch, E. F., and Sambrook, J. (1982) Gel Electrophoresis, in *Molecular Cloning, A Laboratory Manual,* Cold Spring Harbor Laboratory, Cold Spring Harbor, New York.

12. Kedes, L. H., Chang, A. C. Y., Houseman, D., and Cohen, S. N., (1975) Isolation of histone genes from unfractionated sea urchin DNA by subcellular cloning in *E. coli. Nature* **255**, 523–538.

13. Glover, D. M., White, R. L., Finnigan, D. J., and Hogness, D. A. (1975) Characterisation of six cloned DNAs from *Drosophila melanogaster*, including one that contains the genes for rRNA. *Cell* **5**, 144–157.

14. Williams, B. G. and Blattner, F. R. (1979) Construction and characterisation of the hybrid bacteriophage lambda charon vectors for DNA cloning. *J. Virol.* **29**, 555–575.

15. Morrow, J. F. (1979) Recombinant DNA Techniques, in *Methods in Enzymology* (Wu, R., ed.) Academic, New York.

16. Christiansen, C. (1984) Molecular Cloning in Bacteriophage lambda and in Cosmids, in *Methods in Molecular Biology* Vol. 2 (Walker, J. M., ed.) Humana, New Jersey.

Chapter 17

Construction of Mammalian Genomic Libraries Using λ Replacement Vectors

Alan N. Bateson and Jeffrey W. Pollard

1. Introduction

The ideal genomic library should consist of a series of clones containing overlapping sequences that are representative of the entire genome. Such an ideal state can be approached by cutting the DNA randomly and cloning large pieces of this DNA into a suitable vector (1). The DNA can be cleaved either by mechanical shearing, in which case the DNA is fragmented in a truely random fashion, but with the introduction of problems associated with blunt end ligation, or better, by partial digestion with a restriction endonuclease such as Mbo1 or Sau3A. These enzymes recognize four base sequences that are predicted to occur on average (although in practice this is not the case) every 256 bases, and hence digestion results in cleavage of the DNA in a pseudorandom manner.

The simplest vectors to use are the λ replacement vectors, although cosmids are also suitable and are described in Chapter 18. The λ replacement vectors exploit the fact that the central region, consisting of nearly 40% of the genome, contains genes inessential for λ replication (2). This fragment, known as the stuffer fragment, may therefore be replaced with foreign DNA. Since λ will package with high efficiency only in the size range 40–52 kb, religation of the left and right arms alone will result in a molecule too small to package, and thus all plaques should contain recombinant phages. Background packaging, however, does occur owing to the religation of all the pieces. This background can be reduced in a number of ways: (1) the introduction of a unique restriction site(s) into the stuffer fragment such as a SalI site allows for the secondary cleavage of the stuffer fragment, thus reducing the possibility of religation of a packageable DNA molecule from the λ sequences alone, (2) the left and right arms may be separated from the stuffer fragment using a sucrose gradient or agarose gel electrophoresis, (3) phage that contain the stuffer fragment can be selected against using the Spi selection system (3–5). This selection is effective because the *red* and *gam* genes that prevent plating on P2 lysogens (*see* refs. 2,3, for details) are contained in the stuffer fragment and, therefore, are lost in recombinants allowing these recombinants to form plaques on P2 lysogens (*rec* BC⁺ hosts), providing a *chi* sequence is introduced into the λ arms (3,4).

All these features are contained in such vectors as λL47 (5) and EMBL 3 and 4 (4). These vectors also contain a polylinker with a BamH1 cloning site at either end of the stuffer fragment, which allows insertion of fragments generated by Mbo1 of Sau3A digestion. This is because, although BamH1 cuts at G↓GATCC, the core recognition sequence ↓GATC is the recognition site for Mbo1 and Sau3A, thus allowing ligation into this site. In the EMBL vectors, the BamH1 site is also flanked by Sal1 in EMBL 3 (excluding the use of this enzyme to destroy the stuffer fragment) and EcoRI in EMBL 4, which allows for the use of these enzymes to excise the donor fragment. These particular replacement vectors are therefore extremely useful for constructing genomic libraries, although it should be pointed out that they can only be effectively used if they are propagated, after the *spi* selection, on *recBC⁻* hosts.

This chapter will describe the construction of a mammalian genomic library using the EMBL 4 vector and is based on a combination of protocols derived from refs. *1, 3,* and *4.* In this case large genomic DNA is partially digested with Mbo1 and fractionated by sucrose gradient centrifugation to obtain approximately 20-kb fragments. This procedure prevents multiple insertions of donor DNA into a single recombinant clone and therefore facilitates the isolation, without ambiguity, of overlapping clones within the library. Once isolated the 20-kb DNA fragments are ligated to the arms, packaged, and plated onto a P2 lysogen (Q359) to select for the *spi⁻* phenotype. This method should give a mammalian genomic library with a probability of greater than 99% containing the sequence of interest.

2. Materials

1. L-Broth (1 L): 10 g of bacto-tryptone, 5 g of yeast extract, 5 g of NaCl; adjust to pH 7.2 and autoclave. If used for growth of λ and its derivatives, add $MgSO_4$ to 10 mM.
2. CY medium (1 L): 10 g of casamino acids, 5 g of bactoyeast extract, 3 g of NaCl, 2 g of KCl, 2.46 g of $MgSO_4$ • $7H_2O$; adjust to pH 7.5 and autoclave.
3. 10 mM $MgSO_4$ • $7H_2O$.
4. λ Gradient buffer: 10 mM Tris-HCl, 100 mM NaCl, 10 mM $MgSO_4$ • $7H_2O$, pH 7.4.
5. λ Dialysis buffer: 10 mM Tris-HCl, 25 mM NaCl, 1 mM $MgSO_4$ • $7H_2O$, pH 8.0.
6. 0.5M EDTA, pH 7.0.
7. 10% (w/v) sodium dodecyl sulfate (SDS).
8. Proteinase K: 10 mg/mL (w/v) in 10 mM Tris-HCl, 5 mM EDTA, 0.5% (w/v) SDS, pH 7.8.
9. Recrystallized phenol equilibriated with 0.5M Tris-HCl, pH 8.0, and finally 0.1M Tris-HCl, pH 8.0 [Note: equili-

brate with several changes of 0.5*M* Tris to ensure the pH is 8.0. Phenol:chloroform means a 1/1 (v/v) mix of buffered saturated phenol with chloroform:isoamylalcohol in a ratio of 24/1 (v/v). Phenol causes severe burns and should be handled accordingly].

10. Sucrose gradient buffers: make a gradient of 15% (w/v) and 35% (w/v) sucrose in 1*M* NaCl, 20 m*M* Tris-HCl, 5 m*M* EDTA, pH 8.0.
11. TE buffer: 10 m*M* Tris-HCl, 1 m*M* EDTA, pH 7.5.
12. 0.01*M* $MgCl_2$.
13. Loading buffer: 0.25% (w/v) bromophenol blue, 0.25% (w/v) xylene cyanol, 4% (w/v) sucrose in H_2O.
14. Homogenization buffer: 10 m*M* Tris-HCl, 60 m*M* NaCl, 10 m*M* EDTA, 0.5% (v/v) Triton X-100, pH 7.4.
15. 10% (w/v) sarkosyl.
16. Stop solution: 25% (v/v) loading buffer (solution 13, above) containing 28 m*M* EDTA.
17. 0.25*M* EDTA, pH 8.0.
18. Top agar: 7 g of bacto-agar in 1 L of L-broth (plus $MgSO_4$).
19. Bottom agar: 15 g of bacto-agar in 1 L of L-broth (plus $MgSO_4$).
20. SM medium (1 L): 5.8 g of NaCl, 2 g of $MgSO_4 \cdot 7H_2O$, 50 mL of 1*M* Tris-HCl, pH 7.5, 2% (w/v) gelatin. Autoclave.
21. Enzymes: Mbo1, Sal1, BamH1, Xhol, Hind III, T4 ligase, DNaseI are supplied with the appropriate buffers by all manufacturers. Follow their instructions for storage and shelf-life.
22. Bacterial strains and bacteriophages: λ EMBL 4 (4) accepts DNA between 9 and 22 kb. QR48: A, *rec*– host used for growing the parent vector.
   Q358: A *su*+ host used for the growth of the recombinant bacteriophage for secondary screening and titering EMBL 4 parent (6).
   Q359: A *su*+ host used to detect *spi*− recombinants for initial plating of the library (6).
   LE359: Used for the growth and titering of wild-type λ.

All solutions should be made up with double-distilled, sterile deionized water. Solutions 3, 4, 5, 6, 10, 11, 12, 14, and 17 can be autoclaved and stored at room temperature. All media (solutions 1, 2, 20) can be autoclaved and stored indefinitely at room temperature. Phenol should be stored at 4°C in brown glass bottles (do not use bottles that have previously been used for tissue culture medium; residual serum baked onto the glass will be stripped off by the phenol). Proteinase K should be stored at –20°C. All remaining solutions can be stored at room temperature.

3. Method

3.1. Preparation of Phage Stocks

1. From a frozen stock of bacteria QR48 (*see* ref. 1), seed a 10-mL liquid culture in L-broth at 37°C.
2. Inoculate 1 mL of overnight culture into a conical flask (at least 4x the volume) containing 50 mL of CY medium and incubate it, with shaking (250 rpm) for 2–2.5 h at 37°C until the optical density is 0.4 at 630 n*M*.
3. Harvest the cells by centrifugation at 3000*g* for 10 min at 4°C. Pour off the supernatant and resuspend the pellet in 25 mL of 10 m*M* MgSO$_4$. This suspension may be stored for up to 10 d at 4°C, but the highest plating efficiency is achieved with 0–2 d-old cells.
4. To obtain the maximum yield of phage, it is necessary to calculate the optimum ratio of phage with which to infect the bacteria: the infectivity ratio. This is determined in a small-scale experiment prior to the full-scale preparation of phage (*see* note 1 in section 4). The day before the experiment, determine the titer of both the plating bacterial cells, QR48, and the phage lysate (refs. *1,7, see* Note 1 in section 4).
5. Inoculate 0.1 mL containing 2 x 10^8 plating cells with the following titers of phage in 0.1 mL into separate sterile

Eppendorf tubes: $10^8, 5 \times 10^7, 10^7, 5 \times 10^6, 10^6$, and 5×10^5 pfu, respectively. Incubate at room temperature for 15 min to allow the phage particles to absorb.

6. Carefully transfer each of these cultures to separate 100-mL conical flasks containing 10 mL of CY medium, prewarmed to 37°C. Shake (140 rpm) overnight at 37°C on an orbital shaker.

7. Clear the lysates by centrifugation at 12,000g for 10 min at 4°C. Titer the supernatants and determine the optimum infectivity ratio.

8. Continue with the main phage preparation. Inoculate 4 × 10^9 plating cells with the optimum infectivity ratio of phage in a 15-mL sterile plastic tube. Incubate at room temperature for 15 min.

9. Transfer this carefully to a 2-L conical flask containing 200 mL of CY medium prewarmed to 37°C. Shake overnight at 140 rpm in an orbital shaker at 37°C.

10. At the end of this period, cultures should still appear cloudy, but with extensive lysis showing as stringy clumps of bacterial cell debris. If this is not the case, the lysis of the cells can be increased by placing the flasks at 4°C for 1–2 h.

11. Remove the cell debris by centrifugation at 5000g at 4°C for 20 min.

12. Remove the supernatant to 50 mL polycarbonate ultracentrifuge tubes, and centrifuge these at 75,000g for 1 h at 4°C.

13. Pour off the supernatant; the pellet should have a dirty yellow center (debris) surrounded by a faintly blue opalescent ring of phage.

14. To resuspend the pellet, add 0.48 mL of λ gradient buffer at 4°C to each tube, and tilt the tubes so that the pellet is under the buffer. Leave the phage to diffuse out overnight.

15. Transfer *gently* to a 15-mL siliconized glass centrifuge tube and centrifuge at 12,000g for 5 min at 4°C to remove cell debris.

16. Transfer the supernatant to a fresh 15-mL glass centrifuge

tube, add 100 μg of DNaseI, and incubate at room temperature for 1 h.

17. Prepare a three-step cesium chloride gradient in λ gradient buffer with concentrations of 1.7 g/cm³ (95 g to 75 mL; η, 1.3990), 1.5 g/cm³ (67 g to 82 mL; η, 1.3815), and 1.30 g/cm³ (40 g to 90 mL; η, 1.3621) using a peristaltic pump to carefully layer sequentially 2 mL 1.7, 3 mL 1.5, and 2 mL 1.3 into the 12-mL transparent ultracentrifuge tube, marking the interface of each layer.

18. Carefully layer the phage suspension using the pump and top up the centrifuge tube with gradient buffer.

19. Centrifuge at 160,000g for 1 h 40 min at 20°C using the "slow acceleration" and "break off" modes of the ultracentrifuge.

20. The expected banding pattern is shown in Fig. 1 (*see* Note 2 in section 4). Carefully remove the phage band with a 21-gage needle and a syringe. This is done by placing a small piece of Scotch tape over the band and then piercing the tube immediately below the band, bevel edge of the needle up. Be careful that you do not go straight through the tube and into your finger.

21. Dialyze the phage against λ dialysis buffer at 4°C using two changes of 2 L each.

22. Titer the phage preparation (*see* Note 3 in section 4).

3.2. Preparation of Phage DNA

1. To the phage preparation add 1/50 (v/v) of 0.5M EDTA.

2. Add 1/50 (v/v) of a 10% SDS solution and incubate for 15 min at 65°C.

3. Cool to 45°C, add proteinase K to 50 μg/mL, and incubate at 45°C for 1 h.

4. Extract with phenol by carefully inverting the tube several times.

5. Break the phases by centrifugation at 7800g for 5 min at 4°C.

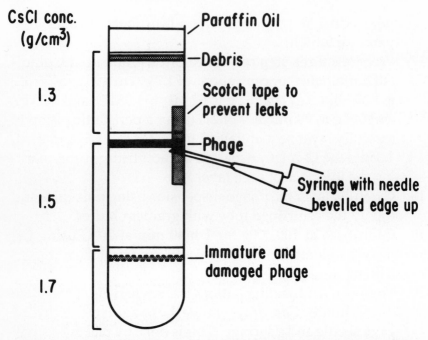

Fig. 1. A diagram showing the expected banding pattern of the phage preparation after ultracentrifugation in a cesium chloride step gradient. The intact phage band appearing immediately below the 1.3 to 1.5 g/cm³ cesium chloride interface is extracted from the gradient with a syringe and needle as shown.

6. Repeat this procedure by extracting the aqueous (top) phase with phenol three times.

7. Extract the aqueous phase two times with chloroform to remove the phenol, each time breaking the phases at 7800g for 5 min.

8. Extract the aqueous phase with water-saturated ether three times to remove the chloroform, separating the phases each time at 1000g for 2 min.

9. Remove residual ether by placing briefly under vacuum.

10. Run a range of dilutions of the resulting supernatant on a 0.5% (w/v) agarose gel using known concentrations of λ as a marker. Stain with ethidium bromide as described in Chapter 7, Vol. 2 of this series (ref. 8).

11. Estimate the DNA concentration from the agarose gel, or better, measure the absorbance at 260 and 280 nm. The OD 260/280 ratio should be 1.8; one OD = 50 μg/mL (*see* Note 4 in section 4).

3.3. Preparation of λ Arms

1. It is advisable to ensure that the DNA can be completely digested with BamH1 on a small scale before proceeding to the large-scale preparation of arms. Take 0.5 μg of DNA and add 2 U of BamH1 in the appropriate buffer (supplied by the manufacturer) and make up to 10 μL and incubate for 1 h at 37°C.
2. Heat-shock the DNA at 65°C for 10 min (to de-anneal cos ends), cool, and analyze by 0.5% (w/v) agarose gel electrophoresis using appropriate Hind III digested λ markers.
3. If the above procedure shows complete digestion, proceed to a large-scale digestion; if not, clean the DNA up with further phenol extractions.
4. Digest 150 μg of vector DNA in a 500-μL volume with 300 U of BamH1 in the appropriate buffer at 37°C overnight.
5. Check for complete digestion by running 0.5 μg of the DNA after heat-shock at 68°C for 10 min on a 0.8% (w/v) agarose gel with appropriate markers. If the reaction is incomplete, add more enzyme, adjust the buffer and incubation volume, and continue the incubation.
6. Once the digestion is complete, add 300 U of Sal1 with appropriate buffer adjustment, to cut the "stuffer" fragment, and leave this to digest at 37°C overnight (*see* Note 5 in section 4).
7. Prepare a 12-mL 15–35% sucrose gradient.
8. Extract the digested DNA with an equal volume of phenol/chloroform once. Separate the phases by a 5-min centrifugation in a microfuge. Extract with an equal volume of chloroform and again separate the phases. Precipitate the DNA with 300 m*M* sodium acetate and 0.6 vol of isopropanol (*see* Note 5 in section 4).

9.   Regain the DNA by centrifugation in a microfuge, dry the pellet in a speed-vac, and resuspend it in 200 µL TE buffer. Remove an aliquot containing 0.5 µg to analyze the DNA after allowing the cohesive ends to reanneal at 42°C for 1 h in the presence of 0.01*M* MgCl$_2$ (*see* Note 6 in section 4).

10. If the DNA is intact, layer the remainder over the gradient and centrifuge at 140,000*g* for 19 h at 15°C using the "slow acceleration" and "break-off" modes of the centrifuge.

11. Collect 0.5-mL fractions through a 21-gage needle inserted at the base of the tube. (To stop the gradient from gushing out, the top may be sealed with parafilm and the flow started by introducing a very small hole in the top. Alternatively a commercial gradient fractionator may be used.)

12. Remove 10 µL of every second fraction, dilute with 35 µL of water, and add 8 µL of loading buffer. Heat to 68°C for 10 min and analyze on a 20 x 20 cm 0.5% (w/v) agarose gel with appropriate restriction digested λ markers adjusted to match the salt and sucrose concentrations in the fractions.

13. Locate fractions containing the arms (20 and 8.9 kb) and pool.

14. Either dialyze this against TE buffer if the volume is large, or dilute the fractions threefold with TE buffer and precipitate overnight with 2 vol of 95% ethanol without additional salt. Regain the pellet by centrifugation, wash it with 70% ethanol, dry it in a speed-vac, and dissolve it in an appropriate volume of TE buffer (to give 300–500 µg/mL).

15. Measure the exact concentration by absorbance at 260 nm and store in 5-µg aliquots at –20°C.

3.4. Preparation of Donor DNA (see Note 7 in section 4)

1.   Immerse up to a gram of freshly dissected tissue in liquid N$_2$, and grind it to a fine powder (under liquid N$_2$) in a pestle and mortar.

2. Transfer the powder to a sterile 50-mL plastic centrifuge tube, and allow it to just thaw before adding 2 mL of ice-cold homogenization buffer.

3. Gently resuspend with a sterile "cut-off" 1-mL pipetman tip, and transfer it to a hand-driven Potter-Elvejen homogenizer. Wash the tube out with 3 mL of homogenization buffer, add it to the homogenizer, and gently homogenize at 4°C until the particulate material is dispersed.

4. Transfer the homogenate to a 10-mL plastic centrifuge tube (Falcon plastics) and centrifuge at 100g for 5 min at 4°C.

5. Remove the supernatant and repeat the homogenization of the pellet, after gentle resuspension, regaining the pellet each time by centrifugation in the same tube. This should be performed twice.

6. Finally resuspend the pellet in a small amount of homogenization buffer to give a total weight of 1.6 g.

7. Add 400 µL of a 10% sarkosyl stock slowly and incubate without shaking for 1 h at 50°C.

8. Add 100 µL of proteinase K and incubate for 3 h (to overnight) without shaking at 37°C.

9. In a separate tube, mix together 3.36 mL of homogenization buffer, 0.84 mL of 10% sarkosyl, and 6.3 g of cesium chloride. Dissolve at 50°C.

10. Gently mix the genomic DNA with the cesium chloride solution and transfer to a 12-mL polyallomer ultracentrifuge tube.

11. Overlay the tube with parafin oil to avoid collapse, and cap the tube. Centrifuge the solution at 250,000g for 18 h at 15°C.

12. After centrifugation, remove the tube, place it in a clamp, and puncture the tube just above the bottom (in order to avoid the RNA pellet) with a 16 to 18-gage needle, and allow the gradient to drip out.

13. Collect the DNA in 0.5-mL fractions (*see* Note 8 in section 4). Those containing DNA should be obvious because of

the increased viscoscity. If low concentrations of DNA were prepared, the fractions containing the DNA can be determined by spotting 5 µL of each fraction onto cling film and adding 5 µL of ethidium bromide at 2 µg/mL. Visualize the spots under a UV transilluminator, and those containing the DNA will be apparent by their fluorescence.

14. Pool the DNA fractions and dialyze against TE buffer using three changes of 2 L.

15. Determine the size of the DNA by electrophoresis on a 0.3% (w/v) agarose gel using size markers such as intact λ ($\cong$50 kb) and intact T4 ($\cong$165 kb) DNA (*see* note 9 in section 4).

16. Determine the DNA content. This may be done by absorbance at 260 nm. One OD $\cong$ 50 µg/mL or by titrations against known concentrations of λ on an agarose gel or as spots on cling film, as described in point 13 in this section. The DNA is usually concentrated enough for use, but if not, it may be concentrated by successive, *gentle* extractions with 2-butanol, allowing the phases to separate by standing (aqueous layer at the bottom). After concentration, dialyze extensively against TE buffer to remove traces of the butanol.

17. Store this at 4°C. Do not freeze.

3.5. Partial Digestion of the Genomic DNA

1. Before starting the large-scale digestion, perform a pilot digestion. Take 10 µg of DNA, 3 µL of Mbo1 buffer (10x stock), and water to give a final volume of 27 µL and place at 37°C to warm.

2. Prepare 13 tubes for time-points ranging from 0 to 60 min and place on ice. Add 5 µL of stop solution to each tube.

3. Dilute the stock Mbo1 enzyme with sterile ice-cold water to give a concentration of 0.5 U/µL and place on ice.

4. Quickly add 3 µL of diluted enzyme to the prewarmed DNA, mix, and start timing.

5. Remove 2 μL immediately and then 2 μL every 5 min for 1 h.
6. Run each time-point on a large (20 x 20 cm) 0.3% (w/v) agarose gel using λ, and λ digested with XhoI or Hind III as markers. Stain and photograph the gel.
7. Determine the time of digestion required to give maximum fluorescence in the 15–22 kb range. Ideally this should be approximately 30 min of digestion.
8. Scale up the digestion to 150 μg of DNA in a total volume of 450 μL. The digestion requires only half the corresponding enzyme concentration, however, to give the maximum number of molecules in the required size range (1). Determine the exact time by digesting for the optimum time of the pilot experiment ±5 min. At each time, stop the experiment by adding 150 μL of the reaction mixture to an ice-cold microfuge tube containing EDTA, pH 8.0, to give a final concentration of 20 mM.
9. Pool the three fractions and load onto a 15–35% (w/v) sucrose gradient as described for the preparation of vector arms.
10. After centrifugation, fractionate the gradient and run 10 μL of every second fraction on a 0.3% (w/v) agarose gel with the appropriate markers.
11. Label the fractions according to the sizes determined from the agarose gel (*see* Note 10 in section 4).
12. Dilute each fraction threefold and ethanol precipitate with 2 vol of 95% (v/v) ethanol, dry it in a speed-vac, and resuspend in 20 μL of sterile TE buffer.
13. Determine the concentration of DNA using the "spot-test" described above and compare to known concentrations of λ DNA. Alternatively estimate the concentration after electrophoresis on a 0.8% (w/v) agarose gel.

3.6. Ligation of Vector to Insert (see note 12 in section 4)

1. Ligations of 1 μg of DNA total should be tested with ratios of 4:1, 2:1, 1:1, and 0.5:1 arms to insert. Carry out reactions

at 150 µg DNA/mL with 1 U of T4 DNA ligase overnight at 4°C. Also ligate 0.5 µg of vector arms only under similar conditions to estimate the background caused by vector contamination with the stuffer fragment (*see* Note 13 in section 4).

2. Package 250 ng of the ligated DNA using a commercial packaging kit or home-made packaging components as described in Chapter 36, Vol. 2 of this series (*1,9*). At this step also package *bona fide* λ to determine the packaging efficiencies of the extract (commercial kits always provide this control).

3. Remove the freeze thaw and sonicated extracts from the –70°C freezer and allow to thaw on ice. The freeze thaw lysate will thaw first. It should be added to the still-frozen sonicated extract. Mix gently, and when they are almost, but not quite, defrosted, add the DNA to be packaged. Mix and incubate for 1 h at room temperature.

4. Add 0.5 mL of SM medium and a drop of chloroform and mix.

5. Remove the debris by centrifugation in a microfuge for 30 s and titer the packaged DNA on the appropriate strain:
    λ —LE392
    arms + insert—Q359 (*spi⁻* selection)
    arms alone—Q358 and Q359

6. Calculate the efficiency of packaging of λ (*see* Note 14 in section 4) and of the arms + insert. λ Should be about 100x more efficient than the religated material.

7. If one of the test ligations give a suitable efficiency in the packaging reaction, then select this and scale up the reaction to produce a full library (*see* Notes 14 and 15 in section 4). Test a small aliquot.

8. Once the library is produced, it should be tested for completeness as outlined in note 15 (section 4) and can be stored under chloroform indefinitely. It can thereafter be screened with or without amplification as required (*1,3,10*).

9. The library should be amplified by growing as a plate stock, making sure that the plaques do not overgrow each other, thereby reducing the possibility of recombination between different recombinant phages. To prepare the plate stock, mix aliquots of the packaging reaction containing about 2×10^4 recombinants in a 50-µL volume with 0.2 mL of plating bacteria and incubate it for 20 min at 37°C.
10. Melt 6.5 mL of top agar with each aliquot and spread on a 150-mm plate of bottom agar, previously dried.
11. Incubate for 8–10 h, making sure the plates are absolutely horizontal so that spread of clones is minimized.
12. Overlay the plates with 12 mL of SM medium and store at 4°C overnight.
13. Recover the bacteriophage suspension and transfer it to a sterile polypropylene tube. Rinse the plate with 4 mL of SM medium and pool with original suspension.
14. Add chloroform to 5% (v/v) and incubate for 15 min at room temperature with occasional shaking.
15. Recover the bacteriophage by removal of the cell and agar debris by centrifugation at 4000g for 5 min at 4°C.
16. Remove the supernatant and add chloroform to 0.3% (v/v), at which point the libraries may be stored in aliquots at 4°C for many years. They can be concentrated if necessary by centrifugation in a CsCl gradient as described above.

4. Notes

1. To obtain a good yield of phage, it is essential to have very clean glassware. This is achieved by soaking with chromic acid and then thoroughly rinsing with distilled water. It is also necessary to maintain good sterility control because the single-stranded *cos⁻* ends of lambda are extremely vulnerable and their integrity is essential. It is always worthwhile testing to see if the *cos⁻* ends will anneal by incubating at 42°C for 1 h in the presence of 10 mM MgCl$_2$ and analyzing the product on a 0.3% (w/v) agarose gel. Obvi-

ously, it is also necessary to have good housekeeping over the storage and verification of bacterial and phage stocks. Details can be obtained from ref. 1.

2. If the lowest band on the cesium chloride gradient is larger than the phage band, this suggests that mixing and pipeting have been too vigorous. If the top band is too large, it will contaminate the phage preparation. It may, therefore, be necessary to rerun the phage preparation on another gradient. If the yield is low, it may be necessary to titer at each step to ascertain where the losses occur so that you improve that part of the preparation.

3. It is necessary to always maintain the Mg^{2+} concentration; otherwise the phage particles will disintegrate.

4. 10^{10} Plaque-forming units of phage gives approximately 0.44 μg of DNA. Therefore a 200-mL cleared lysate with a yield of 6×10^{10} pfu/mL will give 500 μg of DNA. The DNA should be clean by this step. If for any reason the DNA has a low 260/280 ratio, return to step 4 in this section and start again. The presence of sheared DNA usually means vigorous pipeting or poor quality phenol. Cut pipet tips and redistilled phenol are necessary.

5. Both EMBL 3 and 4 were designed such that by digestion with Sal1 the stuffer fragment is cut without the arms being destroyed. With EMBL 4 the Sal1 site lies between the BamH1 site and the stuffer fragment (the converse is true with EMBL 3). Thus cutting with the two enzymes substantially reduces the probability that there can be religation of complete phage genome. This religation, in fact, is made even further unlikely because the small piece of poly linker produced by Sal1 digestion is lost during the isopropanol precipitation because it is too small to be precipitated (see ref. 2). This further improves the quality of the library by preventing false recombinants that may have escaped the genetic selection imposed by the *spi* selection.

6. The procedure for isolating arms can be further improved if the cohesive ends are allowed to anneal at 42°C for 1 h in

the presence of 10 mM MgCl$_2$. This enables the isolation of a single species of DNA from the sucrose gradient and at the same time checks the DNA.

7. This method may also be used for preparing DNA from tissue culture cells. In order to do this, grow about 2×10^8 cells (do not harvest from confluent cultures), harvest them by scraping or trypsinization and washing, and centrifuge them at 400g for 3–5 min. Quick-freeze the cell pellet in liquid N$_2$ and transfer it to a small pestle and mortar and then proceed as described in the text.

8. If the DNA in the pooled fractions has the appearance of a "cloudy clot," then the DNA is contaminated with protein. This DNA will not aliquot uniformly, nor will it be completely digested with restriction endonucleases. It can be cleaned up by phenol extraction, although generally this is not advisable because it can result in shearing of the DNA. If there are continuous problems after increasing the degree of homogenization and the length of the proteinase K digestion, then a phenol extraction may be incorporated into the protocol between the proteinase K and the cesium gradient step. This is performed by transferring the DNA to a 30-mL glass siliconized centrifuge tube and addition of an equal volume of phenol:chloroform (1:1 by volume) equilibriated with TE buffer, pH 7.5, 2% (w/v) sarkosyl. Stopper the tube with a silicone rubber bung, and place the tube on its side. Roll the tube *gently* to mix and leave on ice for 30 min for the phases to separate. Repeat this procedure three times, each time removing the phenol from below. Chloroform extract the resulting DNA solution three times in a similar manner. Remove the trace contamination of chloroform with three ether extractions, and then the residual water-saturated ether by placing under a vacuum. Measure the final volume and proceed to step 10 making appropriate adjustment of the volume.

9. The size of the DNA is the most important determinant in the production of a complete library. Although some

breakage is inevitable, the DNA should be greater than 100 kb, and the bigger the better; otherwise, when a random collection of fragments is generated by partial restriction endonuclease digestion, there will be a large proportion of noncohesive ends. These fragments substantially lower the efficiency of packaging because they join nonproductively to vector DNA since only one of the two λ arms can join to them. Moreover, since concatemers of phage particles are the usual substrates for packaging, these ragged ends cause termination of this process. Thus if the donor DNA preparation appears smeared down the gel, it is better to go back and prepare it again rather than proceeding with the library. The usual causes for the poor quality of DNA are mechanical shearing (too vigorous pipeting, use of uncut pipet tips, or too rough mixing) or DNase contamination caused by poor sterility control over solutions and glassware.

10. Do not throw away these tubes. This material is very precious and may be used later for either different vectors or to repeat packaging with a slightly smaller or larger insert size. Store it and the DNA prior to ligation at −20°C.

11. Both the arms and donor DNA may be fractionated on a low-gelling temperature agarose gel. But these gels have a lower capacity than sucrose gradients and any impurities in the agarose severely inhibit the ligation reaction. In practice, therefore, we have not found this to be a suitable method to fractionate the DNA.

12. To calculate the amount of vector and donor DNA required in g, use the following formula: Insert size equals 20 kb and, therefore, has a molecular weight of approximately 1.27×10^7. Arms are 30 kb and molecular weight of 1.9×10^7. Therefore, for a 1/1 ratio

$$\frac{\mu g \text{ arms}}{1.9 \times 10^7} = \frac{\mu g \text{ insert}}{1.27 \times 10^7}$$

Therefore, μg arms = 1.496 x μg insert. Thus for a 1-μg reaction, 0.6 μg of arms and 0.4 μg of insert are required. This needs to be adjusted according to the average size of the insert and for different ratios of arms to insert (*see* ref. *1*). The theoretical ratio of arms to inserts should be 2/1, but some of the molecules may lack a cohesive terminus (*see* Note 9 in section 4), so that the effective concentrations are different from those calculated from the measured DNA concentration. Therefore, it is necessary to do a range of test ratios before the large-scale packaging, as described in section 3. Prior to the ligation, it is possible to alkaline phosphatase the donor DNA to prevent self-ligation. This increases the effective concentration of available ends and will increase the efficiency of productive ligations. Care should be exercised over the quality of the alkaline phosphatase since many preparations contain some nuclease activity.

13. The size of the DNA may also be checked at this stage. If the ligation of insert plus arms has been successful, the ligated DNA should at least be the size of λ. It is also a good idea to anneal the cos ends of the vector by incubating it in the presence of 10 m*M* MgCl$_2$ at 42°C for 1 h.

14. λ Should give at least 10^8 pfu/μg DNA. If not, the packaging extract (or your handling of it) is defective, and precious sample should not be committed to packaging. Remember when packaging not to defrost many aliquots of packaging extracts and leave them on ice: follow the protocol. We have also found it preferable to do several small-scale packagings rather than a single large-scale one, because of the problems of handling the extracts; we only use 2–3 vials at any one time. The vector arms alone should give less than 10^4 pfu/μg when plated on Q358 and lower when plated on Q359. If, providing the packaging is efficient, the arms + insert give lower than 10^6 pfu/μg, check the ligase by ligating a commercial λ Hind III digest and checking this using agarose gel electrophoresis. If the

ligase is efficient, repeat steps 1 and 2 in section 3.6 using the arms only. If these ligate efficiently as assessed by agarose gel electrophoresis, then the insert DNA is not of suitable quality. As mentioned in note 9 in this section, this is the single most common reason for the failure to prepare a good library. The method described is protocol 2 of ref. *1*, which gives a bias to the packaging reaction to include insert molecules of around 20 kb.

15. The ideal is to construct as complete a library as possible such that every sequence in a mammalian genome is represented at least once. To calculate the number of plaques to achieve a library of a 99% probability of containing the required sequence, the following formula should be used (*1,3*):

$$N = \frac{\ln(1 - P)}{\ln(1 - x/y)}$$

where P is the desired probability, x is the insert size, y is the haploid genome size, and N is the necessary number of recombinants. Therefore, given the 20-kb hypothetical insert in order to get a 99% probability, about 8×10^5 plaques in a single packaging reaction are required. The library should be tested for its validity by picking approximately 100 plaques randomly and spotting them as an array onto plating bacteria in top agar. Blot onto nitrocellulose after growth, and probe this with nick-translated total genomic DNA of the host species. Because of the large numbers of repeat sequences in the mammalian genome, approximately 80% of the clones should give a signal after hybridization to this probe.

16. We have always found some loss of efficiency upon scale up, and this should be compensated for in the final packaging reaction. Remember to add a drop of chloroform and clarify the extract. Packaging extracts can inhibit bacterial growth and lower the titer.

References

1. Maniatis, T., Fritsch, E. F., and Sambrook, J. (1982) *Molecular Cloning, A Laboratory Manual* Cold Spring Harbor Press, Spring Harbor, New York.
2. Hendrix, R. W., Roberts, J. W., Stahl, F., and Weisberg, R. A., eds. (1983) *Lambda 11* Cold Spring Harbor Press, Cold Spring Harbor, New York.
3. Kaiser, K. and Murray, N. (1985) The Use of Phage Lambda Replacement Vectors in the Construction of Representative Genomic DNA Libraries, in *DNA Cloning: A Practical Approach* Vol. 1 (Glover, D. M., ed.) IRL Press, Oxford, Washington.
4. Frischauf, A. M., Lehrach, H., Poustka, A., and Murray, N. (1983) Lambda replacement vectors carrying polylinker sequences. *J. Mol. Biol.* **170**, 827–842.
5. Burt, D. and Brammer, W. J. (1985) Bacteriophage λ as a Vector, in *Basic Cloning Techniques: A Manual of Experimental Procedures* (Pritchard, R. H. and Holland, I. B., eds.) Blackwell Scientific, Oxford.
6. Karn, J., Brenner, S., Barnett, L., and Cesareni, G. (1980) Novel bacteriophage λ cloning vector. *Proc. Natl. Acad. Sci. USA* **77**, 5172–5176.
7. Dale, J. W. and Greenaway, P. J. (1984) Preparation and Assay of Phage λ, in *Methods in Molecular Biology* Vol. 2 *Nucleic Acids* (Walker, J. M., ed.) Humana, New Jersey.
8. Boffey, S. A. (1984) Agarose Gel Electrophoresis of DNA, in *Methods in Molecular Biology* Vol. 2 *Nucleic Acids* (Walker, J. M., ed.) Humana, New Jersey.
9. Dale, J. W. and Greenaway, P. J. (1984) In Vitro Packaging of DNA, in *Methods in Molecular Biology* Vol. 2 *Nucleic Acids* (Walker, J. M., ed.) Humana, New Jersey.
10. Dale, J. W. and Greenaway, P. J. (1984) Identification of Recombinant Phages by Plaque Hybridization, in *Methods in Molecular Biology* Vol. 2 *Nucleic Acids* (Walker, J. M., ed.) Humana, New Jersey.

Chapter 18

Cosmid Library Construction

John D. Haley

1. Introduction

To study the organization and regulation of eucaryotic genes from cloned DNA, vectors capable of maintaining large inserts have been constructed, largely from bacteriophage λ, where nonessential intergenic regions were identified. It had been noted that only λ DNA 37–52 kb in length (or 78–102% of the wild-type length) were efficiently packaged and passaged in vivo. Thus vectors from which the intergenic region had been removed could be positively selected for by the insertion of target DNA, thus restoring a packagable DNA length and producing viable phage.

In the cloning of the mouse β-globin gene, known from hybridization studies to reside on a 7-kb EcoRI fragment, mouse DNA was cut to completion with EcoRI, size fractionated for the desired fragment length, and inserted into phage λ DNA from which nonessential DNA had been removed, and plaques were transferred to nitrocellulose filters (1) and probed with labeled globin sequences. An insert DNA of 7 kb containing the mouse β-globin gene and the flanking sequences was isolated and characterized (2), although this approach to the isolation of eucaryotic genes required enrichment of sequences of a discrete length, and consequently the library was not suitable for the selection of genes on large and small restriction fragments. This problem was circumvented by the construction of large random libraries of partially cleaved eucaryotic DNA 10–20 kb in size using newly developed λ vectors capable of stably maintining DNA inserts of up to 20 kb (3–5), which could be introduced into bacterial hosts by in vitro packaging of the recombinant phage DNA into infectious λ particles (6,7), thus raising the efficiency from 10^3 phages/µg ligated insert DNA by Ca^{2+}-mediated transformation to 10^6 phage/µg (8,9). In the construction of a random library representing all genes within a eucaryotic genome, DNA was partially cleaved with Hae 3 and Alu to an average size of 20 kb and fractionated by sucrose density gradient centrifugation. The 20-kb eucaryotic DNA was joined to the λ vector arms via EcoRI linkers attached to the target DNA, and the DNA packaged in vitro. The recombinant phage within the gene libraries constructed by this technique (10) could be amplified, were of sufficient number to statistically represent >99% of the organism genome (11), and could be maintained as a high titer phage stock.

The development of a new class of plasmid:λ hybrid vectors termed cosmids came from work on the specificity and identification of sequences involved in the packaging of λ DNA and have made it possible to clone insert DNAs of 33–45 kb in length, package the cosmids in vitro into infectious phage particles, yet stably and nonlytically propagate the recombinant molecules as plasmids.

Though the bacteriophage λ chromosome is a linear duplex (48.5 kb), upon infection of a suitable host, complementary 12-bp single-stranded termini (designated cohesive ends, ref. 12) are sealed by *Escherichia coli* DNA ligase, and the λ chromosome replicates bidirectionally as a ring. Late in infection, the change to rolling circle replication generates long DNA concatamers containing multiple cos sites (13–15), which are cleavage sequences recognized by the A gene product (7,8). This reaction is coupled to the packaging of phage DNA (~250-fold condensation) in the head capsid (16) and phage maturation.

During investigation into the role of cos sequences during morphogenesis, it was shown that DNA 39–52 kb in length containing the cos sequences could be packaged (17,18) in vitro, regardless of origin (19,20). Small plasmid vectors (cosmids) 6–10 kb in length were subsequently developed (21–23), capable of harboring DNA inserts 33–45 kb in length, and contained the replication origin, multiple antibiotic markers, single restriction enzyme clevage sites, and λ cos sequence (24). These chimeric molecules act as substrate in vitro λ packaging extracts, and transduction of E. coli with cosmid-containing infective particles with the subsequent circularization of the injected DNA allow it to be maintained in the plasmid state and confer antibiotic resistance. The use of cosmids has been advantageous in the study of gene clusters (25–28) and in the transformation of yeast (29) and mammalian cell lines (30–33) with large DNA segments.

The approach to cosmid cloning (Fig. 1) taken by the author was to partially digest high-molecular weight DNA with Sau3A and fractionate those molecules 35–50 kb in length by velocity sedimentation. The cosmid vector pHC79 (22) was cleaved with BamH1 phosphatase to prevent the formation of packagable vector multimers and ligated to the partial Sau3A DNA by virtue of the complementary nature of BamH1 (GGATCC) and Sau3A (GATC) termini. Those molecules 38–52 kilobases in length were efficiently packaged, in vitro, into λ phage heads, to which the tail protein may be added, yielding infectious lamboid particles able to transfer the high molecular weight cosmid molecules across the cell membrane at efficiencies considerably higher than Ca^{2+}-mediated transformation techniques. A porcine genomic library of approximately 200,000 cosmids was constructed by this protocol, from which two overlapping clones coding for the peptide hormone relaxin were isolated (47).

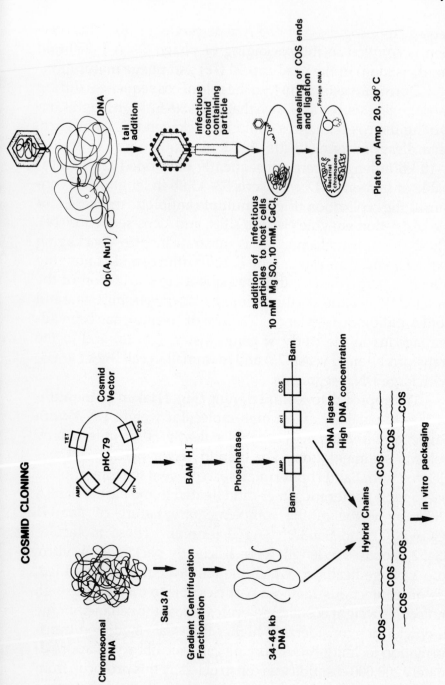

Fig. 1. Overall procedures in the construction of genomic libraries using cosmid vector pHC79.

The following procedures detail the isolation of genomic target DNA, the preparation of vector DNA, the ligation conditions, the preparation of efficient λ packaging mixes, and, finally, the infection of the *E. coli* host, creating a bacterial bank containing large genomic sequences autonomously replicating as a plasmid.

2. Materials

2.1. Stock Solutions

1. 5M NaCl.
2. Sucrose buffer: 20% sucrose, TE buffer, pH 8.
3. Triton lysis buffer: 2% triton X-100, 50 mM Tris-HCl, 10 mM EDTA, pH 7.8.
4. 40% Polyethylene glycol 6000 (PEG-6000).
5. TE: buffer 10 mM Tris-HCl, pH 8, 1 mM EDTA.
6. Buffered phenol.
7. Phenol/chloroform.
8. 3M Sodium acetate, pH 5.5.
9. CsCl solution: 0.95 g/mL solution in TE buffer.
10. TE buffer, 0.1% SDS.
11. 10% *n*-Lauryl sarcosine.
12. 10 mg/mL Ethidium bromide (weigh in fume cupboard wearing gloves; this compound is a known mutagen).
13. NaCl-saturated isopropanol: mix equal volumes of 5M NaCl and isopropanol; shake and allow phases to form; isopropanol is the upper phase.
14. 10x BamH1 restriction enzyme buffer: 100 mM Tris-HCl, pH 7.8, 100 mM MgCl$_2$, 10 mM DTT, and 1.5M NaCl.
15. Low TE: 10 mM Tris-HCl, pH 8, 0.1 mM EDTA.
16. Phenol lysis buffer: 25 mL of phenol with 0.1 mL of β-mercaptoethanol and 25 mL of aqueous phase that is 20 mM Tris, pH 7.8, 0.5% SDS, 1.0M NaCl, 1 mM EDTA.
17. SB3 gel buffer: 50 mM Tris-acetate, pH 7.9, 1 mM EDTA.
18. 10x Sau3A restriction enzyme buffer: 100 mM Tris-HCl, pH 7.4, 100 mM MgCl$_2$, 10 mM DTT, 0.4 M NaCl.

19. 10 and 30% Glycerol step solutions: either 10 or 30% glycerol in 0.2M sodium acetate, pH 7.8, 10 mM EDTA, 10 mM Tris-acetate, pH 7.8.

20. 10x DNA ligation buffer: 500 mM Tris-HCl, pH 7.4, 100 mM MgCl$_2$, 10 mM DTT.

21. ATP stock: prepared to 10 mM final concentration with 10 mM Tris, pH 7.4.

22. Sonication buffer: 20mM Tris-HCl, pH 8, 50 mM putrescine, 20 mM MgCl$_2$, 30 mM β-mercaptoethanol: pH to 8.

23. Packaging buffer: 20 mM Tris-HCl, pH 8, 1 mM EDTA, 3 mM MgCl$_2$, 5mM β-mercaptoethanol.

24. Sucrose buffer: 10% sucrose, 50 mM Tris, pH 7.5.

25. Lysozyme mix: 2 mg/mL lysozyme in 0.25M Tris, pH 7.5.

26. λ Dilution buffer: 10 mM Tris-HCl, pH 7.4, 10 mM MgCl$_2$, 100 mM NaCl, 0.1% gelatin.

27. Hybridization buffer: 50% formamide (deionized), 0.1% SDS, 50 mM sodium phosphate, pH 6.8, 5x SSC, 1 mM sodium pyrophosphate, 0.02% polyvinyl pyrrolidone, 0.02% bovine serum albumin, and 100 μg/mL sheared (through a 19-gage needle) boiled herring sperm DNA.

28. 20x SSC buffer: 3M NaCl, 0.3M sodium citrate.

29. Buffered phenol: Melt at 60°C. Add hydroxyquinoline to 0.1%. Add 1/5 of the volume of 2M Tris-HCl, pH 8.0. Mix and add half of the original volume of distilled water. Mix and decant the upper aqueous phase. Add 1/10 of the volume of 10 mM Tris-HCl, pH 8.0.

30. Phenol-chloroform: 25/24/1, buffered phenol:chloroform:isoamyl alcohol (v/v/v).

31. 10x TBE buffer: 120 g of Tris, 60 g of boric acid, 8 g of Na$_2$ EDTA.

2.2. Enzymes

1. BamH1.
2. Sau3AI.

3. DNA polymerase I (Klenow fragment; Boehringer Mannhein).
4. T4 DNA ligase (Amersham).
5. Alkaline phosphatase [calf intestinal; Boehringer; $(NH_4)_2SO_4$ precipitate; or molecular biology grade].
6. Lysozyme (Boehringer).

2.3. Nucleic Acids

Prepared in sterile water as 10 mM stocks and store frozen at –20°C.

1. dATP, dGTP, dCTP, dTTP (PL Biochemicals).
2. ATP (Boehringer).

2.4. Bacterial Strains and Media

1. DH1: F⁻, recA1, endA1, gyrA96, thi-1, hsdR17, (r_k^-, m_k^+), supE44 (Duttanaham).
2. LE392: F⁻, hsdR514 (r_k^-, m_k^-), supE44, supF58, lac/Y1, or Δ (lacIZY), galK2, galT22, metB1, trpR55, λ⁻ (44,45).
3. HB101: recA⁻, r_k^-, m_k^-, suII, leu⁻, B1⁻, thr⁻, pro⁻, lacZ⁻, Sm⁻ (46).
4. BHB2690: N205 recA⁻ (imm⁴³⁴, cIts, b2, red⁻, Dam, Sam/λ; B. Hohn).
5. BHB2688: N205 recA⁻ (imm⁴³⁴, cIts, b2, red⁻, Eam, Sam/λ; B. Hohn).
6. L broth (per L): 10 g of bacto-tryptone, 5 g of yeast extract, 5 g of NaCl. Adjust pH to 7.5 with 500 μL of 5M NaOH/ L of broth and autoclave.
7. Antibiotics: ampicillin, 20–50 μg/mL, tetracycline, 10 μg/ mL.
8. Soft agar: 0.7% bacto-agar; plates: 1.3% bacto-agar.
9. NZ broth (per L): 10 g of NA-amine, 5 g of NaCl, 2 g of $MgCl_2$ • $6H_2O$. Adjust pH to 7.5 with NaOH and autoclave.

10. LAM plates: L broth, 1% bacto-agar, 10 mM MgSO$_4$.
11. NZM plates: 10 g of NZ amine, 5 g of yeast extract, 5 g of NaCl, 1 mM of MgSO$_4$, 0.4% maltose, 1.3% bacto-agar. Adjust pH to 7.5.

2.5. Vector pHC79

Cosmid pHC79 has a length of 6.4 kb and was derived from pBR322 by the insertion of a 1.65-kb fragment from λ phage Charon 4A containing the cos or cohesive end sequences, which confer the ability to act as a DNA substrate in the packaging of λ both in vivo and in vitro (22).

3. Methods

"Ethanol precipitation" is carried out throughout these protocols by adjusting the solution to be precipitated to 0.3M sodium acetate, pH 5.5, then adding 2.5 vol of ethanol.

3.1. Purification of Cosmid DNA

Plasmid DNA is isolated by a modification of the Triton-lysis procedure (34,35). Amplification of plasmid DNA by chloramphenicol is adapted from the original procedure in minimal media (36) for the isolation of pHC79, which contains the ColEl replicon. Plasmid-containing host strains are verified by selection of antibiotic resistance markers (e.g., ampicillin), with a single colony being picked and used to inoculate a 10-mL overnight culture supplemented with the ampicillin (50 µg/mL). One liter of LB broth is inoculated with an aliquot of the fresh saturated culture at a 1:100 dilution. The culture is grown with aeration at 37°C to OD$_{600nm}$ of 0.3 (approximately 3–5 hours), at which point chloamphenicol is added to 170 µg/mL and the incubation continued an additional 16 h. For cosmids with inserted genomic DNA, the LB broth is supplemented with 50 µg/mL ampicillin/aeration is at 30°C, with no chloramphenicol amplification, with growth until saturation. The following procedure is used to extract plasmid or cosmid DNA from its bacterial host.

1. Chill the culture on ice.
2. Spin at 6000g 10 min. Pour off the supernatant and resuspend in 10 mL of cold sucrose buffer.
3. Transfer to a 50-mL Oakridge tube. Vortex.
4. Add 2 mL of 0.5M EDTA, pH 8.0. Vortex.
5. Add 2 mL of 10 mg/mL lysozyme prepared in water. Mix gently. Keep on ice for 60 min.
6. Add 15 mL of cold Triton lysis solution. Keep on ice for 15 min.
7. Spin at 100,000g for 1 h in a Beckman SW28 rotor or an equivalent.
8. Remove the supernatant and add 40% PEG-6000 to 10% and 5M NaCl to 1M. Mix and store at 4°C overnight.
9. Spin at 12,000g for 20 min.
10. Resuspend the pellet in 10 mL of TE buffer with 0.1% SDS. Extract with 5 mL of buffered phenol/CHCl$_3$ [1/1 (v/v)].
11. Remove the aqueous phase and reextract with ether.
12. Add sodium acetate, pH 5.5, to 0.3M and 2.5 vol of ethanol. Mix and chill at –70°C for 15 min. Spin at 12,000 g for 10 min.
13. Wash the pellet and sides of the tube with ethanol. Remove residual ethanol by desiccation, and resuspend the pellet in 10 mL of 0.95 µg/mL total solution CsCl in TE buffer.
14. Add 10% *n*-lauryl sarcosine to 0.1% and 1 mg ethidium bromide solution in reduced light. Spin at 120,000g in polyallomer tubes at 20°C for 40 h or the equivalent (e.g., 40,000 rpm for 40 h in a Ti 70.1 or SW55).
15. Illuminate the centrifuge tube with long-wave UV light (i.e., 354 nm) and remove the *lower* band with a large-bore needle, taking care not to take the upper open circle/

chromosomal band. RNA is found at the very bottom of the tube.

16. Quickly extract the supercoiled plasmid DNA with an equal volume of NaCl-saturated isopropanol with a brief centrifugation to better separate the phases, until no ethidium color is seen in the organic upper phase.

17. Extract once more with NaCl-saturated isopropanol.

18. Prepare dialysis tubing (Spectropore 12,000–14,000 cutoff) by boiling for 5 min in 10 mM EDTA, pH 7.8. Wash with distilled water. Add plasmid DNA to the tube and dialyze against 2 L of TE buffer for at least several hours or overnight at 4°C.

19. Recover the DNA from the dialysis membrane, adjust to 0.3M in sodium acetate, pH 5.5, add 2.5 vol of ethanol, mix, and store at –70°C.

20. Recover by centrifugation at 12,000g for 10 min, wash with ethanol, and desiccate.

21. Resuspend in 0.5 mL of TE buffer. Store at 4°C, or for long-term storage, freeze at –80°C.

3.2. Preparation of BamH1-Cut Phosphatased Cosmid Vector pHC79

1. The cosmid vector pHC79, isolated from *E. coli* BHB3000/pHC79 by the Triton-lysis method, is additionally extracted with an equal volume of buffered phenol followed by ether extraction and re-ethanol precipitation, to remove any exonuclease activity that might damage the vector termini generated during the subsequent BamH1, digestion.

2. Fifty micrograms of the purified vector are cleaved at its single BamH1 restriction site in 10 mM Tris, pH 7.8, 10 mM MgCl$_2$, 1 mM DTT, 150 mM NaCl, with 50 U of BamH1, at 37°C for 2.5 h in a 250-μL reaction volume.

3. The reaction is chilled on ice and quickly assayed for complete cutting on a 0.8% mini-agarose gel. The reaction

is terminated by addition of 0.5M EDTA, pH 7.5, to 10 mM, followed by extraction with phenol/chlorofom, ether extraction twice, and ethanol precipitation by addition of sodium acetate, pH 5.5, to 0.3M and 2.5 vol of ethanol, at −70°C for 30 min.

4. The vector is pelleted by centrifugation, washed with ethanol, dried *in vacuo,* and resuspended in 225 µL of 10 mM Tris, pH 7.4, 0.2% SDS.

5. The vector is preheated to 65°C for 3 min to separate any annealed BamH1 termini and quick cooled; 4.0 U of purified calf intestinal phosphatase (25 µL) is added (*see below* for purification), and the reaction mixture is incubated at 45°C for 1h.

6. The dephosphorylation reaction is terminated by three cycles of extraction with phenol/chloroform and centrifugation at 12,000g for 10 min, two cycles of ether extraction, followed by addition of 3M sodium acetate (pH 5.5) to 0.3M and 2.5 vol of ethanol, and stored at −20°C overnight.

7. The BamH1-cut, phosphatased vector is pelleted by centrifugation at 12,000g for 10 min, washed extensively with ethanol, dried *in vacuo,* and resuspended in 100 µL of 10 mM Tris, pH 7.4, 0.1 mM EDTA at a final concentration of 0.9 µg/µL.

3.3. Dephosphorylation of 5' Termini

The recircularization of linearized DNA cloning vectors (cut at a single restriction site) during ligation may be prevented by removal of the terminal 5' phosphate (37). This greatly improves the efficiency with which foreign DNA may compete in the ligation to complementary or blunt-end restriction enzyme sites at the termini of the linearized vector DNA. Calf intestinal alkaline phosphatase is used in preference to bacterial alkaline phosphatase, since the former enzyme is somewhat more easily inactivated by phenol extraction. Most commercial preparations of alkaline phosphatase contain nuclease activity(s), however, which upon prolonged exposure may nick or degrade the vector termini, preventing optimal ligation to staggered or blunt-end duplex termini of foreign DNA inserts. Therefore the following schematic procedure (38) is employed to remove nuclease activity from calf intestinal phosphatase, commercially available as an ammonium sulfate suspension.

1. Pellet 0.4 mg (160 U) of phosphatase by 5 min of centrifugation in a microcentrifuge.
2. Resuspend in 100 μL of 20 mM Tris, pH 8.4, and 100 mM KCl, and load on a Sephadex G-75 (1.0 x 30 cm) column equilibrated with the same buffer.
3. Monitor the elution profile spectophotometrically at 280 nm and collect 150-μL fractions.
4. Pool peak fractions, add an equal volume of sterile glycerol, and store at –20°C.
5. Assay for activity by hydrolysis of gamma-^{32}P-ATP (10 min, 37°C, 10 mM Tris, pH 7.8). Analyze the reaction by chromatography on PEI-cellulose plates in 0.7M KCl (one dimension: see note 1 in section 4) autoradiography.
6. Nuclease activity is analyzed by phosphatase treatment (16 h, 37°C, 10 mM Tris, pH 7.8) of a radiochemically labeled DNA fragment of specific length with subsequent electrophoresis (8% polyacrylamide) and autoradiography (see note 2 in section 4).

The alkaline phosphatase reaction is optimized using the following conditions:

1. Treat the restriction enzyme-digested vector (100 ng) with varying concentrations of purified alkaline phosphatase (10 mM Tris, pH 7.8, 0.2% SDS, 45°C 1 h). Stop the reaction by phenol/CHCl$_3$ extraction (3x), ether (2x), and ethanol precipitate.
2. Ligate the BamH1-cleaved, phosphatased vectors both to themselves and to 1 kb Sau3A-cut test DNA inserts (DNA

concentration: 100–200 μg/mL; 50 mM Tris, pH 7.4, 10 mM MgCl$_2$, 1 mM DTT, 1 mM ATP, pH 7.4, 0.5 U of T4 ligase; 10°C overnight). Ligations are analyzed by transformation and antibiotic selection (amp and tet).

Conditions that limit self ligation (recircularization of the vector), yet maximize joining to test DNA inserts, are chosen for large-scale vector preparation. Approximately 0.12 U of purified phosphatase/μg vector is optimal for BamH1-cut pHC79. In preliminary tests, equivalent units of "molecular biology grade" calf intestinal alkaline phosphatase (Boehringer Mannheim) have also been found effective at similar concentrations.

3.4. Preparation of High Molecular Weight DNA

1. High molecular weight DNA is prepared (39 and Chapter 16) from 5 g of frozen tissue in liquid N$_2$ and finely ground in a mortar and pestle chilled with additional aliquots of liquid N$_2$. The ground, frozen tissue is slowly added to phenol lysis buffer while stirring gently on a magnetic stirrer for approximately 15 min.

2. Organic and aqueous phases are separated by centrifugation at 5,000g for 10 min in a Sorvall HB-4 rotor, and the aqueous solution is reextracted with 25 mL of fresh buffered phenol as before.

3. Residual phenol is removed by extraction with 2 vol of ether (repeated three times), and the remaining ether bubbled off with filtered nitrogen gas.

4. RNA is removed by enzymatic digestion with 50 μg/mL of RNAse A (dissolved and made DNAse-free by boiling for 20 min in 20 mM sodium acetate, pH 5.5) at 37°C for 16 h while dialyzing against 20 mM Tris, pH 7.8, 10 mM NaCl, and 1 mM EDTA.

5. The solution is adjusted to 50 mM EDTA and 0.5% SDS (SDS activates proteinase K).

6. Crystalline Proteinase K is added to 100 µg/mL and the solution gently swirled at 37°C for 3 h.

7. The solution is made 1*M* with 5*M* NaCl and extracted once with buffered phenol and centrifuged as before.

8. The final DNA solution is extensively dialyzed (with three changes of 4 L of dialysis buffer) against 20 m*M* Tris, pH 7.8, 10 m*M* NaCl, and 1 m*M* EDTA. This procedure is similar to that described by Blin and Stafford (*40*), with the exception that phenol extraction precedes the proteinase K digestion. Final DNA concentration is measured by absorbance at 254 nm, with an expected yield of 40 mL of a 200-µg/mL genomic DNA solution.

9. DNA is sized on a 0.2% agarose gel with 2 µg/mL of ethidium bromide layed on a 1% agarose support, run in SB3 buffer at 20 mA overnight. Care must be taken to avoid shearing the DNA, especially during the phenol extractions, to maximize the length of isolated DNA and ensure proper concatamer formation during the ligation of vector and insert (requiring two sticky ends per genomic DNA molecule) and high-efficiency packaging of the DNA into infectious particles.

3.5. Preparation of Target DNA

1. High molecular weight genomic DNA is prepared and sized on 0.2% agarose gels as described. To select the proper concentration of Sau3A required to digest the DNA to an average size of 40 kb, the following equation is used (*41*):

$$40,000 \text{ (bp)}/256 \text{ (average four base cleavage frequency)} = 156.25 \text{ µg DNA per unit enzyme}$$

2. Consequently three samples of 16 µg of genomic DNA each are digested with 0.04, 0.40, and 4.00 U of enzyme in a 50-µL reaction with 10 m*M* Tris, pH 7.4, 10 mM MgCl$_2$, 1 m*M* DTT, 40 mM NaCl, at 37°C for 15 min, and the prod-

ucts analyzed on a 0.3% gel with λ DNA markers (uncut, HinD III, and XhoI), and conditions generating 35–50 kb DNA estimated.

3. In preparative digests two aliquots of 278 μg genomic DNA each are added to 2 and 3 U of Sau3AI, respectively, in 10 mM Tris, pH 7.4, 10 mM MgCl$_2$, 1 mM DTT, 40 mM NaCl in a 1.5-mL reaction volume at 37°C for 15 min. The reactions are terminated by the addition of 10% *n*-lauryl sarcosine to 0.1%, 0.5M EDTA (pH 7.5) to 15 mM, heated to 65°C for 5 min, and the products placed on ice.

4. Glycerol gradients for velocity sedimentation of the Sau-3A-cleaved DNA are prepared from autoclaved 10 and 30% glycerol buffers in SW28 polyallomer tubes using a gradient maker. The Sau3A-cleaved DNA is gently layered onto the surface of the gradient and centrifuged in a Beckman SW28 rotor at 70,000g for 16 h at 10°C. The gradients are fractionated with a 100-μL sterile glass capillary tube attached to a peristaltic pump. The glass micropipet is gently lowered to the bottom of the SW28 tube and held in place with a clamp. One-milliliter fractions are collected at a flow rate of 1 mL/min in sterile 1.5-mL Eppendorf tubes.

5. Ten microliters from every fifth fraction are removed, bromphenol blue and xylene cyanol are added to 0.1%, and the samples are loaded on a 0.3% agarose gel in 1x SB3 and run relatively quickly at 100 mA. Fractions containing DNA approximately 30–50 kb in length are reanalyzed on a 0.2% agarose gel in SB3, but at 20 mA overnight.

6. DNA fractions of 35–50 kb in length are pooled and diluted 1:2 with TE buffer, 2.5 vol of ethanol are added, and the DNA is precipitated at –70°C, overnight in a SW28 polyallomer tube.

7. The precipitated DNA is pelleted by centrifugation in a SW28 rotor at 80,000g for 30 min at 4°C.

8. The pellet is *gently* resuspended (again care must be taken to avoid shearing the DNA even after partial cleavage) in

4 x 100-μL aliquots of 10 m*M* Tris, pH 7.4, 0.1 m*M* EDTA, 50 m*M* Na acetate, pH 5.5, and transferred to a sterile Eppendorf tube, 2.5 vol of ethanol is added, and stored at –70°C overnight.

9. The DNA is repelleted by centrifugation at 12,000*g* for 10 min in an Eppendorf centrifuge, washed well with ethanol, and gently resuspended in low TE buffer at a final concentration of 0.3–0.7 μg/μL, and stored at 4°C. Expected yields are 10–20 μg of Sau3A-cleaved 35–50 kb DNA per 250 μg of starting genomic DNA.

10. The DNA concentration is determined by adding 1 μL of the sample to 19 μL of a 5 μg/mL ethidium bromide solution (in water) in a microtiter plate, estimating the concentration visually by transilluminating the sample with UV light (254 nm) and comparing to the fluorescence of similarly ethidium complexed genomic DNA standards previously determined spectrophotometrically at high concentration.

3.6. Ligation of Vector and Target DNA

1. From experiments designed to optimize the vector/insert ratio for maximal packaging, 0.45 μg of BamH1-phosphatased pHC79 (0.5 μL) is added to 0.3 μg of Sau3A-cut fractionated chromosomal DNA (1 μL) (1.5/1.0 vector/insert by weight) in 50 m*M* Tris, pH 7.4, 10 m*M* MgCl$_2$, 1 m*M* DTT, heated to 65°C for 3 min to separate any annealed molecules, and allowed to cool briefly at room temperature. ATP (pH 7.4) is added to 1 m*M*, followed by H$_2$O to 4.5 μL, and 0.5 μL of T4 DNA ligase (0.5 Weiss units), and the reaction mixed gently and incubated at 10°C overnight.

3.7. Preparation of In Vitro Packaging Extracts

The procedure was developed by V. Pirotta (unpublished), modified by B. Hohn, and appears here with only slight alteration.

3.7.1. Prehead Extract

1. Strain BHB2690 is streaked out on duplicate LAM plates and incubated at 30 and 42°C, respectively, overnight. The 42°C plate should show little or no growth.
2. The bacteria grown at 30°C are used to inoculate 750 mL of NZ broth in a 2-L Erlenmeyer flask prewarmed to 30°C, at an $OD_{600\,nm}$ of 0.1.
3. The culture is shaken at 30°C to an OD_{600nm} of 0.3 (approximately 3–5 h).
4. The flask is quickly transferred to a 45°C waterbath for 20 min without shaking (to induce the ts prophage).
5. The flask is again quickly moved to a 39°C incubator and grown under vigorous aeration for an additional 2 h.
6. Induction of the prophage is tested by adding a drop of chlorofom to several milliliters of culture and checking for clearing of the culture after several minutes.
7. If successful induction is observed, the cells are chilled on ice water and spun (6000*g*, 10 min in a Sorvall GS3 rotor) in ice-cold centrifuge bottles, the supernatant thoroughly removed, and the pellet suspended in 2.5 mL of cold sonication buffer.
8. The cell suspension is transferred to a 10-mL corex tube and placed in a salted-ice bath. The culture is lysed with 10-s blasts (Braun Bronsonic sonicator with a 2-mm microprobe at 100 W) until the solution clears and the viscosity drops (about 3 min).
9. Bacterial debris is removed by centrifugation in a prechilled Sorvall HS-4 rotor at 6000*g* for 10 min. A one-sixth volume of packaging buffer is added to the supernatant (about 2 mL), and the sonicate extract is distributed in 50-µL aliquots in precooled sterile microfuge tubes and frozen, open, in liquid nitrogen. The extract may be stored in liquid nitrogen indefinitely or at –70°C for at least 6 mo.

3.7.2. Freeze-Thaw Extract

1. Strain BHB2688 is streaked out on duplicate LAM plates, and incubated at 30 and 42°C, respectively, overnight. The 42°C plate should show little or no growth.
2. The bacteria grown at 30°C are used to inoculate 2 x 750 mL of NZ broth (in 2-L flasks) prewarmed to 30°C to an OD_{600nm} of 0.3, tested for induction, and the cells spun down as for the prehead extract.
3. The cell pellet is resuspended in 5 mL of cold sucrose buffer in a cold glass corex tube, centrifuged again (6000g for 10 min in a prechilled Sorvall HS-4 rotor), and the supernatant transferred (using prechilled pipets) to a cold Beckman sealable polycarbonate Ti70 tube.
4. One hundred microliters of fresh lysozyme mix is added, and the solution is mixed gently and frozen in liquid nitrogen.
5. The frozen spheroplasts are slowly thawed on ice (for approximately 1 h), 100 μL of packaging buffer is added, and the viscous solution is gently mixed and centrifuged at 250,000g in a chilled Beckman Ti70 for 30 min at 4°C. Fifty-microliter aliquots of the resulting supernatant are distributed as before with similar storage characteristics.

3.8. Packaging of Cosmid DNA

1. The sonicate and freeze-thaw extracts are thawed on ice, quickly spun at 12,000g for 2 min to remove any remaining debris, and returned to ice.
2. Ten microliters of sonicate extract is then quickly added to half of the ligation reaction (2.5 μL) and mixed gently, and 10 μL of freeze-thaw lysate is added, also mixed gently, and the packaging allowed to proceed at room temperature (18–20°C) for 90 min.
3. In later experiments, because the freeze-thaw extract appears to be somewhat unstable, 5 μL of freshly thawed

freeze-thaw lysate is added to the sonicate extract/ligation mix every 30 min for a total of 90 min, with a slightly higher efficiency of packaging. After 90 min, 250 µL of cold λ dilution buffer is added. To control for the efficiency of packaging and transduction, 90 ng of λ DNA (cl857Sam7) is packaged as a control with each cosmid packaging experiment.

3.9. Infection of the E. coli Host

1. *Escherichia coli* (HB101, ED8767, DH1, 490A, and LE392) are grown from a glycerol stock to late log phase ($OD_{600\,nm}$ of 1.0) in L broth supplemented with 0.4% maltose, spun at 5000g for 5 min in a cold 10-mL sterile tube, resuspended in an equal volume of 10 mM $MgCl_2$, 10 mM $CaCl_2$ (5), kept on ice for 10 min, spun at 5000g for 5 min, and resuspended in 0.1 vol of 10 mM $MgCl_2$ solution. Best results are always obtained with freshly prepared cells.

2. Infection is initiated by addition of 10 µL of host bacteria to 250 µL of packaged cosmid. The phages are allowed to absorb to the maltose binding protein (malB) on the surface of the host, and inject their DNA across the cell membrane, by incubation at 37°C for 30 min without shaking.

3. To allow for circularization of the cosmid DNA and expression of the newly acquired ampicillin resistance gene, 3 mL of L broth (for 150-mm plates; 1 mL for 90-mm plates) with 10 mM $MgCl_2$ (to stabilize any free phage) are added and shaken for an additional 30 min at 37°C.

4. The cosmid-containing host bacteria are plated at a density of 10^4–2 x 10^4 /150-mm plate on detergent-free nitrocellulose filters (27) to avoid a reduction in titer, overlaid onto NZM plates with 20 µg/mL ampicillin, and incubated overnight at 30°C (instead of 37°C) to avoid segregation problems likely encountered by bacterial cells replicating 38–52 kb of extrachromosomal DNA within a given doubling time. Indeed, a twofold greater efficiency in

plating is noted at 30°C. The efficiency of cosmid cloning should range from 2×10^4 to 2×10^5 cosmids/µg of insert DNA.

5. The filters are replica plated by the method of Hanahan and Meselson (42; see below), regrown at 30°C on NZM plates supplemented with 50 µg/mL ampicillin, and prepared for colony hybridization, whereas the master plate is stored at 4°C. Cosmid DNA is isolated by Triton-lysis extraction as previously described. Antibiotic selection pressure is always maintained to avoid segregation and loss of the cosmid.

3.10. In Situ *Filter Hybridization*

1. Recombinant *E. coli* are plated directly on sterile detergent-free nitrocellulose filters (washed Schleicher and Schull BA85 or Millipore HAWP) and grown for 16 h or until the colonies are 1–2 mm in diameter. The nitrocellulose is lifted off the agar and placed colony side up on a sterile Whatman No. 1 filter.

2. A second nitrocellulose filter, premoistened by contact with a fresh agar plate, is gently laid on the original nitrocellulose, and another Whatman No. 1 filter is placed on top, forming a sandwich. The colonies are transferred by applying pressure to the sandwich between two glass plates, and the filters are keyed by puncturing with a No. 24 needle (27,42).

3. The two filters are regrown on fresh agar plates for several hours, and the original is stored at 4°C with the agar plate sealed and inverted.

4. The DNA from the bacterial colonies is bound to the nitrocellulose as described by Grunstein and Hogness (17) involving lysis of the bacteria *in situ* on the nitrocellulose filter and denaturation of the DNA by contact with filters soaked in 0.5M NaOH and 0.5M NaCl for 5 min.

5. The filters are individually blotted on dry Whatman 3MM, transferred onto two Whatman No. 1 filters in a Buchner

funnel under vacuum, neutralized by addition of 10 mL of 0.5*M* Tris, pH 7.4, and1.5*M* NaCl, then washed with 10 mL of 2x SSC, 0.1% SDS and, finally, *in vacuo* at 80°C for 2 h.

6. High backgrounds occasionally encountered in screening the cosmid library at a density of 20,000–30,000 colonies/ 150-mm plate are reduced by wiping the surface of the nitrocelllulose filter with a tissue soaked in 2x SSC, 0.1% SDS following the second neutralization in 0.5*M* Tris, pH 7.4, and 1.5*M* NaCl, to remove bacterial debris and/or by prewashing the filters after baking in 2x SSC and 0.1% SDS at 60°C for several hours with three changes of buffer.

7. The baked nitrocellulose filters are prehybridized for >1h in hybridization buffer at 42°C in a plastic bag sealed with a commercial heat sealer. The denatured probe (see below) is added to the bag and incubated at 42°C for 1–3 d.

8. The nitrocellulose filter is carefully removed from the bag, washed in 2x SSC (twice for 5 min each) at room temperature, in 0.1% SSC, 0.1% SDS at 40°C (twice for several hours), or, if a homologous probe is used, in 0.1% SSC, 0.1% SDS at 60°C (twice for several hours). The filters are air dried, covered with plastic wrap, and autoradiographed using an intensifying screen in a film cassette.

3.11. Preparation of Hybridization Probes

Radiolabeled DNA probes for hybridization experiments may be prepared from DNA fragments produced by restriction digestion and purified by native acrylamide gel electrophoresis or by use of random primers prepared from calf thymus DNA (43) or are commercially available as a mixture of all possible 6mers (PL Biochemicals) and DNA polymerase I Klenow fragment.

1. The preparative isolation of DNA fragments by polyacrylamide gel electrophoresis is best achieved using 20 x 20 cm glass plates with spacer widths of 1.5 mm. The gel plates are sealed with Scotch 3 MM light electrical tape. Gel com-

positions range from 6to18% acrylamide and from 1/20 to 1/60 parts *bis*-acrylamide, depending on the size of the DNA fragments being purified. Acrylamide-*bis* stock solutions are deionized with Dowex mixed-bed ion-exchange resin and diluted to the appropriate percentage with Millipore filtered H_2O. Gels are prepared in TBE buffer, 0.5% ammonium persulfate (fresh), using 100 µL TEMED to initiate polymerization.

2. Gels are run in 1x TBE, at 150–200 V, where the current is fairly low and the gel plates remain cool, thus preventing thermal denaturation of double-stranded DNA. To avoid contact of the sample with the ammonium persulfate front, which often results in the loss of the low moleculare weight DNA fragments, the gels are pre-electrophoresed for at least 30 min before loading. Samples are generally resuspended from an ethanol precipitate in TE buffer and mixed with an equal volume of 10% Ficoll 400, 0.1% SDS, and 0.1% bromophenol blue dye, and are applied to the wells with an automatic pipet.

3. Following electrophoresis, the DNA is visualized by soaking the gel for 10 min in TBE buffer with 10 µg/mL ethidium bromide and illuminating the gel with a 254 nm or preferably, a 354 nm UV transilluminator to fluoresce the ethidium bromide–DNA complex. A photographic record of the experiment is obtained using a Polaroid camera mounted on a tripod equipped with both a Wratten 25 and UV filter, using Polaroid-type 107 (ASA 3000) film.

4. The electroelution of DNA fragments into dialysis bags is used to isolate specific restriction fragments electrophoresed in polyacrylamide gels. The fragments of interest are localized on the gel by ethidium bromide staining and UV transillumination or autoradiography, cut out with a single-edged razor blade, and placed on a sheet of wax paper (Nexcofilm).

5. Dialysis membrane (mol wt cutoff, 12,000–14,000) is prepared by boiling (5 min) in 10 mM EDTA, pH 7.6.

6. The gel slice is placed in the membrane with 200 µL of 100 m*M* Tris-borate and 2 m*M* EDTA, and the bag is sealed with a Spectropore clip.

7. The bag is placed in a horizontal electrophoresis tank of the type used for agarose gel electrophoresis, and the DNA is eluted in gel running buffer overnight at 40 V or for several hours at 150 V.

8. The buffer is removed, the bag is washed with 2x 100 µL of 10 m*M* Tris, pH 7.8, and 1 m*M* EDTA, and the pooled fractions are made 0.3*M* in sodium acetate, pH 5.5, and precipitated by addition of 2.5 vol of ethanol, –70°C, for 30 min, or –20°C overnight. For DNA isolated from agarose, the dialysate is phenol extracted and ether washed twice prior to ethanol precipitation. Randomly primed probes (43) are prepared by the following typical protocol.

1. Two hundred nanograms of DNA are added to 1 µg of the single-stranded random primers (gift of Dr. John Shine; or from PL Biochemicals), heated to 90°C for 3 min to thermally denature the double-stranded DNA, and slowly cooled to promote annealing of the primers to the single-stranded template.

2. The reaction mixture is adjusted to 50 m*M* Tris, pH 7.4, 10 m*M* MgCl$_2$, 5 m*M* DTT, 600 µ*M* dGTP, dATP, dTTP, 0.64 µ*M* alpha-^{32}-P-dCTP (100 µCi; Amersham 3000 Ci/mmol), and 10 U of DNA polymerase (Klenow fragment), followed by incubation at 37°C for 60 min, and finally chased with 600 µ*M* dCTP for an additional 15 min.

3. The reaction is terminated by the addition of EDTA (pH 7.6) to 10 m*M*.

4. Free mononucleotide is removed by chromatography on a Sephadex G-50 column (8 x 0.5 cm) in 10 m*M* Tris-HCl, pH 7.8, 1 m*M* EDTA, 100 m*M* NaCl.

5. The void volume fractions are detected with a hand Geiger-Muller monitor and pooled, and the DNA is precipitated by the addition of 10 µg of tRNA and 2.5 vol of

ethanol. The probe is resuspended in 100 μL of hybridization buffer and denatured at 90°C for 5 min prior to use.

4. Notes

1. Free phosphate migrates considrably more quickly than nonhydrolyzed ATP in one-dimensional chromatography on PEI-cellulose in $0.7M$ KCl.
2. Residual nuclease activity in calf intestinal phosphatase may be detected by extended digestion of a 5' extended restriction endonuclease terminus prelabeled with alpha-^{32}P -dNTP at an internal position, insensitive to the 5' phosphatase activity, yet readily hydrolyzed by exonuclease.

Acknowledgments

I would like to express thanks to Drs. Barbara and Thomas Hohn and Dr. Elizabeth Dennis for advice on cosmid cloning, Dr. Robert Crawford for helpful comment during a library construction, and Drs. Hugh Niall and Michael Waterfield.

References

1.   Benton, W.D. and Davis, R.W. (1977) Screening lambda gt recombinant clones by hybridization to single plaques in situ. *Science* **196**, 180–182.
2.   Tilghman, S.M., Tiemeier, D.C., Polsky, F., Edgell, M.H., Seidman, J.G., Leder, A., Enquist, L.W., Norman, B., and Leder, P. (1977) Cloning specific segments of the mammalian genome: phage lambda containing mouse globin and surrounding gene sequences. *Proc. Natl. Acad. Sci. USA* **74**, 4406–4441.
3.   Enquist, L., Tiemeier, D., Leder, P., Weisburg, R., and Sternberg, N. (1976) Safter derivatives of bacteriophage lambda gt. lambda C for use in cloning of recombinant DNA molecules. *Nature* **259**, 596–598.
4.   Blattner, F.R., Williams, B. G., Blechl, A.E., Faber, H.E., and Smithies, O. (1977) Charon phages: Safer derivatives of phage lambda for DNA cloning. *Science* **196**, 161–169.

5. Williams, B.G. and Blattner, F.R. (1979) Construction and characterization of the hybrid bacteriophage lambda charon vectors for DNA cloning. *Virology* **29**, 555–575.
6. Hohn, B. and Hohn, T. (1974) Activity of empty, headlike particles for packaging of DNA of bacteriophage lambda in vitro. *Proc. Natl. Acad. Sci. USA* **71**, 2372–2376.
7. Becker, A. and Gold, M. (1975) Isolation of the bacteriophage lambda A-gene protein. *Proc. Natl. Acad. Sci. USA* **72**, 581–585.
8. Sternberg, N. and Weisberg, R. (1977) Packaging of coliphage lambda DNA. I. The role of the cohesive end site and the gene A protein. *J. Mol. Biol.* **117**, 717–731.
9. Hohn, B. and Murray, K. (1977) Packaging recombinant DNA molecules into bacteriophage particles in vitro. *Proc. Natl. Acad. Sci. USA* **74**, 3259–3263.
10. Maniatis, T., Hardison, R., Lacy, E., Lauer, J., O'Connell, C., Quon, D., Sim, G., and Efstratiadis, A. (1978) The isolation of structural genes from libraries of eucaryotic DNA. *Cell* **15**, 687–701.
11. Clarke, L. and Carbon, J. (1976) A colony bank containing synthetic ColEl hybrid plasmids representative of the entire *E. coli* genome. *Cell* **9**, 91–99.
12. Sato, K. and Campbell, A. (1970) Specialized transduction of galactose by lambda phage from a deletion lysogen. *Virology* **41**, 474–487.
13. Feiss, M. and Margulies, T. (1973) On maturation of the bacteriophage lambda chromosome. *Mol. Gen. Genet.* **127**, 285–295.
14. Hohn, B. and Katsura, I. (1977) Structure and assembly of bacteriophage lambda. *Curr. Top. Microb. Immunol.* **78**, 69–110.
15. Feiss, M., Fisher, R.A., Siegele, D.A., Nichols, B.P., and Donelson, J.E. (1979) Packaging of the bacteriophage lambda chromosome: A role for base sequence outside cos. *Virology* **92**, 57–67.
16. Earnshaw, W. and Harrison, S. (1977) DNA arrangement in isometric phage heads. *Nature* **268**, 598–602.
17. Grunstein, M. and Hogness, D.S. (1975) Colony hybridization: A method for the isolation of cloned DNAs that contain a specific gene. *Proc. Natl. Acad. Sci. USA* **72**, 3961–3965.
18. Hohn, B. (1975) DNA as substrate for packaging into bacteriophage lambda, in vitro. *J. Mol. Biol.* **98**, 93–106.
19. Collins, J. and Hohn, B. (1978) Cosmids: A type of plasmid gene-cloning vector that is packageable in vitro in bacteriophage lambda heads. *Proc. Natl. Acad. Sci. USA* **75**, 4242–4246.
20. Umene, K., Shimada, K., and Takagi, Y. (1978) Packaging of ColEl DNA having a lambda phage cohesive end site. *Mol. Gen. Genet.* **159**, 39–45.
21. Collins, J. and Bruning, H. (1978) Plasmids useable as gene-cloning vectors in an in vitro packaging by coliphage lambda: Cosmids. *Gene* **4**, 85–107.

22. Hohn, B. and Collins, J. (1980) A small cosmid for efficient cloning of large DNA fragments. *Gene* **11**, 291–298.

23. Meyerowitz, E., Guild, G., Prestidge, L., and Hogness, D. (1980) A new high capacity cosmid vector and its use. *Gene* **11**, 271–282.

24. Hohn, B. (1983) DNA sequences necessary for packaging of bacteriophage lambda DNA. *Proc. Natl. Acad. Sci. USA* **80**, 7456–7460.

25. Royal, A., Garapin, A., Cami, B., Perrin, F., Mandel, J.L., LeMeur, M., Bregegegre, F., Gannon, G., Le Pennec, J.P., Chambon, P., and Kourilsky, P. (1979) The ovalbumin gene region: Common features in the organization of three genes expressed in chicken oviduct under hormonal control. *Nature* **279**, 125–131.

26. Cattaneo, R., Gorski, J., and Mach, G. (1981) Cloning and multiple copies of immunoglobulin variable kappa genes in cosmid vectors. *Nucl. Acids Res.* **9**, 2777–2790.

27. Grosveld, F., Dahl, H., DeBoer, E., and Flavell, R.A. (1981) Isolation of β-globin-related genes from a human cosmid library. *Gene* **13**, 227–237.

28. Steinmetz, M., Minard, K., Horvath, S., McNicholas, J., Srelinger, J., Wake, C., Long, E., Mach, B., and Hood, L. (1982) A molecular map of the immune response region from the major histocopatability complex of the mouse. *Nature* **300**, 35–42.

29. Morris, D., Noti, J., Osborne, F., and Szalay, A. (1981) Plasmid vectors capable of transferring large DNA fragments to yeast. *DNA* **1**, 27–36.

30. Grosveld, F., Lund, T., Murray, E., Mellow, A., Dahl, H., and Flavell, R.A. (1982) The construction of cosmid libraries which can be used to transform eukaryotic cells. *Nucl. Acids Res.* **10**, 6715–6732.

31. Goodenow, R., McMillan, M., Nicholson, M., Taylor Sher, B., Eakle, K., Davidson, N., and Hood, L. (1982) Identification of the class I genes fo the mouse major histocompatability complex by DNA-mediated gene transfer. *Nature* **300**, 231–237.

32. Lund, T., Grosveld, F.G., and Flavell, R.A. (1982) Isolation of transforming DNA by cosmid rescue. *Proc. Natl. Acad. Sci. USA* **79**, 520–524.

33. Lau, Y-F. and Kan, Y.W. (1983) Versatile cosmid vectors for the isolation, expression, and rescue of gene sequences: Studies with the human alpha-globin gene cluster. *Proc. Natl. Acad. Sci. USA* **80**, 5225–5229.

34. Tanaka, T. and Weisblum, B. (1975) Construction of a colicin E1-R factor composite plasmid in vitro: Means for amplication of deoxyribonucleic acid. *J. Bacteriol.* **121**, 354–362.

35. Davis, R.W., Botstein, D., and Roth, J.R. (1980) *A Manual for Genetic Engineering: Advanced Bacterial Genetics* Cold Spring Harbor Laboratory, Cold Spring Harbor, New York.

36. Clewell, D.B. and Helinski, D.R. (1969) Supercoiled circular DNA-protein complex in *Escherichia coli:* Purification and induced conver-

sion to an open circular DNA form. *Proc. Natl. Acad. Sci. USA* **62**, 1159–1166.

37. Seeburg, P.H., Shine, J., Martial, J.A., Baxter, J.D., and Goodman, H.M. (1977) Nucleotide sequence and amplification in bacteria of structural gene for rat growth hormone. *Nature* **270**, 486–494.

38. Efstratiadis, A., Dafatos, F.C., and Maniatis, T. (1977) The primary structure of rabbit β-globin mRNA as determined from cloned DNA. *Cell* **10**, 571–585.

39. Polaky, F., Edgell, M.H., Seidman, J.G., and Leder, P. (unpublished).

40. Blin, N. and Stafford, D.W. (1976) Ageneral method for isolation of high molecular weight DNA from eukaryotes. *Nucl. Acids Res.* **3**, 2303–2308.

41. Seed, B., Parker, R.C., and Davidson, N. (1982) Representation of DNA sequences in recombinant DNA libraries prepared by restriction enzyme partial digestion. *Gene* **19**, 201–209.

42. Hanahan, D. and Meselson, M. (1980) Plasmid screening at high colony density. *Gene* **10**, 63–67.

43. Taylor, J.M., Illmersee, R., and Summers, J. (1976) Efficient transcription of RNA into DNA by avian sarcoma virus polymerase. *J. Biochem. Biophys. Meth.* **442**, 324–330.

44. Enquist, L., Madden, M.J., Schiop-Stazly, P., and Vande-Woude, G.F. (1979) Cloning of herpes simplex type I DNA fragments in a bacteriophage lambda vector. *Science* **203**, 541–543.

45. Wood, W.B. (1966) Host specificity of DNA produced by *Escherichia coli:* Bacterial mutations affecting the restriction and modification of DNA. *J. Mol. Biol.* **16**, 118–133.

46. Boyer, H.W. and Roulland-Dussoix, D. (1969) A complementation analysis of the restriciton and modification of DNA in *Escherichia coli*. *J. Mol. Biol.* **41**, 459–472.

47. Haley, J., Crawford, R., Hudson, P., Scanlon, D., Tregear, G., Shune, J., and Niall, H. (1987) *J. Biol. Chem.*, in press.

Chapter 19

Construction of cDNA Libraries in λgt10 or λgt11

Michael M. Burrell

1. Introduction

To obtain a cDNA clone of an mRNA, the mRNA must be copied faithfully into DNA, and the cDNA library must be large enough to represent the abundance class that contains the mRNA of interest. For example, in tobacco, Goldberg (1) has shown that it is possible to divide the mRNA population into three classes with most of the mRNAs (11,300) being in the lowest abundance class and making up 39% of the polysomal mRNA. To obtain a cDNA library that contains at least one clone for each mRNA of this class will require about 2×10^5 clones (2). This can be achieved with a few micrograms of mRNA by using the efficient RNase H method of making double-stranded cDNA (3) and a bacterioghage λ vector that exploits the high efficiency with which in vitro packaged phage can be introduced into *Escherichia coli*. The choice of λ vector is important because if the DNA to be inserted makes the λ genome greater than 105% of the wild-type length, the packaged phage will have a low viability.

The length of mRNA molecules ranges from several hundred bases to a few kilobases (kb). A suitable vector for this length of DNA is λgt10, because it will accept DNA fragments of up to 7.6 kb. If the mRNA of interest is known to approach or exceed this length, then a different vector should be chosen.

It is desirable in a cloning strategy for parent phage to be suppressed when the library is plated out. This is achieved with λgt10 because the parent phage is cI$^+$imm^{434}, and these are efficiently repressed on a hf1A^{150} strain of *E. coli* (e.g., C600 hfl). Insertion of cDNA into the EcoRI site of this gene produces a cI$^-$ phenotype. These phages will form plaques. On a non-Cgl strain (e.g., C600), the cl$^+$ phenotype gives a turbid plaque, and the cI$^-$, a clear plaque.

If it is desired to screen a cDNA library with antibodies, then another suitable vector is λgt11. λgt11 has several valuable assets as an expression vector. The EcoRI cloning site is toward the 3' end of the lacZ gene, which is thought to stabilize the expressed protein. The phage can accept up to 7.2-kb inserts, produces a temperature-sensitive lysis repressor (cI857) that is ineffective at 42°C, and contains the amber mutation S100, which makes it lysis-defective in hosts that do not contain the supressor SupF. These features may then be combined with host cells, which carry the plasmid pMC9, which actively represses lacZ expression. The cells are also *lon* protease deficient, and this minimizes the side effects of any protein coded for by the insert during cloning and amplification. When screening with antisera, however, high levels of expression can still be achieved by using IPTG to inactivate the *lac* repressor.

Although there are many advantages in using these vectors, there are several points that should be considered before they are used. λgt10 is 43.34 kb, and therefore a 1-kb insert is a small proportion of the total DNA. Thus much more DNA in total must be prepared from a selected λ clone than from a selected plasmid clone to obtain the same amount of insert DNA. High efficiency of cloning requires good packaging extracts, which are time consuming to prepare, although they can be purchased. It is not considered advisable to maintain clones in λgt10 and λgt11 unless it is known that they are stable. Screening with antibodies sounds attractive, but requires specific antibodies that will recognize the peptide of interest as a fusion product and not necessarily folded in its native form. In addition, many of the correct cDNA clones will be inserted into λgt11 in the wrong reading frame and, therefore, not produce the correct protein.

There are many steps in making a cDNA library in λgt10 or λgt11. This chapter starts with methods for the growth of λgt10 and λgt11 and with the preparation of vector DNA from the phage. These steps are followed by methods of preparing λ arms from the vector DNA and the ligation of the cDNA to these arms. Therefore it is assumed that the reader has previously prepared suitable cDNA (*see* Vol. 2 in this series). The chapter finishes with methods of packaging the DNA into viable phage and infection of *E. coli* to produce the library. The procedures described below have worked well in our hands. There are probably many other methods and variations that will prove equally successful. For further information, the reader is directed to Glover (4), Maniatis et al. (2), this volume, and Vol. 2 in this series.

2. Materials

See Note 1 in section 4.

1. Vectors λgt10 (cIimm^{434}b527) and λgt11 (lac5, S100, cI857, nin5).
2. Host bacteria for λgt10: C600 (hsdR$^-$, hsdm$^+$, supE44, thr, leu, thi, lacY1, tonA21).
3. Host bacteria for λgt11 Y1088 [American type culture collection (ATCC) 37195]: ΔlacU169, supE, supF, hsdR$^-$, hsdM$^+$, metB, trpR, tonA21, proC: tn5(pMC9). Y1089 (ATCC37196): ΔlacU, arad139, strA, hflA150. Y1090 (ATCC37197): ΔlacU169, proA$^+$ Δlon, araD139, supF.

4. TE buffer: 10 mM Tris-HCl, pH 7.5, 1 mM EDTA.
5. TEN buffer: TE + x mM NaCl, as described in text, e.g., TEN-400 is TE + 400 mM NaCl.
6. ATP: 10 mM stored at –20°C.
7. EDTA: 0.5M adjusted to pH 8.0 with NaOH.
8. Phenol: redistilled phenol equilibrated with 100 mM Tris-HCl, pH 8.0.
9. EcoRI 10x restriction buffer: 0.5M Tris-HCl pH 7.5, 1M NaCl, 0.1M MgCl$_2$.
10. Ligation buffer (10x): 0.5M Tris-HCl, pH 7.5, 0.1M MgCl$_2$, 0.1M dithiothreitol. Store at –20°C.
11. Spermidine: 1 mM heated to 80°C for 10 min. Store at –20°C.
12. λ Diluent: 10 mM Tris-HCl, pH 7.5, 10 mM MgCl$_2$, 0.1 mM Na$_2$EDTA.
13. Ultragel AcA34 (LKB).
14. L broth: 10 g of bactotryptone, 5 g of bacto-yeast extract, 10 g of NaCl/L, adjusted to pH 7.5.
15. Bottom agar: 1.5% agar in L broth.
16. Top agar: 0.7% agar or agarose in L broth.
17. Formamide, recrystallized: Store at –20°C.
18. IPTG: 1M isopropylthiogalactoside in SDW. Store at –20°C.
19. Xgal: 2% 5-bromo-4-chloro-3-indolyl-β-D-galactopyranoside in dimethylformamide. Store at –20°C.
20. SDW: sterile double-distilled water.
21. BPB: Bromophenol blue 0.04% in 50% glycerol.
22. Chloroform: Use chloroform/butan-1-ol (50/1) and store in the dark.
23. Enzymes: Concentrations of enzymes supplied by different manufacturers varies, and even that of the same supplier often changes. Therefore always check and adjust volumes in protocols accordingly to use the same activity of enzyme. If enzymes need to be diluted, use the dilution buffer suggested by the manufacturer.

3. Methods

This section assumes that the experimenter has prepared double-stranded cDNA ready for ligation into EcoRI site of λgt10 or λgt11. This is often achieved by adding EcoRI linkers to the cDNA (after methylation) followed by digestion with EcoRI. It is therefore useful to prepare some test DNA of a suitable length by EcoRI digestion of a suitable plasmid or, failing that, genomic DNA followed by size fractionation.

3.1. Preparation of Plating Cells for Infection with λgt10 or λgt11

1. Streak out on L plates that contain 0.2% maltose the desired strain of *E. coli* (see the protocols below for the correct strain). Incubate overnight at 37°C.
2. Place a single colony in a 250-mL conical flask containing 50 mL of L broth + 0.2% maltose and incubate overnight at 37°C with shaking (220 rpm), but without frothing.
3. Centrifuge 40 mL of culture cells for 10 min at 3000g.
4. Resuspend the cells in 20 mL of 10 mM $MgSO_4$ at 4°C (*see* Note 2 in section 4).
5. Adjust OD_{600} to 2.00 with 10 mM $MgSO_4$.

3.2. Titering and Plaque Purification of λgt10 Stocks

To efficiently prepare good λgt10 DNA for cloning, it is first necessary to determine the titer of the phage stock and check whether it is contaminated with mutant phage-producing clear plaques, which complicate screening the library later.

1. Titer the phage stock by incubating aliquots of phage in 0.1 mL of λ diluent with 0.1 mL of plating cells (C600) at 37°C for 20 min. Gently shake the phage and cells during the incubation.
2. Add 3 mL of top agar, swirl to mix, and pour onto an L plate supplemented with 0.2% maltose (*see* Note 3 in section 4).

3. Incubate overnight at 37°C.

4. If there are clear plaques, remove a turbid plaque with a sterile fine Pasteur pipet or capillary and place in an Eppendorf tube containing 1 mL of λ diluent. Add one drop (50–100 µL) of $CHCl_3$ to stop bacterial growth and leave at room temperature for 1 h if required immediately or place at 4°C overnight.

5. Assume 10^6 phage, and serially dilute to determine the titer.

6. If there are clear plaques, repeat step 4.

7. When there are no clear plaques, plate at 2000 plaque-forming units (pfu) per 82 mm plate to screen 15,000 plaques for clear plaques.

8. If there is more than one clear plaque per 10^4, repeat the purification until the clone is sufficiently pure and stable.

3.3 Preparing λgt10 DNA

1. Prepare sufficient plating stock to set up between 15 and 30 82-mm plates at 1×10^6 pfu/plate (see Note 4 in section 4).

2. Prepare fresh plating cells (C600). Pour 15–30 L plates + 0.2% glucose (see Note 5 in section 4). Use 0.1 mL of plating cells, 1×10^6 pfu in 3 mL of top agar per plate. As controls, omit the λgt10 from one plate and both the cells and the λ from another plate. Fewer larger plates can be used, but they require more practice in handling.

3. After 4–6 h of incubation at 37°C, the plates will appear mottled. If in doubt, do not leave the plates too long or the titer will drop. If the plates become evenly turbid, they have gone too far.

4. Place at 4°C or on ice to cool for 30 min. Overlay with 4.5 mL of ice-cold λ diluent and a few drops of chloroform (from a Pasteur pipet). Leave at 4°C overnight for the phage to diffuse into the λ diluent.

5. Carefully remove the λ diluent from the plates without disturbing the top agar and transfer to polypropylene centrifuge bottles.

6. Centrifuge at 10,000g for 20 min to pellet the cells and any agar. Carefully transfer the supernatant to fresh tubes suitable for a swing out rotor.
7. Centrifuge for 90 min at 70,000g.
8. Resuspend the phage pellet in 5 mL of λ diluent, and add 8.5 g of CsCl. Overlay with 14 mL of 1.47 g/mL density CsCl. Centrifuge overnight at 50,000 rpm (184,000g) in a Beckman 70Ti rotor. Accelerate the centrifuge slowly. Decrease the speed to 40,000 rpm for the last 40 min (*see* Note 6 in section 4).
9. Remove the bluish white phage band in as small a volume as possible. Make the volume to 4 mL with at least an equal volume of λ diluent containing CsCl at a density of 1.5 g/mL. Centrifuge at 60,000 rpm for 4 h in a Kontron 80.4Ti rotor.
10. Remove the band in as small a volume as possible and make to 1 mL with λ diluent.
11. Add 1 mL of 3x recrystallized formamide, mix, and leave to stand for 2 h at room temperature.
12. Add 1 mL of SDW and 6 mL of absolute ethanol. The DNA should precipitate immediately.
13. Briefly centrifuge in a bench-top centrifuge and pipet off the supernatant.
14. Redissolve the DNA in 10 mM Tris, 5 mM EDTA, pH 7.5, and ethanol precipitate twice more (*see* Note 7 in section 4).
15. Finally redissolve DNA in TE buffer.

3.4. Preparation of λgt11 DNA

The first steps in preparing λgt11 DNA differ from the above method for λgt10 because it is possible to exploit the temperature-sensitive repression of lysis. Thus cells are first multiplied at 32°C and then induced to lyse at 44°C.

1. Streak out Y1088 cells with λgt11 at 32 and 42°C to check that lysis only occurs at 42°C.

2. Pick a single colony and place in a sterile McCartney bottle with 10 mL of L broth + 10 mM MgCl$_2$, swirl, loosen the cap, and incubate overnight at 32°C.

3. Place 10 mL of overnight culture in a 3-L conical flask containing 1 L of L broth at 32°C. Shake at 280 rpm in an orbital shaker until OD$_{600}$ = 0.6 (2–3 h).

4. Place a sterile thermometer in the flask and swirl in a water bath at 60°C until the flask contents reach 44°C. Then place at 44°C with good aeration for 15 min.

5. Place at 37°C with good aeration for 3 h. (The temperature must not drop below 37°C or poor lysis will result.)

6. Check for lysis by mixing about 2 mL of culture with a few drops of chloroform in a small test tube. The culture should clear in 3–5 min.

7. Add 10 mL of chloroform to the culture and shake at 37°C for 10 min.

8. Pellet the bacterial debris at 5000 rpm (300g) in a 6 x 300 mL rotor (MSE.21 or equivalent) for 10 min at 4°C. Use polypropylene screw-cap bottles.

9. Pellet the phage overnight at 10,000 rpm (15,500g). If a suitable centrifuge for the overnight run is not available, the phage may be recipitated with polyethylene glycol (PEG). To 250 mL of supernatant at room temperature add 15 g of NaCl, 25 g of PEG-6000, 1.5 mL of 10 mg/mL RNase A, and 1 mg/mL DNase I. Stir at 4°C for at least 3 h, preferably overnight. Collect the phage at 10,000 rpm for 30 min, resuspend in 5 mL of λ diluent and chloroform extract at RT to remove the PEG. We find, however, that the recovery of this phage seems to be very variable.

10. Continue as for λgt10 (step 8).

3.5. EcoRI Digestion of λ

This and the next step involve preparing the arms of λgt10 and λgt11 so that the cDNA can be ligated between them. Both vectors have a unique EcoRI site. In λgt10 this is in the cI gene, and insertion of DNA will produce the cI⁻ phenotype (clear plaques). In λgt11 the restriction site is in the lacZ gene. Therefore when λgt11 phage are plated on media containing Xgal, the phage with inserts will produce clear plaques and the wild type phage, blue plaques.

1. Mix the following to a total volume of 100 μL : 10 μg of λgt10 or λgt11, 10 μL of 10 x EcoRI buffer, and sterile distilled water, allowing for the addition of the first 30 U of enzyme at step 3.
2. Centrifuge any droplets of solution to the bottom of the tube.
3. Add 30 U of EcoRI in less than 20 μL.
4. Mix and incubate at 37°C for 2 h.
5. Add 30 U of EcoRI and incubate for a further 30 min.
6. Add 2 μL of 0.5M EDTA.
7. Remove the protein by adding 50 μL phenol, and mix gently (Don't vortex, *see* Note 7 in section 4). Add 50 μL CHCl₃/butan-1-ol (50/1).
8. Centrifuge and remove aqueous layer.
9. Back extract the phenol/CHCl₃ layer with an equal volume of 10 mM Tris-HCl, pH 8.0.
10. Chloroform extract (with chloroform/butan-1-ol; 50/1) the combined aqueous phases.
11. Ethanol precipitate at –20°C by adding 2.5 vol of ethanol and 1/40th of the total volume of 4M ammonium acetate (if in a hurry, use ethanol precooled to –20°C).

3.6. Phosphatasing λ

Strictly, this step is unnecessary for λgt10, but we prefer to include it since there are no detrimental effects and it makes checking the cDNA clones on C600 easier.

1. Add 1 U of bacterial alkaline phophatase (Amersham T.2120Y) (in less than 10 μL) to 50 μL of 0.1M Tris-HCl, pH 8.0. Mix and heat to 80°C for 5 min to inactivate DNAse.

2. Add 10 µg of vector in 200 µL of 0.1M Tris-HCl, pH 8.
3. Incubate at 65°C for one-half hour.
4. Add 2.5 µL of 0.5M EDTA, mix, and heat to 50°C for 10 min.
5. Phenol extract with 125 µL of phenol, mix, add 125 µL of CHCl$_3$/butan-1-ol (50/1), mix gently, and incubate at 37°C for 10 min.
6. Re-extract the aqueous phase as above.
7. Back-extract the organic phase with 125 µL of TE buffer, and incubate at 37°C for 10 min.
8. Combine the aqueous phases.
9. Extract (x2) with CHCl$_3$/butan-1-ol (50/1).
10. Ethanol precipitate at –20°C with 2.5 vol of absolute ethanol and 1/40th of the total volume of 4M ammonium acetate. The DNA precipitates in a few minutes.

3.7. Purification of cDNA Linkers on Ultragel

A common way to clone cDNA into an EcoRI site of a vector is to add linkers and then digest with EcoRI to provide ligateable ends. It is most important, however, to remove linker fragments prior to cloning. This can be achieved as described below (*see* Note 8 in section 4).

1. Plug a sterile siliconized Pasteur pipet or a 1-mL disposable plastic syringe with a small amount of siliconized glass wool.
2. Wet column with 2.0 mL of TEN-400.
3. Pour in a slurry of Ultragel AcA34 in TEN-400 to provide a 1.0-mL column.
4. Wash through with 4.0 mL of TEN-400.
5. Add 250 ng of appropriately cut vector in 100 µL of TEN 400, and wash through with 4 mL of TEN-400.
6. Apply between 5 and 40 ng of double-stranded cDNA in a maximum volume of 5 µL of TEN-400 + 2 µL BPB and collect two drops. Add 50 µL of TEN-400 to the column and collect two drops (= fraction 1).

7. Thereafter collect drop fractions equivalent to approximately 100 μL. Collect fractions until the BPB begins to elute.

8. Determine the fractions that contain the cDNA by measuring the radioactivity in each fraction. If there is sufficient radioactivity present in the cDNA, it may be measured directly with a hand monitor. If this is not possible, spot 2 μL from each fraction onto GF/C filters and count in scintillation fluid. The cDNA usually elutes in fractions 4 to 7. The Bromophenol blue gives a guide as to where the cut linkers elute.

9. Pool appropriate fractions avoiding any linker contamination. Add an equal volume of TE buffer to bring the salt concentration to 0.2*M*. Sample 2 μL for TCA precipitation to calculate the yield for the column.

10. Spin on a microfuge for 2 min to pellet any ultragel that has come through. Combine the appropriate amount of ds cDNA from the column in TEN-200 buffer (*see* Note 9 in section 4), and 1 μg of appropriate vector. Add 2.1 vol of absolute ethanol. Mix, and coprecipitate at –20°C overnight. Any unused double-stranded cDNA may be stored in TEN-200 at –20°C.

3.8. Ligation of cDNA and Vector

This step brings together the cDNA and vector DNA ready for packaging into viable phage. It is important, however, to be able to assess after packaging and infection which step if any has not worked as well as expected. Therefore, set up the following ligations:

i.   Uncut λ
ii. EcoRI cut, not phosphatased λ
iii. EcoRI cut, phosphatased λ
iv. EcoRI cut, phosphatased λ, plus test EcoRI cut DNA of 2–3 kb
v.   EcoRI cut, phosphatased λ, plus cDNA

1. Centrifuge for 20 min at 4°C to precipitate cDNA and vector. Decant ethanol carefully.
2. Wash the pellet with 0.5 mL of absolute ethanol (precooled to –20°C). Spin on a microfuge for 5 min.
3. Take off the ethanol and repeat the absolute ethanol wash (precooled to –20°C). Freeze-dry the pellet for 5–10 min.
4. Take up the pellet in 3 µL of sterile double-distilled water. Add 1 µL of a mix of 5 µL of 10x ligation buffer + 5 µL of 1 m*M* spermidine.
5. Heat at 65°C for 5 min. Spin on a microfuge for 30 s.
6. Incubate at 46°C for 20 min minimum submerged.
7. Add 0.5 µL of 5 m*M* rATP and 1.0 µL of T4 DNA ligase (3 U; *see* Note 10 in section 4). Mix well, but do not vortex. Spin for 30 s, then seal the tube in a small plastic bag.
8. Incubate *submerged* at 12°C for 24 h.

3.9. In Vitro Packaging

To obtain a large cDNA library from a minimal amount of cDNA, the packaging of the ligated DNA into phage must work well. It often works inefficiently in unskilled hands, however. Therefore always check the packaging extracts with some control DNA before committing the cDNA.

Packaging extracts are prepared according to protocol II in ref. 2. Commercial packaging kits are now available, however, and although expensive, they are worthwhile if only a few libraries are to be made. The packaging procedure that we have used differs slightly from that described in Chapter 36 in Vol. 2 of this series.

1. Add 5 µL of sonicated extract to the DNA, and mix with the tip of pipet.
2. Leave on ice for 15 min.
3. Add 10 µL of just-thawed freeze-thaw lysate.
4. Incubate for 1 h at RT.
5. Add 180 µL of λ diluent and mix.
6. If not plating immediately, add one drop of CHCl$_3$, store at 4°C.

3.10. Plating Out λgt10 Libraries for Screening

Assess the titer and quality of the library by using 0.1% of the in-vitro packaged library with C600 plating cells as described above. Assess the proportion of clear (recombinant) plaques.

It is probably best to screen a λgt10 library without amplification to avoid selecting against poorly growing recombinant phage. The large plaque size of λgt10 means that only about 1000 plaques can be screened on an 82-mm plate. Doubling the plating cell density reduces the plaque size and therefore can help if large libraries are to be screened.

To screen the library, prepare C600 hfl plating cells, and plate out at the desired density following the procedure in Chapters 43 and 42 of Vol. 2 in this series. Remember to use agarose for the top layer and well-dried or 2-d-old plates. We usually use Biodyne A and sterilize it. The lifts can be done after 6 h rather than overnight, and we always perform duplicate lifts. The hybridization conditions are dictated by the probe. Suitable conditions are described in refs. 5–7.

3.11. Screening λgt11 Libraries

1. Assess the titer of the library by plating with Y1088 cells, but add 40 µL of Xgal and 20 µL of IPTG to the 3 mL of top agar. Parent plaques appear blue, and those carrying inserts will either be clear or faintly blue.
2. To amplify the library for immunoscreening, as in Chapter 20 in this volume, plate at 10^4 plaque-forming units per 82-mm plate with Y1088 cells at 42°C without IPTG or Xgal (*see* note 11 in section 4).
3. Overlay the plates with λ diluent and store as described above.
4. Screen the amplified library as described in Chapter 20 in this volume.

4. Notes

1. All solutions used for nucleic acid work should be sterile and nuclease free. Unless specified, solutions are stored at room temperature. All glassware and plasticware should be siliconized.
2. When resuspending bacteria, always use a small volume of liquid initially until a creamy paste is obtained, then add the rest of the resuspension medium.
3. When plating phage, use well-dried or 2-d-old plates. When adding the top agar + phage, have the top agar at 45–50°C, do each sample in turn, avoid air bubbles, and leave the plates to set on a flat bench for 30 min. Incubate plates upside down.
4. λgt10 can be prepared from liquid lysates, but the presence of cI⁻ phage revertants will not be rocognized and will confuse the assessing of phage-carrying inserts.
5. When preparing λgt10 DNA, the presence of glucose in the L plates increases the titer by 10-fold.
6. For the background to this gradient, consult ref. 8.
7. Always treat high molecular weight DNA gently. Avoid shear forces or it will easily be degraded.
8. To remove linker fragments prior to cloning, the ultragel column is washed with either λgt10 or λgt11. Therefore do not use more of your precious cDNA than necessary. If 40 ng are used on the column, it should provide a cDNA library of at least 10^5 recombinants.
9. The amount of cDNA used depends entirely on its length and quality; 1–5 ng has been used by us to yield large libraries.
10. Beware of the ways different manufactures express DNA ligase units. The ligation units used here are those quoted by Pharmacia.
11. Do not amplify more than required, because even without IPTG, plaque size varies quite a lot. Remember to use 50 µg/mL ampicillin to maintain the plasmid containing repressor.

References

1. Goldberg, R.B., Hoschek, G., and Kamalay, J.C. (1978) Sequence complexity of nuclear and polysomal RNA in leaves of the tobacco plant. *Cell* **14**, 123–131.
2. Maniatis, T., Fritsch, E.F., and Sambrook, J. (1982) *Molecular Cloning. A Laboratory Manual* Cold Spring Harbor Laboratory, Cold Spring Harbor, New York.
3. Gubler, E. and Hoffman, B.J. (1983) A simple and very efficient method for generating cDNA libraries. *Gene* **25**, 263–269.
4. Glover, D.M. (1985) *DNA Cloning* Vol. 1 *A Practical Approach* IRL Press, Oxford.
5. Burrell, M.M., Twell, D., Karp, A., and Ooms, G. (1985) Expression of shoot inducing Ti TL-DNA in differentiated tissues of potato (*Solanum tuberosum* cv. Maris Bard). *Plant Mol. Biol.* **5**, 213–223.
6. Clayton, C.E. (1985) Structure and regulation of genes encoding fructose bisphosphate aldolase in *Trypanosoma brucei*. *EMBO J.* **4**, 2997–3003.
7. Wood, W.I., Gitschier, J., Lasky, L.A., and Lawn, R.M. (1985) Base composition independent hybridisation in tetramethyl-ammonium chloride: A method for oligonucleotide screening of highly complex gene libraries. *Proc. Natl. Acad. Sci. USA* **82**, 1585–1588.
8. Garger, S.J., Griffith, O.M., and Grill, L.K. (1983) Rapid purification of plasmid DNA by a single centrifugation in a two step cesium chloride-ethidium bromide gradient. *Biochem. Biophys. Res. Commun.* **117**, 835–842.

Chapter 20

Immunoscreening of λgt11 Expression Libraries Using an Avidin–Biotin Detection System

Robert McGookin

1. Introduction

Although the avidin–biotin detection system has been used for immunochemical staining of tissues for some time, its use for amplification in immunoassays was not described until 1979 (1). The biological basis for the system is the very tight binding of four molecules of biotin to one of avidin. This has an amplification ability that can be used to increase the sensitivity of detection of antibodies or other biotin-labeled molecules. Using antibodies raised to a purified peptide and an efficient expression system, such as λgtll (2), the clone(s) containing the cDNA for that peptide can usually be identified by detection of expressed antigens.

The method described here uses the maximum amplification available with this system. A complex of biotin-labeled horseradish peroxidase (HRP) and avidin is prepared and reacted with the biotin-labeled secondary antibody. The secondary antibody is used to detect the primary antibody on the nitrocellulose filter. The method is modified from that provided in the Clontech Laboratories Inc. immunoscreening kit.

2. Materials

1. 10 mM Isopropyl β-D-thiogalactopyranoside (IPTG). Prepare fresh solution in sterile distilled water for each set of filters.
2. 10x TBST: 0.5M Tris-HCl, pH 7.9, 1.5M NaCl, 0.5% (v/v) Tween-20. Store at room temperature; dilute as required.
3. 10x TBS: 0.5M Tris-HCl, pH 7.9, 1.5M NaCl. Store at room temperature and dilute as required.
4. Blocking solution: TBST containing 3% (v/v) fetal calf serum and 3% (v/v) rabbit serum. Prepare from stock 10x TBST and undiluted sera just before use.
5. Staining solution: 0.5 mg/mL 4-chloro-1-napthol, 8.3 mM imidazole, 42 mM Tris-HCl, pH 7.9, 125 mM NaCl, 0.042% (v/v) 30% H_2O_2. Prepare a fresh solution of 3 mg/mL 4-chloro-1-napthol in methanol, and mix with TBS containing 10 mM imidazole to give a final concentration of 0.5 mg/mL. Immediately before use, add 5 μL of 30% H_2O_2 per 12 mL of solution.
6. Antibody solutions (primary antibody, biotinylated anti-IgG as secondary antibody).
   a. Primary antibody. This can either be a polyclonal or monoclonal antiserum raised in any suitable species—usually rabbits for the former and mice for the latter. *Escherichia coli*-specific antibodies should be removed by diluting the serum 1/1 with a sonicate of late log phase *E. coli*, incubating at 37°C for 60 min and centrifuging at 100,000g for 60 min to remove complexes. Store at –20°C in aliquots.
   b. Secondary antibody. Commercial biotinylated anti-IgG against the species in which the primary antiserum was raised. This also should have any *E. coli*-specific antibodies removed as in (a). Store at +4 or –70°C in aliquots.

7. Avidin: 1 mg/mL in TBS. Store in aliquots at –20°C.
8. Biotinylated horseradish peroxidase: 1 mg/mL in TBS. Store in aliquots at –70°C. Stable at –20°C for short periods (less than 4 wk).

3. Method

For details of phage growth, *see* chapter 29 in Vol. 2 of this series. (Also, *see* Chapter 19 in this volume.)

1. Plate the phage onto LB plates containing 50 µg/mL of ampicillin using *E. coli* Y1090 at 5000–10,000 plaques per 90-mm Petri dish and incubate at 42°C for 3.5 h.
2. Meanwhile soak the required number of nitrocellulose filters in 10 m*M* IPTG and leave to air dry in laminar flow cabinet. Place the filters onto the plates after the initial 3.5 h growth, and mark the filter and the dish assymetrically to allow for realignment. Place the plates plus filters at 37°C for 3.5 h. The IPTG in the filters induces mazimum expression of the cloned DNAs (2).
3. All incubations during this detection are carried out at room temperature. Remove the filters, place the plates at 4°C for storage, and rinse in TBST (2.5 mL/filter) for 5 min with gentle agitation. Place the filters in blocking solution to saturate the nitrocellulose with nonreactive protein. They can conveniently be left overnight at 4°C at this stage or agitated for 30 min at room temperature. Use 5 mL of blocking solution per filter.
4. Rinse the filters twice for 5 min each in TBST as above, and incubate for 1 h with appropriate dilution of primary antiserum in TBST (*see* Note 1 in section 4) using 5 mL per filter. Rinse the filters three times in TBST as above.
5. Incubate the filters for 1 h with the appropriate dilution of secondary antibody in TBST using 5 mL per filter as before. Prepare the avidin-biotinylated HRP complex by incubating 29 µL of each of the two components per filter together

for 30 min. Rinse the filters three times with TBST as above and incubate with the complex diluted inTBST to give 5 mL/filter, for 30 min.

6. Before staining, rinse the filters three times in TBS (no detergent). Add the H_2O_2 to the staining solution, and incubate with the filters for 20–30 min. Positive results are seen as blue-purple spots or rings (see Note 4 in section 4). Use 5 mL per filter. Finally, rinse the filters three times in distilled water and leave to air dry.

7. Align the filters with the plates using the marks as a guide. Usually there is enough background to align the two exactly. Remove the agar from the positively staining area into 1 mL of 10 mM Tris-HCl, pH 7.5, and 10 mM MgSO$_4$ using the blunt end of a Pasteur pipet. Add one drop of chloroform to kill the bacteria, and leave the phage to diffuse out of the agar overnight at 4°C. Titer the phage as described in Chapter 29 of Vol. 2 in this series and replate at 100–200 plaques/plate for rescreening.

4. Notes

1. This is a complex procedure, and many of the steps can cause problems. It is important, therefore, to include as many controls as possible. The optimum dilution of the primary antiserum and the other biological constituents should be determined beforehand by detecting antigen spotted directly onto nitrocellulose membrane, using serial dilutions of each component in turn. Include a positive control strip containing an antigen spot and an *E. coli* lysate spot with each set of filters screened.

2. Avidin may react nonspecifically with some proteins. Therefore, after the initial screening, any positive plaques should be checked by including a filter during the rescreening that does not get treated with antiserum. If reactivity with avidin is a problem, streptavidin is a useful alternative (3). Note also that ovalbumin must not be used

as a blocking agent since most preparations are contaminated with avidin and lead to very high background problems.

3. There are many variations on this type of detection system from biotinylated primary antibody reacting with avidin-labeled horseradish peroxidase to the full amplification system described here (3). In cases in which background problems are encountered, the improved signal-to-noise ratio of the simpler system may be preferable to maximum amplification.

4. True positive plaques usually have a ring of reactivity around a less intense central core. This is because, for the first 3.5 h, no IPTG is present to induce protein expression. False positives are often uniformly stained and tend to be associated with discontinuities in the nitrocellulose.

References

1. Guesdon, J.I., Ternynck, T., and Avrameas, S. (1979) The use of avidin-biotin interaction for immunoenzymatic techniques. *J. Histochem. Cytochem.* **27**, 1131–1139.
2. Huynh, T.V., Young, R.A., Davis, R.W. (1985) Constructing and Screening cDNA Libraries in Lambda gt10 and Lambda gt11, in *DNA Cloning, A Practical Approach* (Glover, D.M., ed.) Vol. 1, IRL Press, Oxford and Washington, DC.
3. Buckland, R.M. (1986) Strong signals from streptavidin-biotin. *Nature* **320**, 557–558.

Chapter 21

Primed Synthesis and Direct Sequencing in the Isolation of cDNA Clones Using Short Oligonucleotides

John D. Haley

1. Introduction

With the development of improved techniques for the construction of cDNA libraries representing rare messenger species has come increased demand for screening techniques to isolate specific cDNAs. A variety of techniques has been developed:

1. The use of exact and degenerate oligonucleotides complementary to RNAs coding for established protein sequence.
2. Bacterial expression of cDNA and immunological detection of the protein products.
3. Selection by hybrid-arrest or hybrid-select translation.
4. Differential hybridization using cDNAs obtained under different physiological conditions.

5. Bacterial expression of cDNA and selection by biological assay.
6. Transfer of cDNA libraries in eukaryotic expression vectors into mammalian cell lines and subsequent biological selection.
7. Expression of cloned cDNAs and mRNA in SP6 vectors with subsequent microinjection of RNA and biological selection of decreasing pools.

A modified selection technique using short oligonucleotides complementary to RNAs coding for known protein sequence is described here, originally developed by Noyes et al. (1) and Agarwal et al. (2), and can be used both as a cDNA selection method (Fig. 1) and as a means of obtaining sequence information directly from RNA, particularly when short oligonucleotides (11–15 nucleotides) are used. Thus specific and nonspecific cDNA transcripts can be identified.

This procedure, as described, has been used to isolate cDNA recombinant coding for porcine relaxin (3), rat relaxin (4), and a stylar glycoprotein associated with self incompatibility in *Nicotiana alata* (5). Oligonucleotide synthesis is described in Chapters 13 and 14.

2. Materials

1. dATP, dGTP, dCTP, dTTP: 10 mM stocks are prepared and stored frozen at –20°C (PL Biochemical).
2. Oligo-dT cellulose: type 7 from PL Biochemicals.
3. Buffered phenol: melt at 60°C, add hydroxyquinoline to 0.1%, add 1/5th the volume of 2M Tris, pH 8, mix, and add 1/2 of the original volume of distilled water; mix and decant the upper aqueous phase; add 1/10th the volume of 10 mM Tris-HCl, pH 8. This reduces the acidity of phenol, and the hydroxyquinoline prevents oxidation.
4. Phenol/chloroform: 25/24/1 mixture of buffered phenol, chloroform, and isoamyl alcohol, respectively.

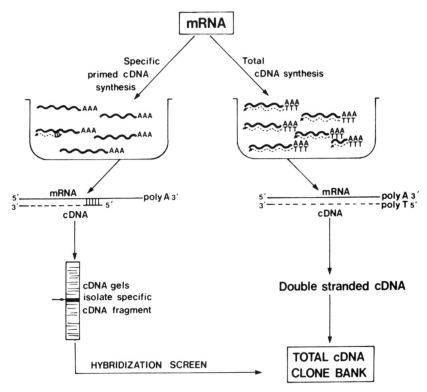

Fig. 1. The use of primed cDNA transcripts in the selection of specific cDNA clones.

5.   3*M* Sodium acetate, pH 5.5.
6.   TE buffer: 10 m*M* Tris-HCl, pH 8, 1 m*M* EDTA.
7.   Reverse transcriptase buffer (10x): 500 m*M* Tris-HCl, pH 8.3, 400 m*M* KCl, 80 m*M* MgCl$_2$, 4 m*M* DTT.
8.   Low TE buffer: 10 m*M* Tris-HCl, pH 8, 0.1 m*M* EDTA.
9.   0.5*M* EDTA: adjust pH to 7.5 with solid NaOH.
10. 10x TBE buffer: 120 g of Tris base, 60 g of boric acid, 6 g of Na$_2$ EDTA.
11. Guanidinium lysis buffer: 5*M* guanidinium thiocyanate, 10 m*M* Tris-HCl, pH 7.4, 1 m*M* EDTA, 1% β-mercaptoethanol, 0.5% sarkosyl, and 0.1% Sigma "antifoam."
12. Extraction buffer: 10 m*M* Tris-HCl, pH 7.4, 10 m*M* EDTA, 0.5% *n*-lauryl sarcosine.

13. High-salt buffer: 0.5M NaCl, 10 mM Tris, pH 7.4, 1 mM EDTA, 0.1% n-lauryl sarcosine.
14. Low-salt buffer: 1 mM Tris-HCl, pH 7.4, 1 mM EDTA.
15. Alkaline denaturing buffer: 100 mM NaOH, 7M deionized urea, 10 mM EDTA, 0.1% bromophenol blue, and xylene cyanol.
16. Urea, analytical grade: 0.45 μm filtered and deionized with mixed-bed ion-exchange resin-AG501-X8(D).
17. Reverse transcriptase (avian myeloblatosis virus): Life Sciences, USA.
18. Human placental ribonuclease inhibitor: Amersham, UK.

3. Methods

3.1. Preparation of Poly (A)⁺ RNA

1. To prepare RNA, frozen tissue (–70°C) is broken into small pieces and quickly homogenized in guanidinium lysis buffer (approximately 1 g/10 mL) using a Bronwell "Ultra-turex" with a medium probe until the tissue is finely dispersed (6). For the isolation of RNA from cells growing attached in culture, media is completely removed, guanidinium lysis buffer added (approximately 2 mL/100 mm dish), and the cell lysate removed with a rubber policeman and sheared by repeated passage through a 19-gage needle.
2. The homogenate is layered over the sterile CsCl step gradient (5.7M; occupying appproximately 20% of the centrifuge tube volume) in polyallomer centrifuge tubes and spun at 150,000g for 18 h, 20°C.
3. The RNA pellet is resuspended in three aliquots of 100 μL of extraction buffer, pooled in a sterile Eppendorf tube, and the combined washes extracted with an equal volume of phenol/chloroform by rapid vortexing.
4. The two phases are restored by centrifugation in an Eppendorf centrifuge for 10 min, the upper aqueous phase is

removed and kept on ice, while the organic phase is re-extracted with an additional 100 µL of extraction buffer, and the aqueous fractions combined. This material is re-extracted with phenol/chloroform as before, washed twice with 1 vol of diethyl ether (in a fume hood!), traces of which are removed under a gentle stream of dry nitrogen.

5. Poly(A)$^+$ RNA is isolated by chromatography on oligo-dT cellulose (7) by virtue of the 3' poly(A)$^+$ extension of 100–200 nucleotides present on most mRNAs (*see* also Vol. 2 in this series). The column is prepared by plugging a sterile Pasteur pipet with a small quantity of glass wool followed by baking (120°C; 2 h), to which a slurry containing ~100–200 mg of oligo-dT cellulose in high-salt buffer is added.

6. The column is rinsed several times with 0.1N NaOH, followed by high-salt buffer until the eluent is of neutral pH.

7. The total RNA sample is heated to 70°C for 3 min to break any inter- or intramolecular hydrogen bonds, quick cooled on ice, and adjusted with 5M NaCl to 0.5M.

8. The RNA is loaded onto the column with the binding of poly(A)$^+$ RNA achieved in high-salt buffer and the eluent is collected, heated to 80°C for 3 min, and reapplied to the column, which is then washed with high salt-buffer until no RNA is detectable as originally monitored with an LKB flow-through spectrophotometer measuring absorbance at 254 nm.

9. Sterile low-salt buffer is applied to the column to elute the bound poly(A)$^+$ RNA, and 400-µL fractions are collected in sterile Eppendorf tubes containing 1 mL of ethanol.

10. Forty µL of sterile 3M Na-acetate, pH 5.5, is added, the tube vortexed, and the RNA precipitated at –70°C for 30 min.

11. The precipitated RNA is pelleted by centrifugation, and the RNA from the first five fractions is resuspended in sterile TE buffer, combined, adjusted to 0.3M sodium acetate, and reprecipitated with 2.5 vol of ethanol.

12. The RNA pellet obtained by centrifugation is resuspended in 10 mM Tris, pH 7.4, 0.1 mM EDTA, and stored at –70°C.

Yields are approximately 50 µg of poly(A)$^+$-enriched RNA/g tissue.

3.2. Purification of Synthetic Oligonucleotides

See also Chapters 13–15.

1. Primers prepared either by phosphotriester (8) or phosphide chemistries are purified by strong ion-exchange HPLC (Whatman Partisil 10-SAX) with 40% formamide and a 0.001M KCl–0.5M KCl salt gradient. The collected peak is dialyzed extensively against Millipore water in a Spectropore 6 membrane (mol wt cutoff, 1000), the optical density of the DNA at 254 nm measured, and the sample lyophilized twice and resuspended at a concentration of 0.5 µg/µL in 10 mM Tris, pH 7.4, 0.1 mM EDTA.

2. Alternatively, primers are purified on 18% acrylamide 7M urea gels, the DNA visualized by UV quenching, and the appropriate band cut out with a razor blade and either passively eluted overnight at 4°C in a Spectropore 1000 mol-wt-cutoff dialysis membrane or transferred to DEAE paper as described (9). Such extensive purification of short oligonucleotides (<20 bases) synthesized using automated phosphide chemistry (*see* ref. 10, e.g., on an Applied Biosystems DNA synthesizer) is usually not required.

3.3. Primed cDNA Synthesis

1. For primed cDNA syntheses, 0.1–1 µg of synthetic primer (in low TE buffer; 1 µg/µL) is added to 1–10 µg of poly(A)$^+$ RNA (in low TE buffer: 1 µg/µL) heated to 90°C for 2 min, and slowly cooled, and the reverse transcription carried out at 42°C in 50 mM Tris, pH 8.3, 40 mM KCl, 8 mM MgCl$_2$, 0.4 mM DTT, 750 µM dATP, dTTP, and dGTP, 20 µM dCTP, 100 µCi alpha-^{32}P-dCTP (Amersham; 3000 Ci/ mmol; 14) with 10 U of human placental ribonuclease

inhibitor (Amersham; *11*) and 10 U of reverse transcriptase (per microgram mRNA) in a typical reaction volume of 50–100 μL.

2. In analytical syntheses the ratio of dCTP to alpha-^{32}P-dCTP is often increased empirically to promote full-length cDNA synthesis, whereas in preparative synthesis, say for probe isolation, the ratio of unlabeled to labeled nucleotide is decreased to maximize the specific activity of the specific cDNA, with the procedure given specifying a typical reaction.

3. These reactions are terminated by addition of 0.5M EDTA to 50 mM, phenol/chlorofom extracted, centrifuged at 12,000g for 10 min, the aqueous phase removed, and the organic phase extracted with an equal volume of TE buffer, the phases separated by centrifugation as before, and the aqueous fractions pooled, washed twice with ether, made 0.3M in sodium acetate, pH 5.5, and precipitated by the addition of 2.5 vol of ethanol at –70°C for 20 min.

4. The RNA/DNA hybrid is pelleted by centrifugation at 12,000g for 10 min, and the pellet is rinsed well with 99% ethanol, dried *in vacuo*, and resuspended in alkaline denaturing buffer.

5. Samples are heated to 70°C for 5 min and quick cooled, and the specific cDNAs analyzed according to molecular weight on 6% acrylamide 7M urea gels (denaturing gels, *see* below) with the appropriate cDNAs cut out and electroeluted.

3.4. Denaturing Gels

1. The analysis and isolation of cDNA fragments by polyacrylamide gel electrophoresis is achieved using 20 x 20 cm glass plates with spacer widths ranging from 1.5 mm for preparative gels to 0.8 mm for analytical gels.

2. The gel plates are sealed with Scotch 3 MM light electrical tape. Gel compositions range from 6 to 10% acrylamide

with 1/20 parts *bis*-acrylamide, depending on the size of the cDNA fragments being run. Acrylamide-*bis* 7M urea solutions are deionized with Dowex mixed-bed ion-exchange resin and diluted to the appropriate percentage with deionized 7M urea. Gels are prepared in 1x TBE buffer, 0.5% ammonium persulfate (fresh), using 100 μL TEMED to initiate polymerization and are run in 1x TBE buffer.

3. Denaturing gels are pre-electrophoresed at 200–400 V and are run hot to the touch (~50°C) to promote complete denaturation of the sample. Alkaline loading buffer is used to hydrolyze RNA and denature single-stranded cDNA in analysis of primed cDNA synthesis reactions. Finished gels are wrapped in plastic wrap and autoradiographed either for a short period at room temperature or frozen at –70°C for longer exposures.

3.5. Direct Sequencing of cDNA Transcripts

Often oligonucleotide primers, even those of a length greater than 14 bases, synthesize a wide variety of cDNA transcripts. Some of these species represent partial transcripts resulting from premature termination of reverse transcription and appear to be sequence related (12), with others reflecting binding of the synthetic primer to other RNA species, either with complete base pairing or with a slight degree of mismatch. The direct sequencing of a radiochemically 5' end-labeled cDNA (1,2) provides a quick assay for the identity of individual cDNA transcripts separated on acrylamide 7M urea gels (Fig. 2). Two methods are shown here, the second of which I have found more reliable, although the first (based on ref. 1) has features that allow for integration with other experiments (for example, the preparation of oligonucleotide probes) and is included for that reason.

3.5.1. Method I

1. Oligonucleotide primer (100–500 ng) (containing a 5'-OH group by virtue of the synthetic method) is added to 200

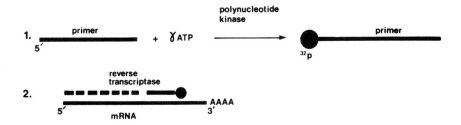

3. NaOH digest of mRNA
4. Run labelled cDNA on 8%PAGE
5. Cut out predominant band, electroelute DNA
6. Sequence by Maxam & Gilbert

Fig. 2. Procedure for the direct sequencing of specific cDNAs using synthetic oligonucleotide primers.

μCi of alpha-^{32}P-ATP (Amersham; 5000 Ci/mmol) in 50 mM Tris, pH 7.4, 10 mM MgCl$_2$, 5 mM DTT, 0.1 mM spermidine, and 10 U of T4 polynucleotide kinase, at 37°C for 90 min.

2. The reaction is made 10 mM with 0.5M EDTA, heat inactivated at 70°C for 5 min, and extracted with phenol/chloroform, the organic phase washed with TE buffer, and the combined aqueous material diethyl ether washed twice and compared to the organic phase with a hand Geiger-Muller monitor. Nearly all the counts should be in the aqueous phase and not still associated with the denatured protein at the phase interface.

3. The unincorporated free label and buffer is removed by chromatography on a Sephadex G-25 fine (8 x 0.6 cm) column using a Pasteur pipet plugged with glass wool, with 1 mM ammonium acetate as the eluent. Three hundred-microliter fractions are collected in sterile Eppendorf tubes, and generally fractions 4–6 are pooled and lyophilized repeatedly.

4. The 5' end-labeled primer is added to ~10 μg of poly(A)$^+$ RNA (in low TE buffer; 1 μg/μL), and the reverse transcriptase reaction is performed as described, except all

four deoxynucleotide triphosphates are present at a concentration of 750 μM and no alpha-^{32}P-dNTP is added (*see* Note 1).

5. The reaction is performed at 42°C for 90 min to ensure complete synthesis.

6. The reaction is terminated with 0.5M EDTA to 20 mM and extracted as before, with ethanol precipitation and resuspension in alkaline denaturing buffer and electrophoresis on 6 or 10% polyacrylamide 7M urea gels. The upper dye is electrophoresed half way down the gel, and the gel wrapped in plastic wrap and autoradiographed for 5–60 min, depending on the degree of incorporation of the labeled primer into the cDNA products.

7. The gel is aligned to the autoradiograph, and the desired cDNA bands are removed with a razor and electroeluted in dialysis membrane (mol wt cutoff, 12,000; preboiled in 10 mM EDTA for 5 min).

8. The cDNA is ethanol precipitated with 80 μg of tRNA carrier and sequenced by the chemical degradation method (*13* and Vol. 2 in this series; *see* Note 2).

3.5.2. Method II

1. Alternatively, the synthetic primer is end-labeled with gamma-^{32}P-ATP in reverse transcriptase buffer supplemented with 0.1 mM spermidine, as described, and the phenol/chloroform, ether extraction steps substituted with thermal inactivation of the reaction at 70°C for 10 min. The Sephadex G-25 chromatography is omitted.

2. Specific reverse transcription of poly(A)$^+$ RNA is initiated by the addition of ~10 μg of RNA (in low TE buffer; 1 μg/μL), the primer/RNA mixture heated to 90°C for 2 min, the reaction supplemented with additional reverse transcriptase buffer, the four deoxynucleotide triphosphates to 750 μM, 15 U of human placental RNase inhibitor, and 54 U of reverse transcriptase at 42°C for 90 min in a 60-μL reaction volume.

3. The reaction is terminated, analyzed, and sequenced as previously described in section 3.5.1. These alterations to the procedure improve the speed and ease of operations and increase the repeatability both in the level of primer incorporation and in the final cDNA banding pattern observed.

4. Notes

1. This produces a cDNA that is end-labeled and suitable for Maxam and Gilbert sequencing. Addition of alpha-^{32}P-dNTP to the reverse transcriptase reaction produces a continuously labeled cDNA that is unsuitable for sequencing by this method.
2. The end-labeled cDNA is suitable *only* for Maxam-Gilbert sequencing. I have had little success with addition of dideoxynucleotides to the reverse transcriptable reaction, producing a staggered ladder, though such approaches have been reported.

Acknowledgments

I would like to thank Dr. Peter Hudson and Dr. Hugh Niall for help in the development of this technique; Dr. Dennis Scanlon and Dr. Geoff Tregear for oligonucleotides; and Dr. Michael Waterfield.

References

1. Noyes, B.E., Mevarech, M., Stein, R., and Agarwal, K.L. (1979) Detection and partial sequence analysis of gastrin mRNA by using an oligonucleotide probe. *Proc. Natl. Acad. Sci. USA* **47**, 1588–1602.
2. Agarwal, K.L., Brunstedt, J., and Noyes, B.E. (1981) A general method for detection and characterization of an mRNA using an oligonucleotide probe. *J. Biol. Chem.* **256**, 1023–1028.

3. Haley, J., Hudson, H., Scanlon, D., John, M., Cronk, M., Shine, J., Tregear, G., and Niall, H. (1982) Porcine relaxin: Molecular cloning and cDNA structure. *DNA* **1**, 155–162.

4. Hudson, P., Haley, J., Cronk, M., Shine, J., and Niall, H. (1981) Molecular cloning and characterization of cDNA sequences coding for rat relaxin. *Nature* **291**, 127–131.

5. Anderson, M.A., Cornish, E.C., Mau, S.-L., Williams, E.G., Hoggart, R., Atkinson,A., Bonig, I., Grego, B., Simpson, R., Roche, P.J., Haley, J.D., Penschow, J.D., Niall, H.D., Tregear, G.W., Coghlan, J.P., Crawford, R.J., and Clarke, A.E. (1986) Cloning of cDNA for a stylar glycoprotein associated with expression of self-incompatibility in *Nicotiana alata*. *Nature* **321**, 38–44.

6. Chirgwin, J.M., Prybyla, A.E., MacDonald, R.J., and Rutter, W.J. (1979) Isolation of biologically active ribonucleic acid from sources enriched in ribonuclease. *Biochemistry* **18**, 5294–5299.

7. Aviv, H. and Leder, P. (1972) Purification of biologically active globin messenger RNA by chromatography on oligothymidylic acid-cellulose. *Proc. Natl. Acad. Sci. USA* **69**, 1408–1412.

8. Itakura, K., Bahl, C.P., Katagiri, N., Michniewicz, J.J., Wrightman, R.H., and Narang, S.A. (1973) A modified triester method for the synthesis of deoxypolynucleotides. *Can. J. Chem.* **51**, 3649–3651.

9. Dretzen, G., Bellard, M., Sassone-Corsi, P., and Chambon, P. (1981) A reliable method for the recovery of DNA fragments from agarose and acrylamide gels. *Anal. Biochem.* **112**, 295–298.

10. Beaucage, S.L. and Caruthers, M.H. (1981) Deoxynucleoside phophoramidites—a new class of key intermediates for deoxypolynucleotide synthesis. *Tetrahedron Lett.* **222**, 1859–1862.

11. Martynoff, G., Pays, E., and Vassart, G. (1980) Synthesis of a full length DNA complementary to thyrobulin 33S messenger RNA. *Biochem. Biophys. Res. Commun.* **93**, 645–653.

12. Efstradiatis, A., Kafatos, F.C., Maxam, A.M., and Maniatis, T. (1976) Enzymatic invitro synthesis of globin genes. *Cell* **7**, 279–288.

13. Maxam, A.M. and Gilbert, W. (1980) Sequencing end-labeling DNA with base-specific cleavages. *Meth. Enzymol.* **65**, 499–560.

14. Huynh, T.V., Young, R.A., and Davis, R.W. (1985) Constructing and Screening cDNA Libraries in λgt10 and λgt11, in *DNA Cloning Techniques, A Practical Approach* Vol. 1 (Glover, D., ed.) IRL Press, Oxford.

Chapter 22

Constructing Expression cDNA Libraries Using Unphosphorylated Adaptors

Keith K. Stanley, Joachim Herz, and Harald Haymerle

1. Introduction

Making libraries of DNA fragments in various cloning vehicles is a basic experimental procedure in molecular biology, yet the methods involved often yield disappointing results, especially for the beginner. When screening for rare DNA molecules, large libraries must be constructed, and the efficiency of each reaction becomes critical. For cDNA libraries, three steps are necessary: cDNA must first be synthesized from poly(A)$^+$ RNA, the double-stranded cDNA must then be inserted into a suitable cloning vehicle, and, finally, cells must be transformed with the chimeric vector/cDNA molecules. Recent improvement in cDNA synthesis protocols (1) and in the quality of commercially available reverse transcriptase enables double-stranded cDNA to be made with an overall yield of about 50% from poly(A)$^+$ RNA. At the other end of the procedure, *Escherichia coli* strains can be made competent for transformation with plasmid DNA with efficiencies of more than 10^8 transformants/µg of supercoiled DNA (2). Frequently, however, the process of cloning DNA fragments into a plasmid vector reduces this potential to disappointing levels. If fragments are cloned directly into the vector by blunt-end ligation or by sticky-end ligation after attaching linkers, the efficiency is reduced by the need to treat the vector with alkaline phosphatase to prevent recircularization without insert. If, on the other hand, the fragments are used in a large molar excess to prevent recircularization of the vector, they must be dephosphorylated to prevent multiple inserts in one plasmid. Even if the alkaline phosphatase treatment is carefully titrated to obtain maximum dephosphorylation with minimum inhibition of ligation or transformation, large libraries are very difficult to construct. In addition, when using kinased linkers, large amounts of restriction endonucleases are required to trim away excess concatenated linker, and internal restriction endonuclease sites must be methylated to prevent cleavage. After many enzymatic reactions on small quantities of cDNA, the losses can be great.

The alternative approach has been to synthesize complementary homopolymer tails on the vector and fragment (3). Since the same base is added on both ends of each fragment, no recircularization or multiple inserts can occur (except for multimers of vector and cDNA). Annealing of (dC)-tailed vector and (dG)-tailed cDNA gives rise to chimeric molecules that transform with high efficiency. Unfortunately, controlling the length of the tail and obtaining DNA that does not tail at internal nicks is a relatively difficult procedure, and many people experience problems with making large libraries by this method. Having synthesized a small amount of cDNA from a precious source of mRNA (e.g., a human tissue), one is reluctant to use a cloning procedure that is not guaranteed to work!

One way around this problem has been to use bacteriophage λ as a cloning vehicle since the superior performance of packaging extracts and infection over transformation over-

1.Adaptor Ligation

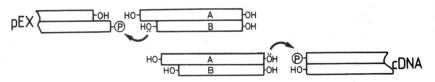

2.Removal of unligated adaptor

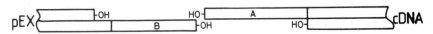

3.Kinasing of 5'-ends

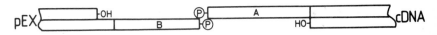

4.Annealing and ligation

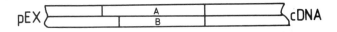

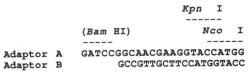

```
                              Kpn  I
                              ------
            (Bam HI)          Nco  I
            -----             ------
Adaptor A   GATCCGGCAACGAAGGTACCATGG
Adaptor B       GCCGTTGCTTCCATGGTACC
```

Fig. 1. Adaptor cloning scheme.

comes the inefficiency of cloning to some extent. λ Clones are cumbersome to handle compared to plasmids, however, because of the high background of vector DNA and increased problems of purification. The following method (4) describes a solution that embraces the reliability of linker ligation with the efficiency of homopolymer tail annealing (Fig. 1). By using unphosphorylated oligonucleotide adaptors with nonidentical ends, the problems of concatamer formation are avoided, and after ligation only one oligonucleotide of each adaptor pair is covalently bonded (to the 5'-phosphates on the vector and cDNA). The second strand of the adaptor is held only by Watson-Crick base pairing and may be removed by precipitation in isopropanol, leaving a perfectly complementary single-stranded extension on every molecule. In this respect the vector and fragments now resemble tailed DNA, except that every fragment has an identical length. This ensures a reproducible annealing reaction and enables the annealed molecules to be ligated together to form closed circular DNA that transforms with high efficiency. In addition, the sequence of the adaptor oligonucleotide may be chosen to give other important features. The adaptors shown in Fig. 1 consist of a 20-base pair overlapping sequence with a blunt end for ligation to cDNA and a BamH1 sticky end for ligation to the vector. In the reading frame of the bacterial expression vector pEX1 (5), a flexible and polar peptide is formed that permits the independent folding of the foreign antigenic determinant and the β-galactosidase encoded by the vector. Care has been taken to use codons for abundant E. coli tRNAs in order to prevent premature termination at this point, and restriction enzyme cleavage sites have been engineered into the sequence to allow for subcloning of the fragments (4).

2. Materials

1. 10x Oligo ligase buffer: 500 mM Tris-HCl, pH 7.6, 500 mM NaCl, 100 mM MgCl$_2$, 50 mM DTT, 5 mM ATP, 10 mM spermidine.
2. 10x Ligase buffer: 500 mM Tris-HCl, pH 7.6, 500 mM NaCl, 100 mM MgCl$_2$, 50 mM DTT, 5 mM ATP.
3. 10x Kinase buffer: 500 mM Tris-HCl, pH 7.6, 100 mM MgCl$_2$, 50 mM DTT, 1 mM spermidine, 1 mM EDTA.
4. TFB (1 L): 7.4 g of KCl (ultrapure), 8.9 g of MnCl • 4H$_2$O, 1.5 g of CaCl$_2$ • 2H$_2$O, 0.8 g of HACoCl$_3$, 20 mL of 0.5M K$^+$MES, pH 6.3. Sterilize 0.5M K$^+$MES by filtration and store in aliquots at –20°C. Dilute to 10 mM using purest

available water, dissolve salts, then sterilize by filtration. Final pH should be 6.20. Stable at 4°C for >1 year. $HACoCl_3$ = hexamine cobalt (III) trichloride.

5. SOB medium: 2% bacto tryptone, 0.5% bacto-yeast extract, 10 mM NaCl, 2.5 mM KCl, 10 mM MgCl$_2$, 10 mM MgSO$_4$. Autoclave tryptone, yeast extract, NaCl, and KCl dissolved in purest water available. Use within 2–3 wk. Make a stock of 1M of each Mg salt, sterile filter, and add just before use.

6. SOC medium: SOC medium is SOB + 20 mM glucose.

7. DMSO/DTT (10 mL): 9 mL of DMSO, 1.53 g of DTT, 100 µL of 1M K$^+$ acetate, pH 7.5. Store frozen in small aliquots.

8. Oligonucleotides: If synthesized by hand, it is important to note that the last base on the 5'-end of the adaptors is essential for ligation in the second ligation reaction. Care should therefore be taken to purify full-length product.

9. S-1000 is obtained from Pharmacia P-L Biochemicals and autoclaved before use.

10. Buffered phenol: 250 g of recrystallized phenol, 37.5 mL of water, 50 mL of 2M Tris base, 0.3 g of 8-hydroxyquinoline.

11. 2x Freezing medium: 40% glycerol in SOB. Sterile filter and store at –20°C.

12. TE buffer: 10 mM Tris-HCl buffer, pH 7.9, 1 mM EDTA.

13. Vector and strains: pEX vectors, oligonucleotide adaptors, and host strains may be obtained from Genofit, Case Postale 239, CH-1212 Grand-Lancy 1, Geneva, Switzerland; Amersham International plc, Lincoln Place Green End, Aylesbury, Buckinghamshire, England, HP20 2TP; or Boehringer Mannheim GmbH, PO Box 310120, D-6800 Mannheim, FRG.

3. Method

This protocol is designed for the cloning of 1 µg of cDNA into the bacterial expression vector pEX1. Three micrograms of pEX1 are used to clone this amount of cDNA giving an approximately equimolar ratio of vector to cDNA after allowing for a

50% recovery of the cDNA on the size fractionation column. This should be sufficient to give a library of 10^6–10^7 clones using the pop 2136 strain, which can give transformation efficiencies of over $10^8/\mu g$ of supercoiled pEX using the Hanahan procedure (2). The protocol described combines vector and cDNA at an early stage to improve the recovery of DNA during precipitations. If libraries containing very low backgrounds of clones without insert are required, the vector can instead be cut in bulk and linear molecules purified on sucrose gradients.

1. Cut 3 µg of pEX1 with BamH1, using a 1.5-fold excess of enzyme over DNA and check for complete linearization on 1.0% agarose gel, while keeping the reaction mix on ice.

2. Stop the reaction by adding 2 µL of 0.5M EDTA, pH 8.0, and 40 µL of buffered phenol. Add 1 µg of blunt-ended cDNA to this mixture and bring to 100 µL using TE buffer. Vortex thoroughly and centrifuge. Remove the aqueous phase into a fresh microfuge tube and back extract the phenol phase with 20 µL of TE. Chloroform extract, and precipitate using 0.1 vol of 5M sodium perchlorate and 0.5 vol of isopropanol. Centrifuge for 10 min, discard the supernatant, and dry the pellet for 1 min at 37°C in a heated dry block with the tube open.

3. Redissolve the pellet in 20 µL of TE buffer and anneal with 250 pmol of each adaptor in a volume of 30 µL by heating to 65°C in a water bath for 5 min and then cooling to room temperature over approximately 15 min. Add 3.5 µL of 10x oligo ligase buffer and 1 µL of T4 DNA ligase (3 U). Incubate overnight at 15°C. The adaptors are in about 100-fold molar excess and prevent recircularization of the vector or cDNA self-ligation.

4. Extract with phenol and chloroform, then precipitate with isopropanol as described above. This removes the bulk of the surplus and unligated oligonucleotides. Adaptor A should now be ligated to the 5'-end of the cDNA, and adaptor B, to the 5'-end of the vector.

If no size fractionation is required, a second isopropanol precipitation may be performed at this point to remove any traces of oligonucleotides, and step 6 skipped.

5. Now the exposed ends of the oligonucleotides are kinased so that they can be ligated later. Dissolve DNA in 20 μL of TE buffer and add: 10x kinase buffer (2.5 μL), 6 mM rATP (1 μL), and 10 U of polynucleotide kinase (1 μL). Incubate 30 min at 37°C.

6. Prepare an 180 x 4 mm column of degassed, autoclaved S-1000 in a 2-mL disposable plastic pipet plugged with siliconized glass wool. Wash extensively with degassed TE while maintaining the temperature at 65°C using a water jacket (this is simply constructed from a piece of polystyrene tubing with the ends sealed using the rubber pistons from disposable syringes). A 10-mL plastic pipet may be connected to the top of the column using silicon rubber tubing and used as a reservoir.

   Load the sample (40 μL) onto the gel bed using a 100-μL glass capillary and allow to soak in. Overlay with 100 μL of TE using the same capillary and allow to soak in, then fill up column and reservoir with TE buffer using a Pasteur pipet. Collect two-drop fractions from the column and measure the radioactivity by Cherenkov radiation. Pool fractions containing cDNA molecules of >1000 base pairs. This can be determined by running 10% of each fraction on an agarose gel, drying down the gel onto DE81 paper, and autoradiographing overnight. As a rough guide, cDNA of >1000 bp elutes in the first 30% of the radioactivity eluted from the column (assuming that the vector is unlabeled).

7. Dilute the pooled fraction to between 1 and 5 μg of vector per mL in TE. Divide into 300 μL portions and add 30 μL of 10x ligase buffer and 1 U T4 DNA ligase. Incubate for 2 h at 37°C.

8. If required, the DNA may be ethanol precipitated at this stage to reduce the volume of the DNA. Competent cells are made and transformed as described by Hanahan (2),

with the modifications shown below for use with pEX DNA in the pop2136 strain because of the temperature-induced expression of fusion proteins above 34°C. Plates should be incubated at 30°C.

9. Scrape the cells off the plates with a sterile spatula, suspend thoroughly in L broth, and dilute with an equal volume of 2x freezing medium. The cell suspension is then snap frozen in liquid nitrogen and stored at –80°C.

3.1. Modified Hanahan Procedure

This protocol is for 50 mL of cells, which gives 4 mL of competent cells, sufficient for 50 ng of DNA.

1. Streak a clump of frozen cells onto SOB agar plate and grow overnight at 37°C.
2. Pick two colonies, 2.5-mm diameter, into 1 mL of SOB in a polypropylene Nunc tube and disperse by vortexing. Grow for 2 h at 37°C. Inoculate a prerinsed 250-mL flask containing 50 mL of prewarmed SOB.
3. Grow at 37°C for about 2.5 h until the cells give an A_{600} value of 0.15 (density = $4 - 7 \times 10^7$ cells/mL).
4. Chill in ice water and transfer into a 50-mL Falcon tube.
5. Centrifuge for 20 min at 1000g at 4°C, then decant supernatant and invert tubes on paper to remove all the liquid.
6. Resuspend the cells in 17 mL of TFB using a precooled plastic pipet, then leave on ice for 10 min.
7. Pellet and drain cells as in step 5. This time the cells should be thinly spread around base of tube.
8. Resuspend in 4 mL of TFB.
9. Add 140 µL of DMSO/DTT solution dropwise while swirling cells. Leave on ice for 10 min.
10. Add a second aliquot of DMSO/DTT as in step 9 to give 7% (v/v). Incubate on ice for 10 min.
11. Add DNA (max, 50 ng in 100 µL) to 1 mL aliquots of the cells in Nunc 17 x 100 mm polypropylene sterile tubes, mix gently, and incubate on ice for 30 min.

12. Heat shock without agitation for 3 min at 37°C in a water bath, then cool on ice for 1–2 min.

13. Add 4 mL of SOC at 20°C, and incubate for 1 h at 30°C (cells should not clump during this procedure), then spread cells onto agar plates containing 30 μg/mL of ampicillin. If desired, cells can be spun down and resuspended in a smaller volume for plating.

4. Notes

If problems arise with the method, they are usually caused by poor ligation of the adaptors onto the vector and cDNA. This is easy to control, and it is recommended that controls 1 and 3 are included in every cloning experiment.

1. Control for good blunt ends on the cDNA: A likely cause of poor ligation is inadequate polishing of the cDNA fragments. To control against this possibility, a small aliquot of cDNA (about 5% of the main reaction) can be ligated in the absence of adaptors in step 3. Run this sample, an equivalent amount of the original cDNA, and 5% of the main reaction mix on a 1.0% agarose gel, dry down onto DE81 paper, and expose overnight. If the cDNA has good blunt ends, it should be completely ligated into high molecular weight DNA. In the main reaction the adaptors should inhibit self ligation of the cDNA. Strong bands of smaller size should, however, be 48 base pairs longer. If the cDNA does not ligate, polish the ends using T4 DNA polymerase, incubating for 5 min at 37°C, and 15 min on ice.

2. Control for good adaptor synthesis: If commercial preparations of adaptors are used, this should not be necessary. If, however, the cDNA has been shown to have good blunt ends, but the adaptors do not inhibit ligation in the above control, then the adaptors or the ligation mix must be at fault. The adaptors are simply checked by cutting some vector DNA with Hae 3 (to test for blunt-end ligation) or

Sau3a (to test the BamH1 sticky-end ligation), labeling the 5'-phosphate by exchange with T4 polynucleotide kinase and ligating with the adaptors under conditions similar to step 3. Run the samples on a thin polyacrylamide sequencing gel and check that the fragments are efficiently converted to ($n + 20$) size.

3. Control for ligation into the vector. Take a sample of the reaction mix in step 7 before and after the ligation, run on a 1.0% agarose gel, dry down on DE81 paper, and autoradiograph. In a successful experiment, most of the radioactive cDNA should move to a size greater than open circular vector after ligation (3). If this condition is achieved, a good library is certain to result provided that good competent cells are available.

4. Although these adaptors may be successfully used with any vector, the advantages for expression cloning are only obtained in the reading frame of pEX1.

5. For expression libraries it is recommended that random primers be used in the first strand cDNA synthesis so that all possible open reading frames are represented in the library (4). The choice of cDNA fragment >1000 base pairs is to exclude artifacts caused by fusions with nonsense reading frames that can be eliminated on the basis of their ORF size.

References

1. Gubler, U. and Hoffman, B. J. (1983) A simple and very efficient method for generating cDNA libraries. *Gene* 25, 263–269.
2. Hanahan, D. (1985) Techniques for Transformation of *E. Coli*, in *DNA Cloning* (Glover, D. M., ed.) Vol. 1, IRL Press, Oxford.
3. Nelson, T. and Brutlay, D. (1979) Addition of homopolymers to the 3'-ends of duplex DNA with terminal transferase. *Meth. Enzymol.* 68, 41–50.
4. Haymerle, H., Herz, J., Bressan, G., Frank, R., and Stanley, K. K. (1986) Efficient construction of cDNA libraries in plasmid expression vectors using an adaptor strategy. *Nucleic Acid Res.* 14, 8615–8624.
5. Stanley, K. K. and Luzio, J. P. (1984) Construction of a new family of high efficiency bacterial expression vectors: Identification of cDNA clones coding for human liver proteins. *EMBO J.* 3, 1429–1434.

Chapter 23

Expression Screening
of cDNA Libraries in pEX

Keith K. Stanley

1. Introduction

Most methods for screening libraries of cDNA clones rely on the selective binding of either nucleic acid probes to cDNA molecules or antibodies to the polypeptide gene product encoded by the cDNA. Although screening of cDNA libraries with synthetic oligonucleotide probes is undoubtedly a powerful procedure, it is not always the easiest way to clone a cDNA, especially if the equipment for amino acid sequencing and oligonucleotide synthesis is not readily available. The oligonucleotide probe synthesized from the back-translated amino acid sequence of the protein must accommodate the degeneracy of the genetic code either by being a mixture of several sequences, or by being sufficiently long to give hybridization at moderate stringency from a single sequence (1). It cannot always be guaranteed that an easy-to-sequence area of a protein (e.g., the amino terminus) will be present in a cDNA library (where the 5'-end is frequently missing), and simply obtaining sufficient pure protein to perform amino acid sequencing can be a problem. On the other hand, polyclonal and monoclonal antibodies are relatively easy reagents to obtain and, in the latter case, have a built-in purity, being themselves derived from a cloning procedure.

Screening procedures based on antibody detection have employed either solid-phase immunoassays (2) or immunoprecipitation of in-vitro translation products after selection of the mRNA on cDNA clones (3). The pEX plasmids (4) employ an immunoassay screening procedure, but differ from earlier systems in the use of plasmids designed for optimal expression (5) and in an efficient detection system called the "colony blot" (6). In this hybrid between a Western blot and an *in situ* colony hybridization (Fig. 1), the inclusion bodies formed during expression of the plasmids are first solubilized in SDS and then bound to cellulose nitrate filters. This screening system allows the full potential of the expression system to be realized.

Optimal expression is achieved by using a strong promoter (the P_R promoter from bacteriophage λ) in combination with the 5'-end of its natural gene (cro). By fusing the promoter and a small piece from this gene coding for nine amino acid residues onto the *Escherichi coli* lac Z gene, high levels of expression were obtained (5), presumably reflecting a favorable folded structure of the hybrid mRNA. Since cDNA is inserted in a cloning linker at the 3'-end of the lac Z (over 3000 bp away) and translation is coupled to transcription in *E. coli*, all fusion proteins are initially made at the same level. This is about 100,000 molecules/cell, representing about 30% of the total *E. coli* protein. A second factor giving high recovery of hybrid protein is their insolubility resulting from the formation of inclusion bodies in the bacteria (6). This helps to protect them against subsequent proteolysis by aqueous phase proteolytic enzymes. Such levels of expression would be lethal to the cells, but by amplifying the plasmid in strains containing the cI857 gene at 30°C, a 300-fold repression of the promoter is obtained and the plasmid is stable against deletions. After incubation of log phase cultures for 1–2 h at 42°C, steady-state levels of fusion protein are obtained.

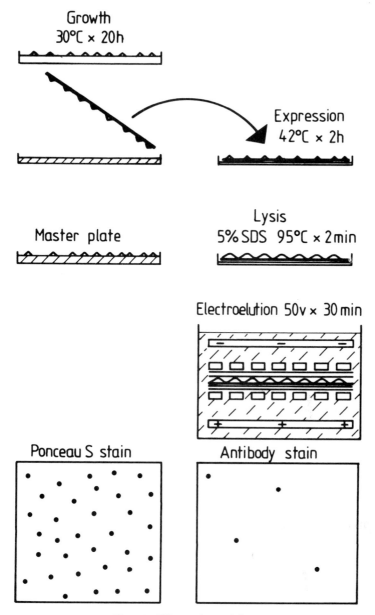

Fig. 1. The colony blot.

The β-galactosidase is not enzymatically active since it has lost essential amino acids at its carboxyterminus, but serves to stabilize the hybrid protein. Thus although pEX vectors with deletions of over 2500 bp in the lac Z still express sufficient fusion protein in colonies for detection with antibodies, it is much more difficult to extract the fusion protein undegraded, and the small size of the protein makes identification difficult in the crowded region of low molecular weight proteins on a polyacrylamide gel.

During the electrophoresis step of the colony blot, it is probable that some renaturation of the fusion protein occurs, since DNA fragments encoding antibody epitopes do not necessarily bind antibody, even though they can bind antibody reacting with an adjacent epitope. Proteins that require disulfide bond formation, however, are unlikely to fold correctly in the absense of protein disulfide isomerase (7), and antibodies that do not react with the original protein on a Western blot usually do not bind to fusion proteins either. Although no N-linked oligosaccharides are added to the fusion proteins, problems have not been reported detecting fusions with glycoproteins.

2. Materials

1.   Filters: Nitrocellulose filters are cheaper and give more signal and a lower background than nylon-based filters in this procedure. BA85 (plain white) filters (0.45 µm pore size) from Schleicher and Schüll work very well. For streaking picked colonies, the BA85/20 filters that are ruled with a grid are convenient.
2.   Plates: Large 24 x 24 cm plates (e. g., biological test plates, Nunc) allow 50,000 colonies to be screened on each plate. The plates may be reused by washing in ethanol and leaving overnight under UV lights. They will also withstand a microwave oven, but slowly buckle with repeated use, and should then be discarded.
3.   Electrophoresis apparatus: 22 x 22 cm filters can be cut in half and processed in standard Western blot apparatus. It is more convenient to build a simple tank with 26 x 26 cm internal dimensions. This can be fitted with stainless-steel

sheet electrodes and perforated perspex shelves on 1-cm perspex legs to support the stack of filters at a short distance from the electrodes. A high-current power pack capable of delivering 50 V at 2 A is required.

4. 10x Electrophoresis buffer: Mix 125 g of Tris base and 600 g of glycine in 5 L of water.

5. Wash buffer: Wash buffer is basically phosphate-buffered saline (PBS) containing 0.1% Triton X-100 or Nonidet P-40 and a protein to quench remaining absorption sites on the cellulose nitrate filters after colony transfer. 0.5% Gelatin (Sigma grade I, catalog no. G2500) quenches adequately for most purposes and is convenient to use. During incubations at 4°C, however, the gelatin concentration should be lowered to 0.1%. 10% Horse or bovine serum is slightly better at quenching the filters, but care should be taken to ensure that there is no cross-reaction between the antibody used and any serum proteins, or between the second antibody and horse/bovine IgG.

In addition the following are sometimes added:

a. Protease inhibitors: These are important when washing filters immediately after the electrophoresis step and during overnight antibody incubations containing *E. coli* extract. 1 m*M* Benzamidine and 40 μg/mL PMSF are suitable.

b. *E. coli* extract (*see* section 3.1) is added to the first antibody incubation to quench cross-reactions with proteins from the *E. coli* colony. This is not normally a problem during the second antibody incubation.

c. DNASe I (Sigma grade II, catalog no. D4527) is added during the filter preparation to digest *E. coli* DNA released by SDS lysis, which would otherwise decrease antibody access to the fusion protein.

6. First antibody: Both monoclonal and polyclonal antibodies have been used successfully in expression screening. Since monoclonal antibodies bind to only one epitope, a higher signal can be obtained by using a mixture of mon-

clonal antibodies. Ascites fluids often have a high background and should be purified for IgG. Polyclonal antibodies should be raised against the purest possible antigen and, if problems arise because of contaminating specificities, should be affinity purified. In both cases antibodies raised against SDS-denatured antigen might be advantageous, but are not essential. It is, however, essential that the antibody works with the antigen in a Western blot. The higher the affinity of the antibody, the better. Good antibodies can detect <1 ng of protein spotted on a filter and processed as for a Western blot. Each colony can synthesize up to 100 ng of fusion protein, so this gives a high signal-to-noise ratio. When screening for rare cDNAs, it is advantageous to remove antibodies binding to *E. coli* proteins on an *E. coli* protein column, as well as by adding *E. coli* extract to the antibody incubation.

7. Second antibodies: Sufficient protein is made to allow detection by horseradish peroxidase (HRP)-coupled anti-immunoglobulin antibodies. These provide a rapid assay that directly stains the cellulose nitrate filter allowing ready identification of positive colonies. Second antibodies can be purchased from the Institute Pasteur, Tago, and Dakko.

8. Developing solution: 50 mg of 3,3-diaminobenzidine are dissolved in 100 mL of 50 mM Tris-HCl, pH 7.4, and 20 μL of 30 vol hydrogen peroxide are added immediately before use. Diaminobenzidine can be made up in 100x solution and frozen in aliquots at –20°C, although it slowly degrades with freeze thawing.

9. 15% Sucrose, 50 mM Tris-HCl, 50 mM EDTA, pH 8.0.

10. 0.1% Triton X-100 in 50 mM Tris-HCl, pH 8.0.

11. Lysozyme: 10 mg/mL in water. Store in aliquots at –20°C.

12. L-Broth (per L, adjusted to pH 7.5 with NaCl):
    a. 10 g of Bactotryptone
    b. 5 g of Bacto-yeast extract
    c. 10 g of NaCl

13. pEX vectors and host strains may be obtained from Genofit, Case Postale 239, Ch-1212 Grand-Lancy 1, Geneva, Switzerland; Amersham International plc, Lincoln Place Green End, Aylesbury, Buckinghamshire, England, HP20 2TP; or Boehringer Mannheim GmbH, PO Box 310120, D-6800 Mannheim, FRG.

3. Method

This protocol is designed for the screening of 500,000 colonies on ten 24 x 24 cm agar plates. This scale of experiment takes two half-days in the initial screening and then 2.5 d for colony purification. In order to save time when screening for several different cDNAs, it is possible to mix antibodies in the first screening and then identity the different sets of clones during colony purification. Obviously it is best to mix antibodies against proteins of similar abundance. Also, if the background on the filters is low after the first screening, they can be reused with another antibody.

1. The condition of agar plates used for spreading the cDNA library is critical to the success of pulling filters. Pour 24 x 24 cm plates with 250 mL of L-broth containing 50 μg/mL of ampicillin and 1.6% agar. Dry inverted with the lids displaced for 15 h at 20°C in still air or for 1 h in a laminar flow hood with the ventilation switched on. Correctly dried plates should have dry lids and a wavy surface on the agar.

2. Thaw an aliquot of cells harboring the cDNA library (*see* Chapter 22) and dilute to approximately 10^8 cells/mL in L-broth. Measure the light scattering at 600 nm in a spectrophotometer and calculate the approximate cell density (0.3 A_{600} = 10^8 cells/mL). Assume that the cell viability is 100% (unless already shown to be otherwise), dilute cells in L-broth to 2.5 x 10^4 cells/mL, and spread 2.0 mL onto each plate (for high-density screening). Even spreading is most

efficiently achieved with the end of a sterile glass pipet raked across the surface at intervals of a few millimeters in orthogonal directions. When all of the cell suspension has been absorbed into the agar, incubate at 30°C for 16–20 h until the colonies are about 1–2 mm in diameter.

3. Displace the lids from the plates and allow the surface to dry for 15–30 min (this is very important for efficient transfer of colonies). Number a dry cellulose nitrate sheet with a water-resistant marker pen, then, holding by diametrically opposite corners, align across diagonal of plate. Lower gently onto the agar surface, making contact first along the diagonal, and then rapidly lowering (but not dropping) the corners. Immediately roller filter onto the plate before the filter is wetted through. Once contact is made, the operation should be completed in <4 s to avoid air bubbles distorting the filter by uneven wetting. If problems are encountered with this technique, the filter can be placed on top of an inverted dish or block of thick polystyrene, and the agar plate then lowered inverted onto the nitrocellulose filter.

4. Stab through the filter and agar at the corners of the plate to facilitate registration of positive clones with colonies. Grip the filter by the top corners and peel off in one movement with colonies attached. Each colony should be entirely transferred to the nitrocellulose sheet, leaving apparently empty depressions on the agar plate. Lay the filter (colonies up) onto a clean surface (e. g., a plate lid), then grip by the two diagonal corners and transfer to a fresh plate. Since this plate will only be used for a short incubation, it is conveniently and cheaply made by adding about 60 mL of L-broth containing 50 µg/mL of ampicillin to two sheets of 3MM paper in a 24 x 24 cm dish. Allow the liquid to soak in for 5 min, then roller out air bubbles with a photographic roller. Discard excess fluid so that paper is moist, but not wet. Establish contact of cellulose nitrate filter along the diagonal, and then lower slowly onto the

paper so as to exclude air bubbles. Incubate for 2 h at 42°C well spread out in an air incubator. It is critical that the temperature does not rise much above 42°C; most incubators benefit from fitting a simple fan on a shelf to prevent local overheating. Replace the master plates into a 30°C incubator for 8 h for colonies to regrow.

5. Prepare another set of polystyrene dishes containing two sheets of 3MM paper impregnated with 5% SDS in the same way as in step 4. Transfer the cellulose nitrate filters to these dishes, taking care to exclude air bubbles as described above. It is particularly critical at this step since colonies above air bubbles will not lyse and will have a higher background in the final stain. Cover the dishes and place in a 95°C incubator for 15–30 min or cook in a microwave oven until almost boiling, e.g., 45 s at 600 W, then 60 s at 200 W. Increase times in proportion to the number of plates in the oven at one time. Wipe the moisture off the lids if there is too much condensation, and stand for a few minutes until colonies are entirely transparent. Do not leave the lid off and allow colonies to dry on the filter. Also avoid boiling; if this happens, the filter may rise in the dish and smear the colonies on the lid.

6. Set up sandwiches consisting of alternate layers of 3MM paper (two sheets) soaked in electrophoresis buffer and the cellulose nitrate filters. Roller each pair of 3MM papers to exclude air bubbles (this is best done slowly), drain until moist but not wet, and add to the stack. Roller gently after each addition. Grip the cellulose nitrate filters across diagonal corners and lower slowly onto 3MM paper so as to exclude air. Do not roller directly on the colonies! Use up to six cellulose nitrate filters in each sandwich, with two 3MM filters at each end. Place stacks of filters onto horizontal perforated shelves in electrophoresis apparatus and cover with buffer (it is best to add the buffer to cover the shelves before adding the sandwich of filters so that air bubbles are not trapped in the perforations). Arrange a

gap of 1–2 cm between shelves so that buffer can circulate and connect with the colony side of the cellulose nitrate filters facing the negative electrode. Electrophoresis should be continued for a minimum of 30 min at 5 V/cm.

7. Remove the cellulose nitrate filters from the electrophoresis apparatus into a box containing wash buffer + 1 mM benzamidine. Agitate on a rocking platform for 5 min, turning the filters over if they show a tendency to adhere to each other. Discard the buffer and add wash buffer containing 1 mM benzamidine and 10 µg/mL DNAse. Agitate for 5 min as above and discard. Wash once again with wash buffer + 1 mM benzamidine, and then remove the filters onto 3MM paper, cover with a second sheet, and roller vigorously. If DNAse digest has been adequate, the paper should peel off with no tendency to adhere. If necessary, filters can be stored at this stage sandwiched between 3MM paper in a plastic wallet at 4°C. Place the cellulose nitrate filters into low gelatin (i.e., 0.1%) wash buffer containing 10% E. coli extract, 1 mM benzamidine, 40 µg/mL PMSF, and antibody at about 0.5 µg/mL (e.g., a polyclonal serum at 1:1000). Preincubate antibody with E. coli protein in wash buffer for a few minutes before adding to filters. Agitate filters with occasional turning on a rocking platform overnight at 4°C.

8. Discard the antibody and wash the filters four times for 5 min in wash buffer. Add horseradish peroxidase-coupled second antibody, and incubate for 60 min at room temperature.

9. Wash a further three times for 5 min in wash buffer, then once in PBS, and finally incubate in developing solution. Positive colonies should appear well ahead of background (in 0.5–20 min) and have a dark rim and characteristic shapes caused by overlapping colonies (Fig. 2). Development of stain may be arrested by thorough washing in H_2O and a brief treatment with 3% trichloroacetic acid. Rinse again in PBS to restore the pH and brown color of the stain.

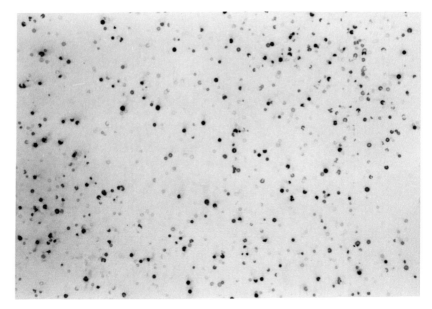

Fig. 2. Immune-positive clones. A library of complement C9 cDNA fragments screened with polyclonal antibody showing the characteristic range of intensity of positive colonies and various shapes caused by overlapping positive and negative colonies.

10. Before the cellulose nitrate filter dries (and shrinks), transfer the positions of positive clones and registration marks accurately onto a plastic sheet with a marker pen. This transparent sheet may then be placed underneath the master plate on a light box, and positive colonies identified. Further help can sometimes be obtained by staining the cellulose nitrate filter for a few seconds in Ponceau S and destaining in deionized water so that the constellations of colonies on the plate may be recognized. At high density the best strategy is to pick the exact colony above the mark with a tooth pick into the first position in a row on a master plate, and then pick a number of colonies surrounding the first into the same row. These mixed colonies should also be streaked onto a nitrocellulose filter, grown at 30°C for 8–16 h, and processed again with the same antibody. Usually the first streak is positive, and this may then be

colony-purified by spreading on a plate, picking 10 individual colonies, and performing a third screen. At high density this method is less work since only 15–20 streaks are made per clone (if a wide area is initially taken, 50 or more streaks may be necessary to find the positive colony).

3.1. E. coli Protein Extract for Quenching Antibodies

1. Dilute an overnight preculture of pop2136 strain 1:200 into 2.5-L flasks containing 800 mL of L-broth, and grow at 37°C overnight.
2. Cool the cells, and then harvest by centrifugation for 20 min at 5000g. Resuspend the pellet thoroughly in 8 mL of 15% sucrose, 50 mM Tris-HCl, 50 mM EDTA, pH 8.0, by drawing up into a glass pipet. Transfer to a Sorvall SS34 tube.
3. Add 2 mL of 10 mg/mL lysozyme, mix thoroughly, and leave on ice for 40 min.
4. Add 24 mL of 0.1% Triton X-100 in 50 mM Tris-HCl, pH 8.0, forcibly by blowing out of a pipet. Leave for 15 min on ice, then centrifuge for 30 min at 20,000 rpm in a Sorvall SS34 rotor.
5. If the lysis has worked well, the pellet should be a homogeneous brown color. Decant the supernatant containing soluble *E. coli* protein and store in aliquots at –20°C. The protein concentration should be approximately 10 mg/mL.

4. Notes

1. If no expression is obtained, this is almost always because of inaccurate maintenance of filters at 30°C for amplification or 42°C for expression. Expression can be checked using an anti-β-galactosidase antibody. Smearing on the filters results if the SDS plates were too moist.
2. Lysing colonies with chloroform or lysozyme/Triton does not solubilize the inclusion bodies and results in a reduced signal-to-noise ratio.

3. Blotting the SDS onto paper rather than electroelution works, but is more variable and often gives a reduced signal-to-noise ratio.

4. Strategy for verifying positive clones: If an oligonucleotide probe is available, it can be used to confirm candidate clones obtained with an antibody screen. Clones can also be checked by hybrid-selected or hybrid-arrested in vitro translation. This is especially powerful if two-dimension SDS polyacrylamide gels are used (8). In cases in which no sequence data are available, the following strategy is recommended:

a. Selection of ORF clones: In a pEX library, only one in six clones is likely to contain an in-frame cDNA, since fusion can occur in any reading frame or orientation. The majority of clones therefore make non-sense hybrids, which are usually expressed at high levels because they contain only short ORFs. For this reason, weak as well as strong positives should always be picked in the initial screening. Genuine in-frame clones may be distinguished on the basis of the size of the expressed protein, few non-sense frames exceeding 600 bp in length (9). This may be achieved by expressing the clones in culture (5) and running the proteins on SDS PAGE, if necessary with Western blotting to identify the correct band. The fusion protein should be at least 140 kdalton in size.

b. Selection of correct epitope: If the clones were selected with a polyclonal antibody, it is then necessary to show that they express the correct epitope rather than that of a contaminating antibody. This is done by affinity purifying the polyclonal antibody on the fusion protein (*see* Chapter 24) and testing for binding to the original antigen in a Western blot.

c. Immunological proof of identity: To finally confirm that a clone is genuine, it is necessary to cut the cDNA into two pieces, and then show that each half binds different monoclonal antibodies or will select antibodies from a polyclonal serum as described above. The chance of a random cDNA molecule coding for two epitopes of a protein is extremely remote.

References

1. Lathe, R. (1985) Synthetic oligonucleotide probes deduced from amino acid sequence data. *J. Mol. Biol.* **183**, 1–12.
2. Anderson, D., Shapiro, L., and Skalka, A. M. (1979) *In situ* immunoassays for translation products. *Meth. Enzymol.* **68**, 428–436.
3. Miller, J. S., Paterson, B. M., Ricciardi, R. P., Cohen, L., and Roberts, B. F. (1983) Methods utilising cell-free protein-synthesising systems for the identification of recombinant DNA molecules. *Meth. Enzymol.* **101**, 650–674.
4. Stanley, K. K. and Luzio, J. P. (1984) Construction of a new family of high efficient bacterial expression vectors: Identification of cDNA clones coding for human liver proteins. *EMBO J.* **3**, 1429–1434.
5. Zabeau, M. and Stanley, K. K. (1982) Enhanced expression of cro-β-galactosidase fusion proteins under the control of the P_R promoter from bacteriophage λ. *EMBO J.* **1**, 1217–1224.
6. Stanley, K. K. (1983) Solubilisation and immune-detection of β-galactosidase hybrid proteins carrying foreign antigenic determinants. *Nucleic Acids Res.* **11**, 4077–4092.
7. Freeman, R. B. (1984) Native disulphide bond formation in protein biosynthesis: Evidence for the role of protein disulphide isomerase. *Trends Biochem. Sci.* **9**, 438–441.
8. Almendral, J. M., Huebsch, D., Blundell, P. A., Macdonald-Bravo, H., and Bravo, R. (1987) Cloning and sequence of the human nuclear protein cyclin. Homology with DNA binding proteins. *Proc. Natl. Acad. Sci.USA* **84**, 1575–1579.
9. Senapathy, P. (1986) Origin of eukaryotic introns: A hypothesis based on codon distribution statistics in genes, and its implications. *Proc. Natl. Acad. Sci. USA* **83**, 2133–2137.

Chapter 24

Producing Antibodies of Predetermined Specificity from *Escherichia coli* Hybrid Proteins

Keith K. Stanley

1. Introduction

A major use of bacterial expression vectors is the expression of fragments of DNA against which monoclonal or polyclonal antibodies may be raised. In some cases this is the only way of generating an antibody, e.g., when investigating an open reading frame of DNA for which the gene product is unknown. By raising an antibody against the in-vitro-expressed DNA fragment, its expression in vivo can be investigated. In other cases, polyclonal or monoclonal antibodies may already be available, but the strategy of raising antibodies against hybrid *Escherichia coli* proteins is chosen because of the advantages that this brings.

The principal advantage of antibodies raised against fusion proteins is that the specificity of the antibody can be manipulated by careful choice of the DNA fragment to be expressed. By presenting only a small region of a protein to an animal, an antibody response against otherwise nonimmunogenic regions of the protein can be elicited (*1*). In this way antibodies may be raised to test the functional importance of various regions of a protein. Care must be taken, however, since antibodies raised against internal parts of the molecule are unlikely to react with the native protein. A better approach is to first map the epitopes of an existing polyclonal antibody against the native protein since the individual antibody species almost always react with surface-located features (2–5). The epitope clones obtained in this process (*see* Chapter 25) may then be used to affinity purify the polyclonal serum or to raise antibodies *de novo*. Affinity purification on fusion proteins is particularly easy since large amounts of the protein are produced, allowing the bacterial colonies to be used directly for the purification. Sufficient antibody can be purified from a few microliters of serum for most analytical experiments.

The second advantage of antibodies raised against fusion proteins is that they lack contaminating activities against related proteins. This is because the antigen has been cloned rather than biochemically purified. Although some bacterial proteins may still contaminate the fusion protein used for immunization, no eukaryotic proteins are present since the bacterial expression vector contains a cloned DNA insert. In the past, monoclonal antibodies have been used to provide this degree of specificity since they too are cloned. On occasions, however, a polyclonal antibody has advantages over a monoclonal antibody because of its heterogeneous composition, giving species reacting under various conditions and with various dissociation constants. Affinity purifying an existing serum is also more rapid than producing monoclonal antibodies (1)!

Antibodies of predetermined specificity were first made using synthetic peptides (1,2,6). By using a fusion protein, however, several advantages are obtained. First, the product is already obtained covalently attached to a carrier molecule. This linkage is also via the α-carbon chain and not via reactive side groups that are normally involved in the immune response. Second, there are less restrictions on the size of the expressed region. Very small pieces (e.g., as little as 11 amino acids) or regions large enough to fold into domains (e.g., 100 amino acids) can be expressed at high levels and successfully used to raise antibodies (7). Finally the relatively insignificant cost of making fusion proteins allows a more thorough random approach to be taken.

The pEX vectors are well suited to antibody production since they produce large quantities of hybrid protein, the hybrid protein may be isolated easily because of its insolubility (8), and vectors with all three translational reading frames are available, allowing restriction fragments of DNA to be expressed. Although it is tempting to subclone convenient restriction fragments of a cDNA clone directly into pEX in order to make antibodies, this strategy may not always yield hybrid proteins that are expressed at high levels since the precise construction made in this way may be particularly susceptible to degradation. A better strategy is to resect the 5'-end of the DNA fragment with Bal 31 before cloning into pEX so that a variety of fusions is made. Suitable fusions for raising antibodies may then be selected on the basis of the level of expression.

The choice of preparation for injecting into an animal depends on the use to which the antibodies will be put. Inclusion bodies prepared simply by precipitation and then washing in a low concentration of guandine hydrochloride (9) are recommended for the first injection because these have never been exposed to SDS and are therefore likely to elicit a response against native structures on the antigen. For similar reasons the boost injections employ fusion proteins purified on SDS gels and then transferred to nitrocellulose (10), which removes much of the SDS. If antibodies that bind to SDS denatured antigens are desired, then electroeluted protein or even gel slices containing the protein may be used.

2. Materials

1. 5x SDS-PAGE sample buffer: 200 mM Tris HCl, pH 8.8, 10% glycerol, 5mM EDTA, 0.1% bromophenol blue, 500 mM DTT.

2. Lysozyme: Dissolve lyophilized powder in water at a concentration of 10 mg/mL and store in aliquots at –20°C. Once thawed, do not refreeze aliquots.
3. L-broth (per L, adjusted to pH 7.5 with NaOH): 10 g of bacto-tryptone, 5 g of bacto-yeast extract, 10 g of NaCl.
4. Buffer A: 25% sucrose and 50 mM Tris-HCl, pH 8.0.
5. Buffer B: 200 mM glycine-HCl, pH 2.5, containing 0.1% gelatin.
6. Detergent mix: 1% Triton X-100, 0.5% deoxycholate, 0.1M NaCl, 10 mM Tris-HCl, pH 7.4.
7. Extraction solution: 1.75M guanidine-HCl, 1.0M NaCl, 1% Triton X-100.

3. Method

1. Dilute 1 mL of a fresh overnight preculture into 100 mL of L broth containing 50 μg/mL of ampicillin and grow at 30°C to a density of 7 x 10^7 cells/mL (A_{600} = 0.2).
2. Transfer the culture to a 42°C water bath, and continue incubation for 2 h.
3. Cool the cells and harvest by centrifugation at 2000g for 15 min.
4. Resuspend the cells in 100 mM NaCl and 50 mM Tris-HCl, pH 8.0, and spin down again as in step 3.
5. Resuspend the cells thoroughly in 0.6 mL of buffer A, and add 150 μL of lysozyme at 10 mg/mL in water, mix, and leave for 15 min on ice.
6. Add 20 μL of 10 mg/mL DNAse 1 and 5 μL of 1M MgCl$_2$, and leave on ice for a further 15 min.
7. Add 250 μL of detergent mix, leave 5 min on ice for cells to lyse, and then spin for 10 min in microfuge.
8. Make a suspension of the inclusion body pellet in extraction solution and spin down for 10 min in a microfuge.
9. Repeat step (8).
10. Wash inclusion bodies in 10 mM Tris-HCl, pH 8.0.
11. Resuspend in 10 mM Tris-HCl, pH 8.0, and check purity and yield on an SDS gel.

12. Dilute to 0.5 mg/mL in PBS for guinea pigs or 1.0 mg/mL for rabbits. Make a 1:1 emulsion with Freunds complete adjuvant and inject 0.5 mL intradermally.
13. For boost injections, solubilize the remainder of the inclusion bodies in 0.5 mL of 20% SDS, 0.25 mL of 0.5M DTT, and 1.75 mL of SDS-PAGE sample buffer. Heat for 5 min at 95°C, then load 20 µL (10–50 µg) per slot of a 7.5% SDS polyacrylamide gel (*see* Volume 1).
14. Run the gel, then transfer the proteins to nitrocellulose by Western blot.
15. Stain the nitrocellulose paper with ponceau S and cut out the fusion protein band carefully.
16. Soak the excised band in liquid nitrogen in a pestle and mortar, then crush until a fine powder. Pour into a precooled 50-mL Falcon tube. Allow any remaining liquid nitrogen to evaporate.
17. Resuspend the powder from 10 slots (containing about 0.2 mg of protein) in 0.5 mL PBS and emulsify 1:1 with incomplete Freunds adjuvant and inject intradermally.

3.1. Affinity Purification of Antibodies on Nitrocellulose

This protocol describes the affinity purification of a polyclonal antibody on a fusion protein using the crude bacterial lysate to absorb the antibody. When using short "epitope" clones the affinity-purified antibody is monospecific and may be used as a monoclonal antibody.

1. Inoculate 5 mL of L-broth with a freshly grown colony of a cloned DNA fragment in pEX using a sterile cocktail stick. A clone should be chosen that is positive in a colony blot (*see* Chapter 23) and contains the desired DNA region.
2. Dilute 1:10 into a second tube of sterile L-broth, then streak 0.3 mL of this dilution directly onto an 82-mm diameter nitrocellulose filter on an ampicillin plate. Grow at 30°C overnight. This should give a filter with about 1000 colonies.

3. Process the filter as for the colony blot (*see* Chapter 23) until the first antibody incubation. It is a good idea to use about 10 times as much antibody as you normally use (e.g., 1 in 100 dilution).
4. Wash the filter four times 10 min in wash buffer (*see* Chapter 23) and then cool on ice in the cold room for 10 min. Arrange each filter face up in its own Petri dish.
5. Elute antibody with ice-cold buffer B. 2.5 mL/filter is sufficient if the elution buffer is kept moving across the surface of the nitrocellulose. After 4 min transfer the eluted antibody with a Pasteur pipet into a tube containing sufficient $2M$ Tris base to restore the pH to between 7.0 and 8.0 (about 100 µL for every 2.5 mL of elution buffer).

4. Notes

1. If using eluted antibody for further immunoblots, dilute 1:1 in wash buffer.
2. If problems are experienced obtaining high-titer antibodies, try pH 2.8 to test if the antibody is particularly pH sensitive or pH 2.2 to test if it is very high affinity.
3. Several epitopes can be absorbed from one dilution of polyclonal antibody at one time provided that the individual antibody species do not react with more than one clone. This is more economical on polyclonal antibody.
4. Occasionally it may be necessary to show that the hybrid protein itself can affinity purify an epitope of a polyclonal antibody, for example when checking the identity of clones. In this case expressed cultures of *E. coli* should be run on standard Western blots and the hybrid bands cut out from the nitrocellulose paper after staining in Ponceau S. The absorption and elution are then carried out as above.
5. The affinity purified antibody is best used immediately, but may be kept at 4°C after sterile filtration for a week or two. It may also be used several times.

References

1. Wilson, I.A., Niman, H.L., Houghton, R.A., Cherenson, A.R., Conolly, M.L., and Lerner, R.A. (1984) The structure of an antigenic determinant in a protein. *Cell* **37**, 767–778.
2. Atazzi, M.Z. (1984) Antigenic structure of proteins. *Eur. J. Biochem.* **145**, 1–20.
3. Benjamin, D.C., Berzofsky, J.A., East, I.J., Gura F.R.N., Hannum, C., Leach, S.J., Margoliash, E., Michael, J.G., Miller, A., Prager, E.M., Reichlin, M., Sercarz, E.E., Smit-Gill, S.J., Todd, P.E., and Wilson, A.C. (19884) The antigenic structure of proteins: a reappraisal. *Ann. Rev. Immunol.* **2**, 67–101.
4. Berzofsky, J.A. (11985) Intrinsic and extrinsic factors in protein antigenic structure. *Science* **229**, 932–940.
5. Fanning, D.W., Smith, J.A., and Rose, G.D.S. (11986) Molecular cartography of globular proteins with application to antigenic sites. *Biopolymers* **25**, 863–883.
6. Anderer, F.A. (1963) Preparation and properties of an artificial antigen immunologically related to tobacco mosaic virus. *Biochem. Biophys. Acta* **71**, 246–248.
7. Tooze, S.A. and Stanley, K.K. (1986) The identification of the epitopes in the carboxyterminal 15 amino acids of the E1 glycoprotein of MHV-A59 using hybrid proteins. *J. Virology* **60**, 928–934.
8. Stanley, K.K. and Luzio, J.P. (1984) Construction of a new family of high efficiency bacterial expression vectors: Identification fo cDNA clones coding for human liver proteins. *EMBO J.* **3**, 1429–1434.
9. Lacal, J.C., Santos, E., Notario V., Barbacid, M., Yamazaki, S., Kung, H-F., Seamans, C., McAndrew, S., and Crowl, R. (1984) Expression of normal and transforming H-ras genes in *Esherichia coli* and purification of their encoded p21 proteins. *Proc. Natl. Acad. Sci. USA* **81**, 5305–5309.
10. Jacob, L., Lety, M-A., Bach, J-F., and Louvard D. (1986) Human systemic lupus erythematosis sera contain antibodies against cell-surface proteins that share epitopes with DNA. *Proc. Natl. Acad. Sci. USA* **83**, 6970–6974.

Chapter 25

Epitope Mapping Using pEX

Keith K. Stanley

1. Introduction

Monoclonal and polyclonal antibodies are important tools in molecular and cell biology, allowing protein molecules to be purified (1), identified in immunoblots of gels (2) and sections of biological material (3), and probed for functionally active regions. Studies on the nature of antibody binding sites show that the molecular interaction between an antibody and its antigen is very specific and localized to a small patch on the surface of a molecule (4). For antibodies that inhibit molecular function, it is clearly of value to be able to map their binding site to a physical location on the polypeptide chain. Even antibodies that do not inhibit binding or catalysis of the protein can give useful information since the epitopes of antibodies are located on the surface of an antigen molecule (5–8). Two classes of antibody binding site have been distinguished. In the first—"continuous" epitopes—synthetic peptides are capable of binding to the antibody (9), suggesting that a single loop of the polypeptide chain is predominantly responsible for the antibody binding. In the second—discontinuous epitopes—linear peptides do not bind to the antibody, but the important amino acids can be mapped by mutational analysis (10). These discontinuous epitopes can sometimes be mimicked by peptides that imitate the spacial orientation of the surface residues (11).

Continuous epitopes have in the major part been mapped by immunoassays using synthetic peptides that cover the whole protein. In order to adequately define the minimum region necessary for binding a large number of small peptides, overlapping along the whole polypeptide chain is required. The approach taken here uses the expression of DNA fragment libraries generated from a cDNA coding for the protein to circumvent this tedious step. Clones isolated from this random library with a monoclonal antibody are then sequenced to identify the region of the cDNA from which they are derived (Fig. 1). Ten to twenty clones of length 100–200 bp are usually sufficient to map continuous epitopes to a stretch of less than 20 amino acid residues, especially if weak positive clones containing DNA fragments coding for the boundary of an epitope are selected. Epitope mapping is therefore easier in pEX (12) vectors than many other expression vectors because of high levels of expression and consequent high signal-to-noise ratio. In order to facilitate sequencing of the epitope clones, a derivative of pEX2 with bases 628–3195 deleted is used. This plasmid (pEX-627) contains the same linker as pEX2, but in the reading frame of pEX1.

Some epitopes that are not perfectly continuous may also be mapped by using libraries containing larger DNA fragments that generate fusion proteins with a more native fold. In this case it is unlikely that the region necessary for antibody binding can be limited to a small stretch of amino acids. This approach for mapping epitopes requiring large areas of flanking sequence would not be possible with a peptide library because of the limitations of peptide synthesis. A further advantage of using expression vectors for epitope mapping is that the peptides expressed from the cDNA are generated as fusion proteins with *E. coli* β-galactosidase, which probably engenders a more native conformation than the peptide alone (13).

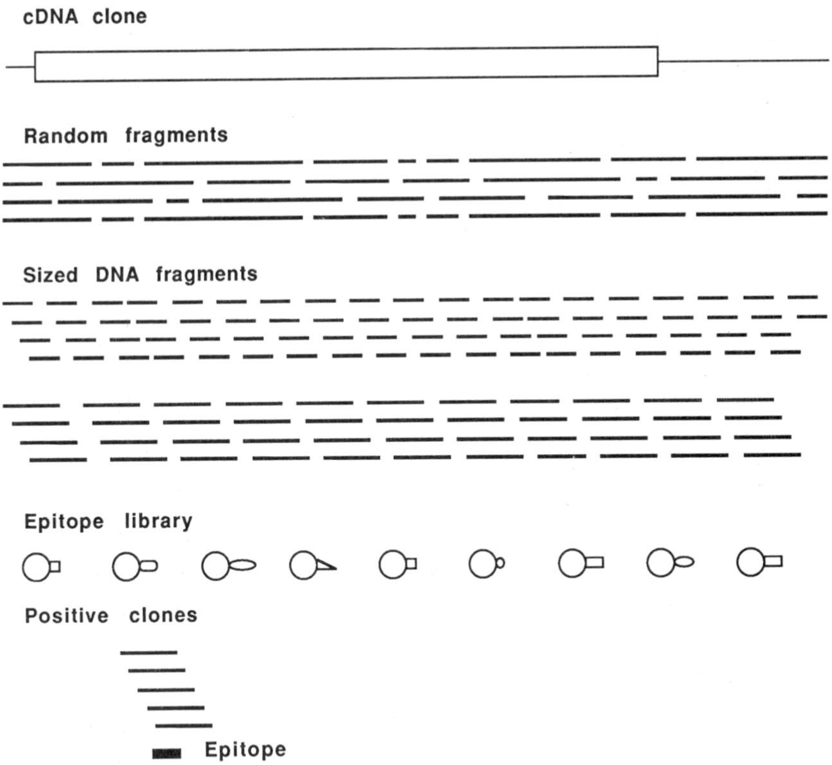

cDNA clone

Random fragments

Sized DNA fragments

Epitope library

Positive clones

Epitope

Fig. 1. Strategy for mapping continuous epitopes.

It is also possible to assign the positive clones obtained with a polyclonal serum to the component antibodies by affinity purifying the serum on individual clones and using these monospecific antibodies to determine the subsets of positive clones with the same epitope. Epitope mapping in pEX therefore consists of a specialized form of adaptor cloning (*see* Chapter 22), followed by screening of an expression library (*see* Chapter 23) and sequence anlysis of the double-stranded plasmid DNA.

2. Materials

1. DNAse I, grade II, can be dissolved in 0.15*M* NaCl and 50% glycerol and stored in aliquots at −20°C.

2. RNAse A: Dissolve in 10 mM Tris-HCl, 15 mM NaCl, pH 7.5, at 1 mg/mL, boil for 10 min, and allow to cool slowly to room temperature. Freeze in aliquots at $-20°C$.

3. Enzymes DNA polynucleotide kinase (10 U/μL), T4 DNA ligase (3 U/μL), *E. coli* DNA polymerase (5 U/μL), and the Klenow fragment of this enzyme (5 U/μL) can all be obtained from Boehringer Mannheim GmbH.

4. (^{32}P-α)-dATP can be used at either high (3000 Ci/mmol) or low (400 Ci/mmol) specific activity.

5. TE Buffer: 10 mM Tris-HCl, 1 mM EDTA, pH 8.0.

6. Buffered phenol: 250 g of recrystallized phenol, 37.5 mL of water, 50 mL of 2M Tris base, 0.3 g of 8-hydroxyquinoline.

7. 10x Nick translation buffer: 500 mM Tris-HCl, pH 7.2, 100 mM MgSO$_4$, 1 mM DTT, 500 μg/mL BSA.

8. 10x Oligo ligase buffer: 500 mM Tris-HCl, pH 7.6, 500 mM NaCl, 100 mM MgCl$_2$, 50 mM DTT, 5 mM ATP, 10 mM spermidine.

9. 10x ligase buffer: 500 mM Tris-HCl, pH 7.6, 500 mM NaCl, 100 mM MgCl$_2$, 50 mM DTT, 5 mM ATP.

10. 10x kinase buffer: 500 mM Tris-HCl, pH 7.6, 100 mM MgCl$_2$, 50 mM DTT, 1 mM spermidine, 1 mM EDTA.

11. Triton lytic mix: 0.5 mL of 20% Triton X-100, 12.5 mL of 0.5M EDTA, pH 8.0, 5.0 mL of 1M Tris-HCl, pH 8.0, 82 mL of water.

12. 20x Gel electrophoresis buffer: 800 mM Tris acetate, 40 mM EDTA, pH 8.0.

13. 10x Agarose gel sample buffer: 0.4% xylene orange in 50% glycerol.

14. 15 mM MnCl$_2$.

15. 200 mM Tris-HCl, pH 7.4.

16. Chloroform:isoamyl alcohol, 24/1 (v/v) (referrred to as "chloroform" here).

17. 3.3M Sodium acetate, pH 6.5.

18. 5M NaClO$_4$.

19. 50 mM Tris-HCl, 100 mM NaCl, pH 8.0.

20. 15% Sucrose, 50 mM Tris-HCl, 50 mM EDTA, pH 8.0.

21. Lysozyme (10 mg/mL) in water.
22. 5*M* Ammonium acetate.
23. Adaptors and primers: The adaptors used in this protocol are described in Chapter 22. Oligonucleotides suitable for use as primers in the double stranded sequencing of pEX627 are as follows:

   5'-primer: 5'GAATTATTTTTGATGGCGTTAACTCGGCG 3'
   3'-primer: 5'CTAGAGCCGGATCGATCCGGTC 3'

3. Method

In this protocol, epitope libraries of various fragment lengths are constructed from 5 µg of purified cDNA. The fragments are cloned into the expression vector pEX1 using oligonucleotide adaptors (see Chapter 22) modified for use with small DNA fragments.

3.1. DNAse Digestion

1. Set up a reaction mix containing: 48 µL of cDNA fragment (5 µg) in TE buffer, 6 µL of 15 m*M* MnCl$_2$, 6 µL of 200 m*M* Tris-HCl, pH 7.4. Prepare a "stop" tube containing 118 µL of TE buffer, 2 µL of 500 m*M* EDTA, and 100 µL of buffered phenol. Then make up a 1:1000 dilution of DNAse at 10 mg/mL and add 1 µL to the reaction mix. Mix the contents of the tube and centrifuge briefly. Take 10-µL samples into the "stop" tube after 0.5, 1, 2, 3, 4, and 5 min.
2. Vortex the "stop" tube after each addition, then finally centrifuge for 2 min in microfuge. Remove the supernatant into a fresh tube, extract with 100 µL of chloroform/ isoamyl alcohol (24/1, v/v), and then precipitate the DNA fragments with 0.1 vol of 3.3*M* sodium acetate, pH 6.5, and 2.5 vol of ethanol.

3.2. Labeling and Polishing

It is essential for adaptor ligation that the DNA fragments have blunt ends. This step gives a short resection of the DNA followed by filling in with ^{32}P-labeled dATP. The reaction is completed to 4°C to give well-polished ends.

1.   Resuspend the pelleted DNA fragments in 20 µL of TE buffer, then add 23 µL of sterile water, 5 µL of 10x nick translation buffer, 1 µL of ^{32}P-α-dATP, 1 µL of *E. coli* DNA polymerase (5 U/µL).
2.   Incubate for 2 min at 37°C, then add 1 mM dNTP mix (5 µL) (100 µM final of each).
3.   Incubate for 30 min at room temperature, then 15 min on ice. Add 125 µL of TE buffer, then phenol extract, chloroform extract, and precipitate in ethanol as described above (step 2). Redissolve polished fragments in 20 µL of TE buffer.

3.3. Adaptor Ligation

Assuming that the recovery is good, 4 µL of the fragments should contain about 1 µg of DNA, which, with an average fragment size of 200 bp, is about 6 pmol. 250 pmol of each adaptor should therefore be a sufficient excess to prevent self ligation.

1.   Mix 4 µL of DNA fragments, 5 µL of adaptor mix (A + B) (250 pmol of each), 8 µL of sterile water. Anneal for 5 min at 65°C, cooling to room temperature over 15 min, then add 2 µL of 10x oligo ligase buffer, and 1 µL of T4 DNA ligase 5 U/µL.
2.   Incubate overnight at 15°C, then add 140 µL of TE buffer, phenol extract, and chloroform extract, and precipitate using 40 µL of 5M NaClO$_4$ and 160 µL of isopropanol. Chill for 10 min on ice and spin for 10 min at 4°C. This should remove the bulk of adaptors, but retain DNA fragments of greater than 100 base pairs.

3.4. Vector Preparation

1.   Cut 5 µg of pEX627 with BamH1, and check on a 1.0% agarose gel for completeness of cut. If satisfactory, phenol

extract, chloroform extract, then preciptiate in sodium acetate and ethanol as described above.

2. Redissolve the pellet in TE buffer and ligate to 40 pmol of adaptors A and B in a total volume of 20 μL as described above, but using 1 U of T4 DNA ligase and incubating for 60 min at 15°C. Phenol extract, chloroform extract, and precipitate in sodium acetate and ethanol as above. Resuspend in 20 μL of TE buffer.

3.5. Kinasing

1. Dissolve fragments in 10 μl of TE buffer and kinase by adding to the DNA 1.5 μL of 10x kinase buffer, 1 μL of 6 mM ATP, 2 μL of sterile water, 1 μL of DNA polynucleotide kinase 10 U/μL. Incubate for 15 min at 37°C, then add 2 μL of Xylene orange sample buffer.

3.6. Gel Fractionation

The aim of this step is to separate the DNA fragments into different size classes.

1. Prepare a 2.0% LMP agarose gel containing 5 μL of 10 mg/mL ethidium bromide in 100 mL of gel. Set the gel in the cold room. Use sterile water and electrophoresis buffer to prevent degradation of adaptor tails during fractionation.

2. Run the whole of the sample in a preparative slot adjacent to molecular weight markers. Examine with 396 nm light, and excise gel slices containing fragments with sizes 50–100, 100–200, and 200–300 bp.

3. Take the gel pieces into weighed microfuge tubes and calculate the volume of gel (assuming density = 1.0). Add TE buffer to make the volume up to 500 μL. Melt the gel slices at 65°C for 5 min, then add 0.1 vol of 10x ligase buffer and 10 μL of the adapted pEX. Cool to 37°C, add 2 μL of T4 DNA ligase (1 U/μL), and incubate for 2 h at 37°C.

4. Phenol extract, chloroform extract, and precipitate with sodium acetate and ethanol as before. Dissolve the pellet

in 200 μL of TE buffer and use for transforming pop2136 cells using the Hanahan procedure with the modifications described in Chapter 22.

3.7. Screening of Epitope Libraries

The libraries of DNA fragments are screened using the colony blot procedure described in Chapter 22 and monoclonal or polyclonal antibodies.

3.8. Sequencing of Epitope Clones

For the double-stranded plasmid sequencing, it is essential to start with a large quantity of clean supercoiled DNA. This is not as easy for pEX as for other plasmids because the copy number of the plasmid is fairly low. It is therefore necessary to start from 40-mL cultures from which 20 μg of DNA may be isolated.

3.9. Preparation of Plasmid DNA

1. Inoculate 40 mL of L-broth containing 100 μg/mL ampicillin in a 200-mL flask with a single colony using a sterile toothpick. Incubate with vigorous shaking overnight at 30°C.
2. Chill the cultures on ice for 10 min, then spin down in SS34 "oak ridge" tubes for 10 min at 6000 rpm.
3. Resuspend the pellet in 1 mL of 50 mM Tris-HCl, 100 mM NaCl, pH 8.0. Transfer to a 2.2-mL microfuge tube and spin for 2 min.
4. Carefully aspirate the supernatant using a drawn-out Pasteur pipet. Resuspend the pellet in 0.4 mL of 15% sucrose, 50 mM Tris-HCl, 50 mM EDTA, pH 8.0, using a Gilson pipet. The yield is dependent on thorough resuspension at this point.
5. Add 100 μL of lysozyme at 10 mg/mL in water, mix thoroughly by vortexing, and leave 30 min on ice. Avoid

vortexing after the incubation has started as this nicks the DNA and prevents efficient separation from chromosomal DNA.

6. Squirt in 1.2 mL of Triton lytic mix, gently invert, then incubate for 15 min on ice. Spin for 30 min at 25,000g, decant the supernatant into a fresh 2.2-mL microfuge tube, and extract with 0.5 mL of buffered phenol. Separate the phases by spinning for 2 min in a microfuge, then remove the supernatant into a fresh tube and extract with 0.5 mL of chloroform.

7. Spin for 2 min, remove the supernatant into a fresh 2.2-mL tube, and precipitate with 0.1 vol of 5M NaClO$_4$ and 0.5 vol of isopropanol.

8. Spin for 5 min, decant the supernatant, spin for another 2 min, and remove the last traces of isopropanol with a drawn-out Pasteur pipet.

9. Resuspend the pellet in 180 µL of TE buffer. Add 2 µL of boiled RNAse (1 mg/mL) and incubate for 15 min at 37°C.

10. Transfer to a 1.5-mL microfuge tube, phenol extract, chloroform extract, and finally precipitate with sodium acetate and ethanol as above. Remove all traces of ethanol from the pellet with a drawn-out pipet, incubate for 1 min with the lids open at 37°C to dry, then resuspend in 50 µL of TE buffer.

3.10. Denaturation of Plasmid DNA

The aim of this procedure is to generate a single-stranded template from the double-stranded plasmid DNA.

1. Mix 10 µL of plasmid DNA (~5 µg), 2 µL of 2N NaOH (fresh), and sterile water to a final volume of 18 µL.

2. Incubate for 5 min at room temperature, then add 8 µL of 5M ammonium acetate, 100 µL of ethanol.

3. Precipitate for 15 min at –70°C, spin in a microfuge for 15 min at 4°C, then wash salt from pellet by mixing with 100 µL of 70% ethanol at –70°C and respinning for 5 min at 4°C.

Remove all traces of ethanol and store at −20°C until ready to sequence.

3.11. Sequencing

Redissolve the denatured DNA in primer annealing mix and proceed with a normal dideoxy sequencing protocol (*14; see* Vol. 2 in this series).

3.12 Sizing of DNA Inserts

It is fastest to sequence one end of the inserts and deduce the position of the other edn by measuring the size of the insert. This also shows that the clone contains only one fragment.

1. Cut 0.5 µg of plasmid DNA with Nco I, then add to the digest mix 1µCi of (^{32}P-γ)-dATP, 1 µL of G+C+T base mix (100 µ*M* final concentration), 0.5 µL of Klenow.
2. Add 6 µL of formamide sample buffer to each tube, heat at 95°C for 3 min, and load on 6% acrylamide sequencing gel with molecular weigh markers.

4. Notes

For controlling the adaptor cloning reaction, follow the notes given in Chapter 22.

References

1. Bailyes, E.M., Richardson, P.J., and Luzio, J.P. (1987) Immunological Methods Applicable to Membranes, in *Biological Membranes: A Practical Approach* (Findlay, J. and Evans, W.H., ed.) IRL, Oxford, Washington.
2. Burnette, W.N. (1981) Western blotting: Electrophoretic transfer of proteins from SDS polyacrylamide gels to unmodified nitrocellulose and radiographic detection with antibody and radioiodinated protein A. *Anal. Biochem.* **112**, 195–203.
3. Griffiths, G., Simons, K., Warren, G., and Tokuyasu, K.T. (1983) Immunoelectron microscopy using thin, frozen sections: Application to studies of the intracellular transport of Semliki Forest virus spike glycoproteins. *Meth. Enzymol.* **96**, 466–485.

4. Amit, A.G., Mariuzza, R.A., Phillips, S.E.V., and Poljak, R.J. (1985) Three-dimensional structure of an antigen-antibody complex at 6 Å resolution. *Nature* **313**, 156–158.
5. Atassi, M.Z. (1984) Antigenic structure of proteins. *Eur. J. Biochem.* **145**, 1–20.
6. Benjamin, D.C., Berzofsky, J.A., East, I.J., Gurd, F.R.N., Hannum, C., Leach, S.J., Margoliash, E., Michael, J.G., Miller, A., Prager, E.M., Reichlin, M., Sercarz, E.E., Smit-Gill, S.J., Todd, P.E., and Wilson, A.C. (1984) The antigenic structure of proteins: A reappraisal. *Ann. Rev. Immunol.* **2**, 67–101.
7. Berzofsky, J.A. (1985) Intrinsic and extrinsic factors in protein antigenic structure. *Science* **229**, 932–940.
8. Fanning, D.W., Smith, J.A., and Rose, G.DS. (1986) Molecular cartography of globular proteins with application to antigenic sites. *Biopolymers* **25**, 863–883.
9. Kazim, A.L. and Atassi, M.Z. (1980) A novel and comprehensive synthetic approach for the elucidation of protein antigenic structures. *Biochem J.* **191**, 261–264.
10. Reichlin, M. (1972) Localising antigenic determinants in human hemoglobin with mutants: Molecular correlation of immune tolerance. *J. Mol. Biol.* **64**, 485–496.
11. Atassi, M.Z. (1978) Precise determination of the entire antigenic structure of lysozyme. Molecular features of protein antigenic structures and potential of "surface-simulation" synthesis. A powerful new concept for protein binding sites. *Immunochemistry* **15**, 909–936.
12. Stanley, K.K. and Luzio, J.P. (1984) Construction of a new family of high efficiency bacterial expression vectors: Identification of cDNA clones coding for human liver proteins. *EMBO J.* **3**, 1429–1434.
13. Shi, P-T., Riehm, J.P., Todd, P.E.E., and Leach, S.J. (1984) The antigenicity of myoglobin-related peptides synthesised on polyacrylamide and polystyrene supports. *Mol. Immunol.* **21**, 489–496.
14. Sanger, F., Nicklens, S., and Coulson, A.R. (1977) DNA sequencing with chain-termination inhibitors. *Proc. Natl. Acad. Sci. USA* **74**, 5463–5467.

Chapter 26

DNA Transfection of Mammalian Cells Using Polybrene

William G. Chaney, Daniel R. Howard,
Jeffrey W. Pollard, Sandra Sallustio,
and Pamela Stanley

1. Introduction

Many techniques have been developed to transfect mammalian cells with DNA (1–2). The most commonly used method is to expose cells to a coprecipitate of DNA and calcium phosphate (3). This technique works very well with both genomic and recombinant DNA sequences for some cell lines, e.g., mouse L-cells, but less well with other cell types such as Chinese hamster ovary (CHO) cells (2–4). Since the ability to rescue mutant phenotypes by exogenous DNA provides a means to identify gene products and to clone their genes (5), the reduced ability to transform CHO cells by the calcium phosphate technique has slowed the use of the large number of CHO cell mutants (6) in transfection and rescue experiments. In this chapter we will describe a simple, reproducible method for DNA transfection (7) that is a modification of a method originally developed for the introduction of viral sequences into chick embryo fibroblasts (8). The method involves exposing the cells to DNA in the presence of the polycation polybrene, followed by a DMSO shock. The polybrene apparently interacts with the charge on the cell and the DNA, allowing the DNA to absorb more easily to the cell membrane. Consequently it gives a high efficiency of gene transfer into CHO cells for both genomic and for recombinant DNA sequences (*see* Fig. 1). In the latter case it is effective over a wide range of plasmid concentrations, even in the absence of carrier DNA. Furthermore it is applicable to a variety of cell lines other than CHO cells for both stable and transient expression of the transfected gene sequences.

Fig. 1 (opposite page). Detection of human DNA sequences in CHO transfectants. Pro⁻5 CHO cells were transfected with a mixture of 100 ng of pSV2-neo, which confers resistance to G418, and 6 μg of human fetal brain DNA by the method described in the text. Colonies resistant to the antibiotic G418 were picked and cultured, and genomic DNA was prepared from them as described previously (7). After digestion with BamH1, 10 μg of each DNA preparation was electrophoresed in 0.7% agarose, transferred to nitrocellulose, prehybridized in the presence of 50 μg/mL CHO DNA, and probed for human DNA sequences using nick-translated genomic DNA from HL60 cells (1.6×10^8 cpm per μg DNA) in the presence of 25 μg/mL CHO DNA. Lanes B–D show the results obtained for a primary transfectant (B) compared with HL60 DNA (C) and CHO DNA (D). These tracks were taken from a single blot that was exposed for different times so that the presence of human DNA sequences in the transfectant could be clearly visualized. Lane (B) was exposed with an intensifying screen for 24 h, whereas lanes (C) and (D) were exposed without an intensifying screen for 3 h. The difference in intensities between the transfectant (B) and HL60 DNA (C) is therefore at least 100-fold. On shorter exposures that gave visible signals for the three human DNA transfectants examined on this gel, no signal was observed in the control lane of CHO DNA (D). Therefore the hybridization shown for lane (B) appears to be specific for human DNA and was typical of each transfectant.

No distinct bands were observed in lane (B), indicating that human DNA sequences representative of the entire genome were integrated in the transfectant. In contrast and as reported previously

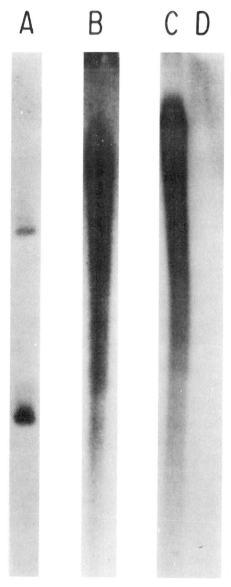

Figure 1 (see facing page for start of caption)

(7), DNA from similar transfectants probed with nick-translated
pBR322 sequences homologous to pSV2-neo gave distinct bands.
An example, in which two bands of molecular weight ~21 and 6.4 kb
were observed, is shown in lane (A). (Molecular weight markers are
not given on the figure because the gels were run for different times.)

2. Materials

1. Growth medium: alpha-minimal essential medium containing 10% (v/v) fetal calf serum, but any appropriate tissue culture medium will suffice.
2. DNA: purified, supercoiled plasmid DNA dissolved in sterile TE buffer (10 mM Tris-HCl, pH 7.6, at 20°C, 1.0 mM EDTA) at approximately 10 µg/mL. Genomic DNA redissolved at 1 mg/mL in TE buffer.
3. Polybrene: 10 mg/mL in glass distilled H_2O and filter sterilized through a 0.2-µm filter.
4. Dimethyl sulfoxide (DMSO): 30% (v/v) in growth medium containing serum. Prepare fresh on the day of use.

Solutions from steps 1 and 3 are stored at 4°C and the solution from step 2 at –20°C.

3. Methods

1. Cells are plated at 5×10^5 cells/100 mm Petri dish in 10 mL of growth medium.
2. 18 h later the medium is removed and replaced with 3 mL of growth medium containing DNA (1–10^3 ng for plasmid DNA, 10–40 µg for genomic DNA) and 30 µg polybrene (3 µL stock solution).
3. The plates are placed in a humidified 37°C incubator in which they are rotated gently every 1.5 h to ensure even exposure of the cells to the DNA/polybrene solution.
4. After 6 h the DNA solution is removed and 5 mL of DMSO containing medium added to each plate at room temperature.
5. *Exactly* 4 min later the DMSO medium is aspirated and the cells quickly washed with 5 mL of growth medium lacking serum, followed by the addition of 10 mL of growth medium containing serum to each plate.
6. After 24–48 h incubation in a humidified incubator, the medium is removed and the appropriate selection ap-

plied. Alternatively the cells can be processed for transient expression of the gene.

4. Notes

1. The major advantages of the polybrene technique are: (1) It involves the simple addition of DNA in a solution of polybrene to cells. This enables the easy handling of large numbers of plasmids in one experiment. (2) The high transformation efficiency allows very small quantities of plasmid DNA to be used. Thus it is relatively easy to introduce only one copy of a sequence per genome. Our experience also suggests that even with quite high plasmid concentrations, only one or a few copies of the plasmid are integrated, and concatemers of integrated plasmids are rarely observed (Fig. 1). Since no carrier DNA is used, the integration of plasmid DNA is directly into the host genome, which should enable analysis of the sequences surrounding the site of integration. (3) Over 95% of the transformants we have selected and 100% of the plasmid transformants contain stable integration of transfected DNA. (4) Different clones of CHO cells give comparable transformation efficiencies.

2. Using any technique of gene transfer, it is necessary to ensure that the transformants are genuine and are not the result of genetic reversion in the recipient cell mutation. The high frequency of transformation with cloned genes in plasmids using the polybrene/DMSO technique essentially ensures that transformants are genuine. For genomic transformation, however, frequencies are often within the range expected for the appearance of revertants. Adequate controls must therefore be built into each experiment to accurately compare transformation with revertant frequencies. It should also be noted that CHO cells are relatively nonadhesive, and during media changes cells will be dislodged and form satellite colonies. Thus clones

picked from the same plate cannot be regarded as independent transformants unless confirmed by showing different sites of integration of transfected gene sequences by Southern blotting. Because of the simplicity of the technique, however, independent clones can be ensured simply by transfecting several plates with low concentrations of plasmid and picking one colony per plate.

3. The transformation frequency appears to be independent of whether the plasmid is supercoiled or linear or if the insert has been removed from the vector. The technique also works equally well for plasmids up to at least 45 kb if transformation efficiencies are calculated on a molar basis. Frequencies also appear linear over the measurable range.

4. The concentration of, and time of exposure to, DMSO are critical. Thus, for CHO cells, treatment with 27.5% (v/v) DMSO results in a 10-fold reduction in transformation efficiency. Similarly exposure of cells treated with polybrene/DNA, as described above, to 10% (v/v) DMSO for 18 h, a treatment that enhances calcium phosphate-mediated chromosome transfer, gives no transformants with the polybrene technique. It should be noted that some cells (e.g., L-cells and HeLa) have lower transformation efficiencies than CHO cells for equivalent plasmids with the polybrene/DMSO technique described here. It is necessary to adjust the DMSO concentration for each cell line to produce optimal transformation efficiencies.

5. The calcium phosphate technique is greatly enhanced by posttreatment of cells with amphotericin (9). With the polybrene technique, however, we have not found that replacement or supplementation of DMSO with amphotericin increases the transformation efficiency. In fact, in the former case few transformants were obtained.

6. Transformation efficiencies are difficult to compare between different reports since different baselines for calculation are used. In our hands, however, for dominantly selectable markers in CHO cells the polybrene/DMSO

technique is superior to others. The efficiencies obtained depend on the promoter used and the selection system employed (7). They are reduced in the presence of excess carrier DNA. This point should be taken into account in any coselection experiment.

7. The polybrene/DMSO method has worked for both stable and transient expression. To date it has been employed effectively for transfecting rat, mouse, chicken, monkey, human, and hamster cell lines and in a recent publication for human diploid fibroblasts (10).

Acknowledgments

To maintain tradition, on this paper the authorship is arranged alphabetically. This technique was developed while the authors' work was supported by NCI grants ROI 30645 and ROI 36434 to P.S., a British MRC grant to J.W.P., and faculty awards from the American Cancer Society and the Irma T. Hirschl Memorial Trust to P.S. Training grant support to D.H. and S.S. was provided by NIH T32 GM7288. Partial support for the laboratory at Albert Einstein was from the NCI Cancer Core grant CA 13330.

References

1. Kucherlapati, R.S. and Skoultchi, A.I. (1984) Introduction of purified genes into mammalian cells. *CRC Crit. Rev. Biochem.* **16**, 349–379.
2. Pollard, J.W., Luqmani, Y., Bateson, A., and Chotai, K. (1984) DNA Transformation of Mammalian Cells, in *Methods in Molecular Biology* Vol. 2 *Nucleic Acids* (Walker, J.M., ed.) Humana, Clifton, New Jersey.
3. Graham, F.L. and Van der Eb, A.J. (1973) New technique for assay of infectivity of human adenovirus DNA. *Virology* **52**, 456–467.
4. Abraham, I., Tyagi, J., and Gottesman, M. (1982) Transfer of genes to Chinese hamster ovary cells by DNA-mediated transformation. *Somat. Cell Genet.* **8**, 23–29.
5. Pellicer, A., Robins, D., Wold, B., Sweet, R., Jackson, J., Lowy, I., Roberts, J.M., Sim, G.K., Silverstein, S., and Axel, R. (1980) Altering genotype and phenotype by DNA-mediated gene transfer. *Science* **209**, 1414–1422.

6.   Gottesman, M. (1985) *Molecular Cell Genetics*, Wiley, New York.
7.   Chaney, W.G., Howard, D.R., Pollard, J.W., Sallustio, S., and Stanley, P. (1986) High frequency transfection of CHO cells using polybrene. *Somat. Cell Mol. Genet.* **12**, 237–244.
8.   Kawai, S. and Nishizawa, M. (1984) New procedure for DNA transfection with polycation and dimethyl sulphoxide. *Mol. Cell. Biol.* **4**, 1172–1174.
9.   Hidaka, K., An, G., Ip, P., Kuwana, M., and Siminovitch, L. (1985) Amphotericin-B enhances the efficiency of DNA mediated gene-transfer in mammalian cells. *Somat Cell Mol. Genet.* **11**, 109–115.
10. Morgan, T.L., Maher, V.M., and McCormick, J.J. (1986) Optimal parameters for the polybrene-induced DNA transfection of diploid human fibroblasts. In vitro cell. *Devel. Biol.* **22**, 317–319.

Chapter 27

Expression of Foreign Genes in Mammalian Cells

Martin J. Page

1. Introduction

Many laboratories are currently interested in cloning cellular genes that code for polypeptides that may have academic or therapeutic use. Once the intact gene is isolated, usually in the form of a cDNA clone, it has to be engineered into an expression cassette that is capable of functioning when introduced into a particular host cell. The choice of using a prokaryotic (e.g., bacterial), lower eukaryotic (e.g., yeast), or higher eukaryotic (e.g., mammalian) expression system is frequently dictated by the nature of the polypeptide to be expressed. It is becoming increasingly evident that for the majority of cloned eukaryotic genes there can be considerable problems in achieving reliable expression of biologically active products in bacterial or yeast systems. Although the product may be expressed in high amounts in these systems, often it is not folded correctly and is therefore biologically inactive or is produced in an insoluble, denatured form within occlusion bodies.

In contrast, the expression of cloned gene products in mammalian cells offers many advantages. First, and most importantly, mammalian expression systems are able to correctly fold complex polypeptides and achieve the necessary tertiary conformational structure for biological activity. This may involve a multiplicity of very precise interactions such as disulfide, ionic, covalent, and van der Waal bonds. Second, some polypeptides require posttranslational modifications (e.g., amidation, phosphorylation, carboxylation, glycosylation) for activity. These modifications can usually only be performed in a suitable mammalian host. Third, if the gene product is naturally secreted, then the cloned gene will contain information for a secretory signal sequence. This sequence is widely recognized by the majority of higher eukaryotic cell types and results in the efficient secretion of the gene product into the culture medium, thereby facilitating isolation and purification.

The rapid development of mammalian expression sytems over the last few years has meant that, at least in principle, it is possible to introduce any cloned gene into any cell type. It is not the purpose here to review all the expression constructs and procedures that are available; instead two methods, involving transient or long-term expression of gene products in mammalian cells, will be described. Both methods have proved to be particularly reliable and reproducible in our laboratory.

2. Materials

Since this chapter is concerned primarily with the design and methodology of expression, no attempt is made to detail the procedures for generating the recombinant expression vectors. It is assumed that the reader is either already familiar with molecular cloning techniques or is able to follow the relevant chapters covered previously in this series and in ref. 1.

1. 5x DEAE-dextran stock solution: dissolve DEAE-dextran (Pharmacia, molecular weight 500,000) at 5 mg/mL in serum-free Dulbecco's medium. Add 1M Hepes, pH 7.3 (use BDH tissue culture grade Hepes), to a final concentration of 50 mM, then filter sterilize through a 0.22-μm filter and keep as a 5x sterile stock at 4°C.

2. Glycerol shock solution: take 70 mL of serum-free Dulbecco's medium, add 25 mL of glycerol and 5 mL of 1M Hepes, pH 7.3, mix well, then filter sterilize through a 0.22-μm filter, and keep as a sterile working solution at 4°C.

3. 100x Chloroquine stock solution: prepare a 10 mM solution of chloroquine (Sigma) in serum-free Dulbecco's medium. Sterilize through a 0.22-μm filter and store at 4°C.

4. COS cells are maintained in a 5% CO_2 incubator using Dulbecco's medium supplemented with 10% fetal calf serum. Take care to keep the cells in log phase growth; do not be concerned about the "ragged" appearance of the cells, since this is their normal morphology.

5. Recombinant DNA constructs: these are usually dissolved in sterile distilled water at a final concentration of 0.5 μg/μL and stored at –20°C.

6. DUK dhfr⁻ (dihydrofolate reductase) CHO cells are maintained in a 5% CO_2 incubator using Dulbecco's medium supplemented with 10% fetal calf serum, 1x nonessential amino acids (Flow Laboratories, cat. no. 16-810-49), and 4 μg/mL each of hypoxanthine and thymidine.

7. 2x HBS solution: 270 mM NaCl, 40 mM Hepes (BDH, tissue culture grade), pH 7.1 ± 0.05. The pH is very critical. Autoclave the solution and store at room temperature. Do not use after 6 wk.

8. Transfection phosphate solutions: prepare 70 mM solutions of Na_2HPO_4 and NaH_2PO_4, filter sterilize through a 0.22-μm filter, and store separately at room temperature.

9. 2M $CaCl_2$ solution: autoclave and store at room temperature.

10. DUK dhfr⁺ selection medium: Dulbecco's medium supplemented with 10% dialyzed fetal calf serum (500 mL serum dialyzed twice overnight at 4°C against 5 L of PBS A) and 1x nonessential amino acids. Do not add nucleosides.

11. Methotrexate (Mtx) stock: Make up a 1-mM stock solution of Mtx (Sigma) by dissolving 5 mg in 10 mL of 0.05M

Na_2CO_3. Filter sterilize through a 0.22-μm filter and store in the dark at −20°C. Dilute in culture medium and add to cells as a 100- or 1000-fold concentrate.

3. Methods

3.1. Transient Expression of Cloned Genes in Monkey COS Cells

In this procedure, a transformed line of monkey CV-1 cells called COS cells (2) are recipients for the expression of the cloned gene. These cells constitutively synthesize SV40 T antigen and are therefore able to support the rapid replication of an incoming plasmid bearing an SV40 origin of replication. This generates a very high plasmid copy number in the cells that have taken up the DNA and results in transient high levels of gene expression.

3.1.1. Vector Design for Transient Expression

Essentially, the expression construct for use with COS cells must have the following features:

1. pBR322 vector sequences containing the bacterial origin of replication and a bacterial selectable marker gene (i.e., the ampicillin-resistance gene), but lacking the so-called "poison sequences" (3). Suitable vectors are pAT153 (4) and pML-2 (3).
2. An SV40 origin of replication.
3. The gene to be expressed in the form of an expression cassette. This involves placing the gene under the control of a strong eukaryotic promoter, such as the SV40 early or late promoters (5), the Rous sarcoma virus LTR promoter (6), or the mouse metallothionein promoter (7). To complete the expression cassette, it is advisable to clone efficient viral or cellular polyadenylation and termination signals, such as those from SV40, polyoma, or rabbit β-globin downstream of the gene to be expressed.

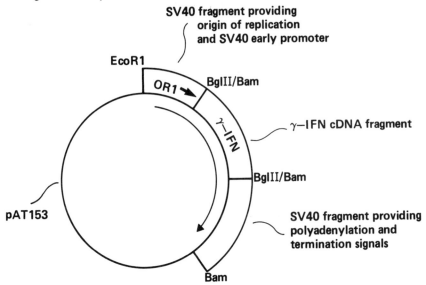

Fig. 1. An example of the essential features for a transient expression construct.

To illustrate the design of a transient expression cassette, Fig. 1 shows a construct used for the expression of the human γ-IFN cDNA in COS cells. This construct consists of four separate DNA components, which are:

1. Bacterial vector sequences from the large EcoRI to BamH1 fragment (3283 bp) of pAT153 to enable the necessary construction of the expression cassette and isolation of large amounts of recombinant DNA for cell transfection.

2. A 342-bp SV40 fragment cloned upstream (i.e., 5') to the γ-IFN cDNA. This fragment was taken from pSV2 dhfr (8) as the small PvuII to Hind III fragment (SV40 coordinates 270-5171) and recloned, using linkers, as an EcoRI to BglII fragment. This fragment is very convenient since it fulfills two of the basic requirements listed above. First, it encompasses the SV40 origin of replication and, second, any gene cloned downstream of the BglII site is placed under control of the strong SV40 early promoter.

3. A human γ-IFN cDNA clone as an 852-bp BamH1 fragment.

4.   An 858-bp SV40 fragment cloned downstream (i.e., 3') to
     the γ-IFN cDNA. This fragment was also taken from pSV2
     dhfr as the small BglII to BamH1 SV40 fragment and used
     in this form to provide efficient viral polyadenylation and
     termination signals.

3.1.2. Transfection Procedure for Transient Expression

1.   From an almost confluent stock 9-cm dish of COS cells,
     harvest and dilute the cells 1/5 into new 9-cm dishes so
     that after a further 24 h they will be about 50% confluent
     and ready for transfection.
2.   Remove the medium and replace with 10 mL of serum-free
     Dulbecco's medium (prewarmed to 37°C); leave for 5 min
     in a 37°C CO_2 incubator. Remove the medium and replace
     with a further 10 mL of serum-free medium and again
     leave for 5 min in the incubator. These steps are important
     to ensure removal of serum, which could interfere with the
     transfection procedure.
3.   During the 5-min intervals above, prepare the transfection
     mixes. Into a sterile Universal tube, place 4 mL of serum-
     free medium, add 1 mL of 5x DEAE-dextran stock solution
     followed by 2 µg of the DNA constuct containing the ex-
     pression cassette. Pipet up and down to complete the
     transfection mix.
4.   Remove the medium from the COS cells and replace with
     the transfection mix, return dishes to the 37°C CO_2 incuba-
     tor, and leave for 30 min.
5.   Remove the medium and add 3 mL of glycerol shock solu-
     tion (prewarmed to 37°C) to the monolayer; this is done by
     pipeting the solution against the side of the dish and
     allowing it to flow over the cell monolayer. Leave in
     contact with the cells for exactly 1.5 min, then rapidly add
     15 mL of complete medium (i.e., containing serum), mix
     well, and remove. Wash the monolayer twice more with
     10 mL of complete medium each time.

6. Add 10 mL of complete medium supplemented with 0.1 mL of the 100x chloroquine stock, leave in the 37°C CO_2 incubator for 4 h, remove the medium, and wash the monolayer twice with complete medium. Finally, add a further 10 mL of fresh complete medium and leave the dishes in the 37°C CO_2 incubator for 2–3 d for expression of the transfected DNA to occur.

7. After this period, examination for expression of the gene product can be performed. If the gene codes for a secreted protein, then the product is likely to be found in the culture medium. Alternatively, if the product is an intracellular protein, the cells can be harvested, washed, and lysed or sonicated to obtain the product. It is also possible to isolate mRNA for Northern blot analysis of gene-specific transcripts or examine whole cells for immunofluorescence if an antibody for the gene product is available.

3.2. Long-Term Expression of Cloned Genes in CHO Cells

This procedure will generate permanent transformed Chinese hamster ovary (CHO) cells, which constitutively (or inducibly if the correct promoter is used) produce the required recombinant gene product. A mutant line of CHO cells deficient in dhfr, called DUK cells (9), is the recipient for transfection, and these have to be maintained in the presence of nucleosides (hypoxanthine and thymidine) to bypass the genetic defect. In the absence of added nucleosides, the dhfr⁻ cells will die, and this forms the basis of an excellent selection procedure to isolate transformed dhfr⁺ CHO cells.

This system for long-term gene expression offers three distinct advantages over other commonly used systems. First, the transformation frequency of dhfr⁻ to dhfr⁺ CHO cells is very high (1 in 10^4 cells), thus enabling the easy isolation of a very large number of transformants. Second, at least in our experience, the reproducibility of the procedure and success of expression of faithful gene products is very good. Third, by further selection of the transformants in the presence of methotrexate (Mtx), which is a direct inhibitor of dhfr, it is possible to amplify the gene copy number of the integrated DNA construct. This results in the isolation of Mtx-resistant cell lines that are capable of producing very large quantities of recombinant gene products.

Procedures for obtaining first-round transformants (i.e., low level expression cell lines) and Mtx-resistant cell lines (i.e., high-level expression cell lines) are described below.

3.2.1. Vector Design for Long-Term Expression

To obtain recombinant gene products from CHO cells, it is necessary to construct two separate expression cassettes. One must be capable of producing dhfr to act as the selectable marker, whereas the other cassette produces the required product. Most of the published reports with this system frequently use the two cassettes in different DNA constructs and cotransfect these together into dhfr⁻ CHO cells. This is perfectly acceptable for obtaining first-round dhfr⁺ transformants, but in our experience, in some instances, there is a tendency for loss of the nonselectable gene (i.e., that directing synthesis of the required gene product) to occur following rounds of amplification in the presence of Mtx. For this reason, to ensure very tight linkage of the selectable and nonselectable expression cassettes, we prefer to clone these both into the same DNA construct.

The requirements for a long-term expression construct are outlined here.

1. pBR322 vector sequences containing the bacterial origin of replication and a bacterial selectable marker gene (i.e., the ampicillin-resistance gene). It is not necessary to use "poison" minus pBR derivatives for long-term expression constructs.

2. A dhfr-selectable marker expression cassette. This is most conveniently taken from the vector pSV2 dhfr (*8*).

3. An expression cassette directing the synthesis of the gene product required (*see* section 3.1.1, item 3 for full details).

4. Ideally the expression cassette in item 3 should be cloned such that the transcriptional orientation of the gene is opposite to that of the dhfr gene. This is to safeguard against the possibility that, if any read-through of the gene product occurred, then the opposite strand (i.e., noncoding strand) of the dhfr expression cassette would be read. This should eliminate anomalous expression products.

5. The simplest way to generate the final required construct is to start with pSV2 dhfr and clone the required gene expression cassette as an EcoRI to BamH1 fragment directly into the EcoRI-BamH1 position of pSV2 dhfr. Alternatively, the desired expresssion cassette could be cloned into either the unique EcoRI or BamH1 sites in pSV2 dhfr as homologous ended fragments (i.e., EcoRI-EcoRI). In this situation the recombinants should be screened to identify constructs in which the expression cassette has been cloned in the clockwise orientation (the dhfr expression cassette is already oriented in pSV2 dhfr in the anticlockwise orientation).

To illustrate the design of a long-term expression cassette, Fig. 2 shows a construct used for the expression of human γ-IFN in CHO cells. This construct consists of three separate DNA segments, which are:

1. Bacterial vector sequences from the large EcoRI to PvuII fragment (2295 bp) of pBR322.

2. A PvuII to BamH1 dhfr expression cassette, which uses the same SV40 control signals as detailed earlier for the design of the transient γ-IFN expression cassette.

The above two components constitute the single large EcoRI to BamH1 fragment from pSV2 dhfr.

3. An EcoRI to BamH1 γ-IFN expression cassette. This uses the same SV40 promoter fragment as shown in Fig. 1 to direct expression of the γ-IFN cDNA. Downstream from the γ-IFN cDNA, however, the polyadenylation and termi-

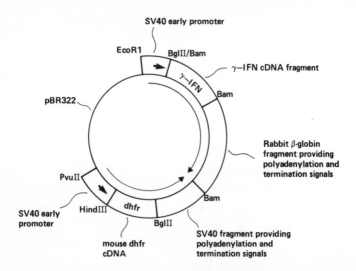

Fig. 2. An example of the essential features for a long-term expression construct for use with dhfr⁻ CHO cells.

nation signal are provided by a rabbit β-globin fragment. This fragment constitutes the 1.9-kb BamH1 to NruI small fragment taken from pKCR (10), in which the NruI site has been converted to a BamH1 site using linkers.

4. It is important to realize that the same γ-IFN expression cassette shown in Fig. 1 cannot be simply excised as an EcoRI to BamH1 fragment and subcloned directly into the corresponding sites in pSV2 dhfr. This would juxtapose identical cloned DNA fragments (e.g., that providing SV-40 polyadenylation and termination signals) in a "back-to-back" arrangement. From a molecular cloning viewpoint, it is extremely difficult to clone this arrangement without extensive rearrangement or deletion events occurring. Consequently, nonhomologous fragments bearing information for polyadenylation and termination are used.

3.2.2. Transfection Procedure and Isolation of dhfr⁺ CHO Transformants

1. One day prior to transfection, plate dhfr⁻ CHO cells at a density of 0.75 x 10⁶ cells/9-cm dish.

2. The next day, remove the medium and replace with exactly 9 mL of fresh medium. The cells are now ready for transfection.

3. Prepare transfection solution A in one tube by adding 1 mL of 2x HBS and 20 µL of a 1:1 mix of the transfection phosphate solutions. Prepare transfection solution B in another tube by making 30 µg of the recombinant DNA expression construct (usually dissolved in distilled water at a concentration of 0.5 µg/µL) up to 0.88 mL with distilled water, then add 120 µL of $2M$ $CaCl_2$ solution. Both solutions are usually prepared in plastic sterile Universal tubes.

4. To form the transfection precipitate, solution B is added to solution A by a "bubble and drip" procedure. This is done by using two electronic pipet aids, each holding a 1-mL pipet. One pipet is placed into solution A, and a slow stream of bubbling is maintained (about two bubbles per s), whereas the other pipet slowly drips solution B (at the rate of about one drop/2 s) directly onto the rising bubbles.

5. The precipitate (2 mL) is left to form at room temperature for 30 min and should appear as a very fine, slightly cloudy precipitate. If there is no cloudiness at all, or if large lumps of precipitate are present, then the "bubble and drip" procedure has not been done correctly.

6. One milliliter of the precipitate is added drop-wise to the 9 mL of medium in the culture dish and left overnight at 37°C in the CO_2 incubator.

7. The following day, the medium is removed and replaced with fresh medium and left for a further 24 h.

8. The next day, the cells are harvested and plated at dilutions of 1/5, 1/20, and 1/100 into T-75 flasks containing dhfr$^+$ selection medium.

9. Every third day the medium is changed with fresh selection medium; dhfr$^+$ colonies start to appear by 7–10 d.

10. Choose the T-75 flask, which has about 100 colonies, harvest the cells, and pool and plate into a single T-75 flask.

Grow the mixed population for two to three generations (e.g., about 1 wk) to establish a mixed basal population. Freeze aliquots in liquid nitrogen as a master stock.

11. Establish cloned lines of the dhfr$^+$ CHO transformants by dilution cloning. This is done by plating the cells at a density of one cell per five wells into 96-microwell plates. Leave the plates for 2–3 wk then harvest individual established clones and generate into cell lines.

12. Test at least 12 such dhfr$^+$ transformed CHO cell lines for the expression of the recombinant gene product (*see* step 7 of section 3.1.2).

3.2.3. Isolation of Amplified dhfr$^+$ Transformants Expressing High Levels of Recombinant Product

1. Complete steps 1–10 of the previous section.
2. From an almost confluent T-75 flask of mixed basal population cells, plate the cells at dilutions of 1/5, 1/5, and 1/3 using dhfr$^+$ selection medium containing $3 \times 10^{-8}M$ methotrexate (Mtx), $10^{-7}M$ Mtx, and $3 \times 10^{-7}M$ Mtx, respectively.
3. Change the medium every third day using medium containing the indicated Mtx levels. After about 2 wk, Mtx-resistant colonies will start to appear in the flasks. Choose the flask that has 20–50 resistant colonies (usually the $3 \times 10^{-8}M$ or $10^{-7}M$ Mtx flask), and proceed as for steps 10, 11, and 12 of the previous section.
4. The Mtx-resistant cell lines expressing the highest levels of recombinant gene product are chosen and subjected to a further round of amplification. If the cells are resistant to $3 \times 10^{-8}M$ Mtx, then they are selected at 1/5, 1/5, and 1/3 dilutions in medium with $10^{-7}M$, $3 \times 10^{-7}M$, and $10^{-6}M$ Mtx. Again, the flask with the appropriate number of Mtx-resistant colonies is pooled, established for 1 wk as a mixed population, then dilution-cloned to establish cell lines. Usually, most amplification of gene copy number occurs during this second round of selection, and the resulting cell lines can express very high yields of recombinant products.

4. Notes

1. There is, of course, a very large degree of flexibility in the vector designs that can be used for transient or long-term expression with mammalian cells. A large number of cloned promoters are available that, often by virtue of upstream enhancer elements, confer cell type and species preference. The promoters mentioned in the text are broad-range promoters and should function efficiently in the majority of experimental cell types.

2. When constructing the expression cassette, it is important to make the fusion between the promoter and cDNA within their respective 5' untranslated leader sequences. Likewise, the cloned fragment containing strong polyadenylation and termination signals must be cloned into the 3' untranslated sequences of the cDNA. The cDNA clone may contain its own polyadenylation signal, but when placed under the control of a strong promoter, there can be significant read-through of the signal. It is therefore advised to always clone efficient viral signals downstream of the cDNA.

3. Transient expression can also be carried out in cell lines other than COS cells. For instance, human Hela or mouse NIH 3T3 cells are widely used; the levels of expression will often be lower than seen with COS cells, however, because of an unamplified copy number.

4. The DEAE-dextran transfection procedure for transient expression must not be used for long-term transfection work. For unknown reasons, the procedure is very inefficient for the generation of long-term transformants. In contrast, the calcium phosphate DNA precipitation procedure outlined for long-term expression work can be equally successful as the DEAE-dextran procedure for transient expression work. For ease and reproducibility, however, the DEAE-dextran procedure is strongly recommended as the method of choice for transient work.

5. If the amplified procedure for high-level expression of recombinant proteins in CHO cells is used, it is advised that the amplification events be monitored by Southern blotting DNA from the established cell lines at each stage of Mtx resistance. For a stable amplified cell line, the majority of the transfected gene copy number should be intact. Conversely, for an unstable amplified cell line, gross rearrangemnets or deletions may be apparent.

References

1. Maniatis, T., Fritsch, E., and Sambrook, J., eds. (1982) *Molecular Cloning, A Laboratory Manual* Cold Spring Harbor Laboratories, Cold Spring Harbor, New York.
2. Gluzman, Y. (1981) SV40 transformed simian cells support the replication of early SV40 mutants. *Cell* **23**, 175–182.
3. Lusky, M. and Botchan, M. (1981) Inhibition of SV40 replication in simian cells by specific pBR322 DNA sequences. *Nature* **293**, 79–81.
4. Twigg, A.J. and Sheratt, D. (1980) Trans-complementable copy number mutants of plasmid ColEl. *Nature* **283**, 216–218.
5. Fiers, W., Contreras, G., Haegeman, R., Rogiers, A., Van deVoorde, A., Van Heuverswyn, H., Van Herreweghe, J., Volckaert, G., and Ysebaert, M. (1978) Complete nucleotide sequence of SV40 DNA. *Nature* **273**, 113–120.
6. Gorman, C.M., Merlino, G.T., Willingham, M.C., Pastan, I., and Howard, B.H. (1982) The Rous sarcoma virus long terminal repeat is a strong promoter when introduced into a variety of eukaryotic cells by DNA-mediated transfection. *Proc. Natl. Acad. Sci. USA* **79**, 6777–6781.
7. Palmiter, R.D., Brinster, R.L., Hammer, R.E., Trunbauer, M.E., Rosenfeld, M.G., Birnberg, N.C., and Evans, R.M. (1982) Dramatic growth of mice that develop from eggs microinjected with metallothionein-growth hormone fusion genes. *Nature* **300**, 611–615.
8. Subramani, S., Mulligan, R., and Berg, P. (1981) Expression of mouse dhydrofolate reductase complementary deoxyribonucleic acid in Simian Virus 40 vectors. *Mol. Cell. Biol.* **1**, 854–864.
9. Urlaub, G. and Chasin, L.A. (1980) Isolation of Chinese hamster cell mutants deficient in dihydrofolate reductase activity. *Proc. Natl. Acad. Sci. USA* **77**, 4216–4220.
10. O'Hare, K., Benoist, C., and Breathnach, R. (1981) Transformation of mouse fibroblasts to methotrexate resistance by a recombinant plasmid expressing a prokaryotic dihydrofolate reductase. *Proc. Natl. Acad. Sci. USA* **78**, 1527–1531.

Chapter 28

Dot-Blot Hybridization Method

Hans E. N. Bergmans and Wim Gaastra

1. Introduction

Nucleic acid hybridization is a very potent technique that can be used for the identification of DNA and RNA species with varying degree of homology and for the estimation of relative amounts of nucleic acid with known homolgy. In most cases, single-stranded (ss) (denatured) DNA or RNA is bound to a filter support (e.g., nitrocellulose) and incubated with a radioactively labeled ss DNA or RNA fragment ("probe") complementary to the nucleic acids of interest. In subsequent washing steps the nonhybridized probe is removed. Hybridized probe will be retained, but may be washed off in further washing steps, depending on the ion concentration and temperature of the wash.

In the sophisticated hybridization techniques known as Southern and Northern blotting (*see* Vol. 2 of this series), restriction digests of DNA or isolated RNA species are subjected to agarose gel electrophoresis before transfer to the filter support. These techniques yield information about the molecular composition of the nucleic acids that bind probe material, but they are also time consuming and expensive. If such detailed knowledge is not needed, e.g., in screening experiments, a simplified blotting and hybridization protocol like the dot-blot assay (1) can be used. In this technique, DNA or RNA samples are applied to filter supports without previous treatment. Samples may be whole genomic DNA or purified fractions, e.g., plasmid DNA or mRNA.

The dot-blot method described here assumes that purified DNA or RNA samples with known concentration have been prepared, e.g., according to the methods described in Chapters 5 and 13 in Vol. 2 of this series.

2. Materials

1. Nitrocellulose filter (Schleicher and Schuell).
2. $1M$ Sodium hydroxyde solution.
3. $1M$ and $2M$ Sodium acetate solutions.
4. 20x SSC solution: $3M$ sodium chloride, $0.3M$ sodium citrate, pH 7.0.
5. 20x STET buffer: $3M$ sodium chloride, $0.6M$ Tris-HCl, pH 8.0, 20 mM EDTA.
6. 50x Denhardt's solution: 5 g of Ficoll, 5 g of polyvinylpyrolidone, 5 g of bovine serum albumin, to 500 mL with water.
7. Denatured salmon sperm DNA: 10 mg/mL in water. Shear by passing the solution several times through an 18-gage hypodermic needle. Denature by heating to 100°C for 10 min and quickly cooling on ice.
8. Blank hybridization mixture: 50% formamide (deionized by stirring with Dowex ion exchanger, e.g., AG 501-X8 from Bio-Rad), 2x Denhardt's solution, 4x STET, 0.1% SDS, 100 µg/mL of denatured salmon sperm DNA.

3. Method

3.1. Preparation of Filters

1. DNA samples should be in water or low ion strength buffer (e.g., 10 mM Tris/HCl, pH 8.0), concentration approxi-

mately 80 μg/mL. Plasmid DNA should be linearized by restriction enzyme digestion and purified by phenol extraction and ethanol precipitation prior to the following procedure.

2. Add NaOH to a final concentration of 0.4N and incubate for 10 min at ambient temperature.
3. Cool on ice and add 1 vol of 2M sodium acetate.
4. Place 22 mm-diameter nitrocellulose filter discs on top of moist 3MM filter paper on top of several layers of dry paper towels. Rinse the filters by passing several drops of 1M sodium acetate through the filters.
5. 50-μL samples of denatured DNA or RNA are applied by means of a micropipet. The pipet tip is pressed firmly on the filter during application. Application of one sample should take approximately 1 min. Rinse by passing a drop of 1M sodium acetate through the filter after application of each sample. Up to 16 samples can be spotted on one filter.
6. Rinse the filter(s) for 5 min in excess (e.g., 200 mL) 4x SSC and air dry.
7. Incubate the filter(s) for at least 1 h in 2x Denhardt's solution at ambient temperature. Drain and air dry.
8. Bake the filter(s) under vacuum at 80°C for 2 h. Filters can be stored dry at ambient temperature.

3.2. Hybridization

1. Incubate the filter(s) for 1 h in 10x Denhardt's solution, 4x STET, shaking at ambient temperature. Incubate in a tightly closed container that fits the filter(s).
2. Remove the buffer and add a volume of blank hybridization mixture so that filter is not in contact with air (*see* note 7 in section 4). If more than one filter is incubated simultaneously, care should be taken to avoid air bubbles between the filters. A blank filter can be put on top to avoid air contact. Incubate for at least 1 h with shaking at 42°C.
3. Remove the blank hybridization mixture, leaving the minimal amount that still leaves the filter(s) thoroughly

wet. Add the radioactive probe, and ensure that it is mixed thoroughly throughout the hybridization mixture. Incubate with shaking at 42°C for 16–48 h.

3.3 Washing

1. Wash the filter(s) twice for 15 min in excess blank hybridization mixture at 42°C.
2. Shake the filter(s) at ambient temperature for 30 min in 50% formamide 2x STET. The filter can now be mounted for autoradiography at this stage.
3. Further melting of hybridized probe from the filter can be done at desired stringency, in different concentrations of SSC and at different temperature (*see* Note 6 in section 4). All probes can be removed by heating the filter for 10 min in distilled water. The filters can be reused after the probe has been removed.

3.4. Autoradiography

1. The filter is placed on wax paper on a glass plate, covered with Saran Wrap, and mounted for autoradiography. If further melting of the probe by washing under different conditions is desired (*see* Note 6 in section 4) care should be taken that the filter does not dry during autoradiography since this will result in irreversible binding of the probe to the filter.

4. Notes

1. The method described above is only one of the many variants on the theme. In fact probably every laboratory has its own variations of general procedure [e.g., *see* Maniatis et al. (1982) in the Further Reading section].
2. Filters should always be handled with gloves to avoid contamination with protein (e.g., DNAse).

3. Although nitrocellulose is a very good filter support for hybridization experiments, new filter materials have been developed with even better performance. Gene Screen (product of DuPont), for instance, is easier to handle (less brittle) than nitrocellulose, and binding of ssDNA to this material does not require baking. A commercially available filter apparatus (Minifold, Schleicher and Schuell) can be used for even distribution of samples to the filter.

4. Double-stranded DNA does not bind on nitrocellulose filters as efficiently as single-stranded DNA; therefore the DNA has to be denatured before application to the filter. This can be done at high base concentrations as described here, or by heating the DNA at 100°C for 10 min, chilling on ice, and adding sodium acetate to a final concentration of 1*M*.

5. When hybridizing to genomic DNA, better results are usually obtained if the genomic DNA is sheared mildly, e.g., by passing through a hypodermic needle, to reduce the length of the DNA fragments and the viscosity of the solution.

6. Washing of filters at different stringencies is a usual procedure to evaluate the degree of homology of the hybridizing nucleic acids. DNA with a higher degree of homology remains hybridized in more stringent conditions, i.e., lower salt concentration and higher temperature. Washing at 65°C in 1x SSC is considered a stringent condition.

7. Complete drying of the filter at any stage in the hybridization procedure may result in irreversible binding of DNA. This would interfere with further removal of probe DNA in further washings at different stringencies or rehybridization with the same filter after all nonirreversibly bound probe has been removed by heating the filter in distilled water at 100°C. Complete removal of probe can be checked by autoradiography between hybridizations.

8. Residual detergent in glassware or the quality of the water used may sometimes be causes for high background bind-

ing of probe DNA. Air bubbles on or between filters
during the hybridization procedure will cause very heavy
background spots that may interfere completely with
hybridization results.

Further Reading

Maniatis, T., Fritsch, E.F., and Sambrook, H. (1982) The Identification of
Recombinant Clones, in *Molecular Cloning, A Laboratory Manual.* Cold
Spring Harbor Laboratories, Cold Spring Harbor, New York.
DuPort Genescreen Plus Brochure (1985) Protocols for electrophoretic and
capillary transfer of DNA and RNA, DNA and RNA hybridization
and DNA and RNA rehybridization.

Reference

1.   Kafatos, F.C., Jones, C.W., and Efstratiadis, A. (1979) Determination of
nucleic acid sequence homologies and relative concentrations by a
dot-blot method. *Nucleic Acids Res.* **7**, 1541–1552.

Chapter 29

Detection of Unique or Low Copy Number DNA Sequences in Complex Genomes

Domingo Gallardo and Albert Boronat

1. Introduction

The characterization of specific DNA sequences by means of hybridization techniques is of great relevance in the development of molecular biology. Because of its simplicity, the method described by Southern in 1975 (1) is widely used for such purposes. This method, also known as the Southern blot, is based on the hybridization of specific probes to DNA fragments immobilized onto a membrane filter after their separation by agarose gel electrophoresis. The identification of specific DNA fragments is possible since their relative positions are maintained during the transfer from the gel to the filter.

The Southern technique is commonly used for the identification and characterization of restriction fragments generated from cloned DNA. Nevertheless, it is also used for the detection of specific DNA fragments in genomic DNA. In the latter approach, some difficulties appear when analyzing unique or low copy number sequences in large genomes. That is the case of mammal and plant genomes, in which size values in the order of 2×10^9 to 7×10^9 base pairs (bp) per haploid genome are normal. Since the amount of DNA loaded in a gel is limited in terms of a good resolution, a specific genomic DNA sequence is diluted among the complexity of fragments generated from the complete genome after its digestion with restriction enzymes. As a consequence, the detection of such sequences is often a problematic and tedious task.

Here we describe a method designed to allow the detection, after 1 d of autoradiographic exposure, of specific DNA sequences corresponding to single or low copy number genes in complex genomes. The method is based on the modification and improvement of conventional transfer and hybridization techniques. The main points of the method are: (1) Treatment of large amounts of genomic DNA with restriction enzymes in conditions ensuring its complete digestion; (2) short-wave UV irradiation of large DNA fragments after agarose gel electrophoresis and extensive transfer to nitrocellulose filter; and (3) optimized hybridization conditions in order to enhance DNA hybridization yield as well as to minimize background.

2. Materials

1. Stock solutions: $2M$ Tris-HCl, pH 7.5; $5M$ NaCl; and $0.5M$ EDTA, pH 8.5. Filter solutions through 0.22-μm membrane filters and sterilize by autoclaving. Store at room temperature.
2. Agarose gel electrophoresis buffer (TAE): 40 mM Tris, 20 mM acetic acid, 2 mM Na_2EDTA, pH 8.1. Prepare a 50x stock dissolving 242 g of Tris, 57.1 mL of glacial acetic acid, and 37.2 g of Na_2EDTA in H_2O to a final volume of 1 L. Adjust pH to 8.1 if necessary. It does not need to be sterilized. Store at room temperature.
3. Ethidium bromide: Prepare a stock of 10 mg/mL in H_2O. Store at 4°C and protected from light. Caution: wear gloves when using ethidium bromide—*it is mutagenic*.

4. 5x Loading buffer: 0.1% SDS, 40 mM, pH 8.5, 30% Ficoll 400 (Pharmacia), and 0.25% (w/v) bromophenol blue. Store at room temperature.
5. Denaturing solution: 1.5M NaCl, 0.5M NaOH. Prepare from 10M NaOH and 5M NaCl stocks. Store at room temperature. It can be reused several times.
6. Neutralizing solution: 1.5M NaCl, 0.5M Tris-HCl, pH 7.5. Prepare from stock solutions, store at room temperature. It can be reused several times.
7. 20x SSC (1x SSC is 0.15M sodium chloride and 0.015M sodium citrate, pH 7.0): Dissolve 175.4 g of NaCl and 88.2 g of trisodium citrate · 2H$_2$O in 900 mL of H$_2$O and adjust to pH 7.0 with HCl. Complete volume to 1 L. Filter through a 0.22-μm membrane filter and sterilize by autoclaving. Store at room temperature.
8. Denatured salmon sperm DNA: Dissolve salmon sperm DNA (sodium salt type III, Sigma) in sterile water at a concentration of 10 mg/mL using a magnetic stirrer. Pass the DNA preparation through an 18-gage hypodermic needle four or five times. Heat the DNA at 100°C for 10 min and cool it down on ice. Store in aliquots at –20°C.
9. Denhardt's solution: Prepare a 100x stock by mixing together 2 g of Ficoll 400 (Pharmacia), 2 g of polyvinylpyrrolidone (PVP-360, Sigma), and 2 g of bovine serum albumin (BSA-fraction V, Sigma) in 100 mL of H$_2$O. Store in aliquots at –20°C.
10. Deionized formamide: Mix 100 mL of formamide with 10 g of mixed bed resin AG501-X8 (D) (Bio-Rad), stir for 30–60 min at room temperature; filter twice through Whatman No. 1 filter paper. Store in small aliquots at –20°C.
11. 1M Sodium phosphate buffer, pH 6.5: Dissolve 11.8 g of Na$_2$HPO$_4$ · 12H$_2$O and 9.25 g of NH$_2$PO$_4$ · H$_2$O in a final volume of 100 mL of water and verify pH. Sterilize by autoclaving and store at room temperature.
12. Prehybridization solution: 5x Denhardt, 300 μg/mL denatured salmon sperm DNA and SSC at the appropriate con-

centration (*see* section 3.4, step 1). Prepare from the stocks (heat salmon sperm DNA at 100°C for 10 min before use).

13. Hybridization solution: 50% Deionized formamide, 20 m*M* sodium phosphate, pH 6.5, 5x Denhardt, 6x SSC, 300 µg/mL salmon sperm DNA, and ^{32}P-labeled DNA probe. Mix salmon sperm DNA and the DNA probe in water (0.5 mL final volume), denature by heating at 100°C for 10 min, and add to a prewarmed solution containing the rest of the components (*see* section 3.4, step 2).

14. 10% SDS: Prepare a 10% (w/v) solution of SDS (sodium dodecylsulfate) in H_2O (caution: wear a mask when weighing SDS). Heat at 70°C for 30 min. Do not autoclave.

3. Method

3.1. Digestion of Genomic DNA with Restriction Enzymes

1. Perform digestions of DNA with restriction enzymes at the conditions recommended by the manufaturers using a DNA concentration of 0.2 µg/µL in the reaction mixture. Incubate for 6 h adding 5 U of restriction enzyme per microgram of DNA at intervals of 2 h. Stop reactions by adding 5x loading buffer and heating at 65°C for 10 min.

3.2. Agarose Gel Electrophoresis

1. Prepare the gel (15 x 11 x 0.5 cm) at the appropriate agarose concentration (0.7 to 1.2% agarose; type I Low EEO, Sigma) in TAE buffer containing 0.5 µg/mL ethidium bromide. Load samples in the gel (10 µg of the digested DNA per well), including the appropriate DNA markers. Run electrophoresis at 1.5 V/cm for 12–16 h in the same buffer.

2. Visualize the DNA bands using UV light (302 nm), and photograph the gel including a transparent rule alongside it to refer the relative position of the DNA markers.

3.3. Transfer of DNA Fragments to Nitrocellulose Filters

1. Irradiate the DNA fragments larger than 4 kb with short-wave UV light (254 nm) for 5 min using an Ultra-Violet Products Inc. hand lamp UVG-11 (or equivalent) placed 10 cm over the gel (protect the rest of the gel with aluminum foil). This UV treatment nicks large DNA fragments, thus facilitating their transfer to the nitrocellulose filter.
2. Cut the gel alongside the wells to refer the origin of electrophoresis, and make an additional cut in a corner in order to reference the gel position henceforth.
3. Soak the gel in denaturing solution (500 mL) for 60 min at room temperature with gentle shaking.
4. Rinse the gel in distilled water and soak it for two periods of 30 min in 500 mL of neutralizing solution at room temperature with gentle shaking.
5. Place a glass plate (18 x 20 cm approximately) onto a suitable baking tray containing 0.5–1 L of transfer buffer (10x SSC).
6. Wet three sheets of Whatman 3 MM paper (16 x 35 cm approximately) in 10x SSC, and place them on the glass plate ensuring that the ends are in contact with the transfer buffer. Remove any possible air bubble trapped between them using a smooth glass rod.
7. Place the agarose gel on the Whatman 3 MM sheets, avoiding the trapping of air bubbles. Place parafilm strips around the edges of the gel to prevent short-circuiting of the buffer between the transfer bridge and the absorbent papers (*see below*, steps 9 and 10).
8. Cut the nitrocellulose filter (Schleicher & Scheull BA83, 0.2 μm) to the same size of the gel and wet it in 2x SSC for 5 min (float the filter on the surface of the solution until it wets completely from beneath and then immerse it). Place the filter onto the gel, avoiding trapping air bubbles between them (*caution: wear gloves when handling nitrocellulose filters*).

9. Cut three to four sheets of Whatman 3 MM paper at gel size, pre-wet in 2x SSC, and place onto the nitrocellulose filter. Stack 8–10 cm of absorbent paper onto the 3 MM paper.

10. Place a glass plate on top of the stack, cover the whole system with transparent autoadhesive plastic film (to prevent evaporation), and press it down with a weight of about 500 g. Allow transfer to proceed for 24–36 h at room temperature.

11. Disassemble the transfer system, peel off the filter, and soak it in 6x SSC for 5 min.

12. Dry the filter at room temperature and bake it at 80°C under vacuum for 2 h (the filter can be stored at room temperature before its use).

13. Monitor the efficiency of the transfer process by soaking the gel in a solution of 1 µg/mL ethidium bromide (in H_2O) for 30 min with gentle agitation, and visualize under UV light.

3.4. Hybridization

1. Place the filter in a sealable plastic bag. Add the prehybridization solution (0.15 mL/cm² of filter) prepared at the appropriate concentration of SSC (usually 1–2x) to obtain conditions equivalent to –25 T_m (*see* section 4, Note 5). Seal the bag by heat and incubate at 68°C for 12–15 h.

2. Remove the prehybridization solution and add the hybridization solution (0.075 mL/cm² filter) containing 1–2 x 10⁶ cpm/mL of the denatured labeled DNA probe (1–2 x 10⁸ cpm/µg). Seal the bag by heat and incubate for 48 h with gentle agitation at the appropriate temperature (usually 42–45°C; *see* section 4, Note 5).

3.5. Washing of Filters

1. Remove the filter from the bag and rinse it in 100 mL of 2x SSC, 0.1% SDS for 2 min at room temperature.

2. Soak the filter twice in 350 mL of 2x SSC, 0.1% SDS at 45°C for 20 min with gentle agitation.
3. Soak the filter twice in 350 mL of 2x SSC, 0.1% SDS at 65°C for 20 min with gentle agitation.
4. Soak the filter three times in 350 mL of 0.5x SSC, 0.1% SDS at 65°C for 20 min with gentle agitation.
5. Dry the filter on absorbent paper at room temperature.

3.6. Autoradiographic Detection

1. Fix the filter and radioactive marks on a cardboard and cover with transparent autoadhesive plastic film.
2. Expose the filter, for at least 24 h, to an X-ray film using an intensifying screen (Micron R or equivalent) at –70°C.
3. Develop the film according to manufacturer's directions. Position the autoradiographic bands on the filter referring to that of the radioactive marks.

4. Notes

1. The best results are obtained when using highly purified DNA preparations (preferably by cesium chloride gradients). This avoids the presence of impurities that might interfere with the action of restriction enzymes.
2. The completion of the genomic DNA digestions is difficult to assess as a consequence of the extremely complex pattern of fragments obtained after agarose gel electrophoresis. We routinely use internal controls consisting in the addition of small amounts of λ or pBR322 DNA to the genomic DNA (80 ng control DNA/10 μg genomic DNA). The DNA digestion can then be monitored by analyzing a small aliquot by the conventional Southern technique using ^{32}P-labeled control DNA.
3. It is convenient, whenever possible, to use restriction enzyme preparations of high concentration to prevent the "star activity" (caused by the presence of glycerol) described for several restriction endonucleases (2).

4. Satisfactory results are usually obtained when using DNA probes labeled by the conventional nick translation technique, as described by Maniatis et al. (*3*; and *see* Vol.2 of this series). Nevertheless, the use of two ^{32}P-deoxyribonucleosides triphosphate highly improves the results by shortening the exposure time and increasing the signal/noise ratio.

5. Perform hybridization in the presence of formamide at conditions equivalent to $-25\ T_m$ approximately (*4*). Calculate this parameter according to the following equation (*5*):

$$T_m = 16.6 \log [\mathrm{Na^+}] + 0.41\ (\%\ \mathrm{G+C}) + 81.5$$

    assuming that each increase of 1% in the formamide concentration lowers the T_m of a DNA duplex by about 0.65°C (*6,7*).

6. The size of the probe may also influence the hybridization parameters. The T_m of a DNA–DNA duplex is modified according to the following criterion:

$$T_m\ (\mathrm{real}) = T_m\ (\mathrm{theoretical}) - 500/l$$

    where l is the length of the duplex in base pairs (*5*). So, when labeling DNA by nick translation, it is important to monitor the average size of the denatured probe in an alkaline agarose gel as described by Maniatis et al. (*3*). For short probes it is advisable to seal the nicks after the labeling using T4 DNA ligase.

7. When using nonhomologous DNA probes, the hybridization conditions should be modified according to the degree of homology, assuming that each 1% mismatch decreases T_m by 1°C (*8*).

8. Depending on the purposes, the stringency of the washing conditions can be modified by changing the salt concentration and/or the temperature.

References

1. Southern, E.M. (1975) Detection of specific sequences among DNA fragments separated by gel electrophoresis. *J. Mol. Biol.* **98**, 503–517.
2. Fuchs, R. and Blakesley, R. (1983) Guide to the use of type II restriction endonucleases. *Meth. Enzymol.* **100**, 3–38.
3. Maniatis, T., Fritsch, E.F., and Sambrook, J. (1982) *Molecular Cloning. A Laboratory Manual.* Cold Spring Harbor Laboratory, Cold Spring Harbor, New York.
4. Beltz, G.A., Jacobs, K.A., Eickbush, T.H., Cherbas, P.T., and Kafatos, F.C. (1983) Isolation of multigene families and determination of homologies by filter hybridization methods. *Meth. Enzymol.* **100**, 266–285.
5. Davis, R.W., Botstein, D., and Roth, J.R. (1980) *Advanced Bacterial Genetics. A Manual for Genetic Engineering.* Cold Spring Harbor Laboratory, Cold Spring Harbor, New York.
6. McConaughy, B.L., Laird, C.D., and McCarthy, B.J. (1969) Nucleic acid reassociation in formamide. *Biochemistry* **8**, 3289–3295.
7. Casey, J. and Davidson, N. (1977). Rates of formation and thermal stabilities of RNA:DNA and DNA:DNA duplexes at high concentrations of formamide. *Nucleic Acids Res.* **4**, 1539–1552.
8. Bonner, T.I., Brenner, D.J., Neufeld, B.R., and Britten, R.J. (1973) Reduction in the rate of DNA reassociation by sequence divergence. *J. Mol. Biol.* **81**, 123–135.

Chapter 30

Photobiotin-Labeled DNA and RNA Hybridization Probes

James L. McInnes, Anthony C. Forster, and Robert H. Symons

1. Introduction

Nucleic acid hybridization, the formation of a duplex between two complementary nucleotide sequences, is being increasingly utilized in the research laboratory and for the routine diagnosis of disease. Biotin-labeled nucleic acid hybridization probes have advantages over radioactively labeled probes in terms of stability, safety, and time of detection. The hybridization of a biotin-labeled probe to a target nucleic acid is carried out on a nitrocellulose or nylon filter; the bound probe is usually detected by the binding of an avidin–enzyme conjugate or a streptavidin–enzyme complex (both avidin and streptavidin have a high affinity for biotin), followed by a colorimetric reaction in which the bound enzyme converts a colorless substrate into a colored product.

We have developed a rapid chemical method for small- or large-scale labeling of single- or double-stranded DNA or RNA with biotin (*1*). Nucleic acids are mixed with a photoactivatable

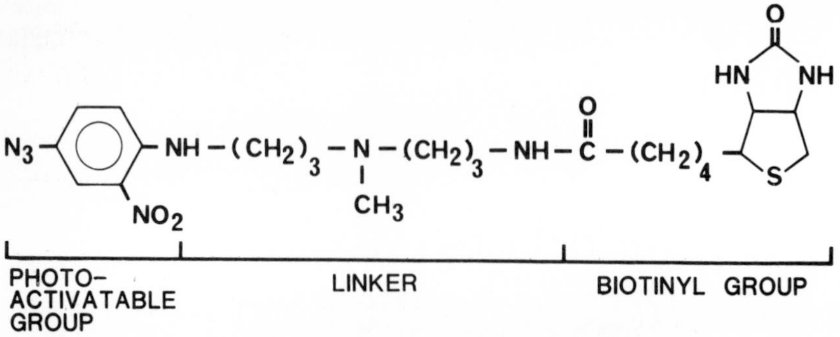

Fig. 1. Structure of photobiotin.

analog of biotin, called photobiotin (Fig. 1), and then irradiated briefly with visible light. This results in the stable linkage of one biotin to every 100–150 nucleotides. Although the precise nature/site of the linkage is unknown, the linkage is stable under the hybridization conditions and is presumably covalent. These probes are as sensitive as radioactive probes in Northern, Southern, and dot blot analyses, and should be particularly suited to hybridization *in situ* and the routine diagnosis of disease.

2. Materials

2.1. Labeling of Nucleic Acids with Photobiotin

1. Photobiotin™ acetate (BRESA, GPO Box 498, Adelaide, South Australia, 5001, Australia).
2. Light source for photoactivation reaction: Lamps suitable for photoactivation must provide a strong source of white light—at least 250 W, preferably 400–500 W. Tungsten lamps are not suitable because of their high heat output. The most practical lamps are mercury vapor discharge lamps containing a tungsten filament (no ballast necessary) with a reflector to direct the light ouput.

Examples: EYE: R57 500 W (Iwasaki Electric Co., Japan); OSRAM: National (MBFTR/V) 500 W; OSRAM: National (MBFTR/V) 250 W.

If such lamps are unavailable, a mercury vapor discharge lamp of 300–500 W with a built-in reflector can be used. Ballast equipment will be required, however.

Examples: OSRAM: GEC (MBFR/U) 400W; PHILIPS: HPLR 400W.

If reflector discharge lamps cannot be easily obtained, nonreflector discharge lamps are suitable. Their main disadvantage, however, is the high light intensity in all directions.

Examples: Mercury vapor/tungsten filament lamps (no ballast required): PHILIPS: ML 500 W; OSRAM: National (MBFT/V) 500 W. Mercury vapor lamps (ballast required): PHILIPS: HPL-N 400 W; HP/T 400 W; OSRAM: GEC (MB/U) 400 W; National (MBF/U) 400 W.

3. Buffer: $0.1M$ Tris-HCl, pH 9.0, 1.0 mM EDTA.
4. Butan-2-ol.
5. $3M$ Sodium acetate solution, pH 5.2.
6. Ethanol.
7. 70% (v/v) Ethanol.
8. Vacuum oven, 80°C.
9. 0.1 mM EDTA solution. This solution is prepared by dilution of $0.5M$ sodium EDTA, pH 8.0. The final pH will be approximately neutral.

Note: Aqueous solutions are prepared with twice glass-distilled water or water of equivalent quality and autoclaved when possible.

2.2. Hybridization Protocol

1. Nitrocellulose filters (Schleicher and Schuell, BA85, pore size 0.45 μm).
2. Deionized formamide.
3. 20x SSC: $3M$ Sodium chloride, $0.3M$ trisodium citrate.
4. 250x Denhardt's solution: 5% (w/v) Ficoll 400 (Pharmacia), 5% (w/v) polyvinylpyrrolidone M_r 40,000 (Sigma),

5% (w/v) bovine serum albumin (BSA) (Fraction V, Sigma).

5. Buffer: 1M sodium phosphate, pH 6.5.

6. Sonicated denatured salmon sperm DNA. Salmon sperm DNA is prepared as a 5 mg/mL stock solution. It is then sonicated to generate DNA fragments in the size range 200–8000 bp, and denatured by heating at 100°C for 10 min, followed by cooling on ice.

7. 0.5M sodium EDTA solution, pH 8.0.

8. Dextran sulfate solution (Pharmacia), 50% (w/v).

9. Sodium dodecyl sulfate (SDS) solution, 10% (w/v).

10. Prehybridization buffer: 50% (v/v) deionized formamide, 5x SSC, 5x Denhardt's solution, 50 mM sodium phosphate buffer, pH 6.5, 0.25 mg/mL sonicated denatured salmon sperm DNA, and 5 mM EDTA.

11. Hybridization buffer: 50% deionized formamide, 5x SSC, 1x Denhardt's solution, 20 mM sodium phosphate buffer, pH 6.5, 0.05 mg/mL sonicated denatured salmon sperm DNA, 5 mM EDTA, and 10% (w/v) dextran sulfate.

Stock solutions of 20x SSC, 1M Na-phosphate (pH 6.5), 0.5M Na-EDTA (pH 8.0), and 10% (w/v) SDS are stored at room temperature. Stock solutions of deionized formamide, 250x Denhardt's, sonicated denatured salmon sperm DNA, and dextran sulfate must be stored at –20°C. Prehybridization and hybridization buffers should be prepared fresh as required.

2.3. Colorimetric Detection of the Biotin-Labeled Probes

1. Buffers: Buffer 1: 100 mM Tris-HCl, pH 7.5, 1.0M NaCl, 2 mM MgCl$_2$, and 0.05% (v/v) Triton X-100. Buffer 2: 100 mM Tris-HCl, pH 9.5, 1.0 M NaCl, and 5 mM MgCl$_2$. Buffer 3: 100 mM Tris-HCl, pH 9.5, 100 mM NaCl, and 5 mM MgCl$_2$. Buffer 4: 10 mM Tris-HCl, pH 7.5, 1.0 mM EDTA. Stock solutions of 1M Tris-HCl (pH 7.5), 1M Tris-HCl (pH 9.5), 1M MgCl$_2$, and 50% (v/v) Triton X-100 are used for the preparation of these buffers.

2. BSA (Fraction V, Sigma) 3% (w/v) in Buffer 1. This is prepared as follows: Dissolve BSA in water—leave adequate volume for other additions. Adjust to pH 3 with concentrated HCl. Heat at 100°C for 15–20 min, then cool to room temperature. Adjust to pH 7.5 with 10M NaOH. Add 0.1 vol 10x concentrated Buffer 1 (minus NaCl). Add NaCl to give a final molarity of 1.0. Make to final volume. This solution can be stored for 1–2 mo at 4°C.

3. Stock (500x) avidin–alkaline phosphatase (Sigma) solution, 0.5 mg/mL. Dissolve 500 μg avidin–alkaline phosphatase in 1 mL of water. Store in small aliquots (e.g., 50 μL) at –20°C to avoid gradual loss of activity by repeated freezing and thawing of solution.

4. Stock (250x) nitro blue tetrazolium (NBT, Sigma) solution, 75 mg/mL. Dissolve 30 mg of NBT in 400 μL of 70% dimethylformamide. Store at –20°C.

5. Stock (250x) 5-bromo-4-chloro-3-indolyl phosphate (BCIP, Sigma) solution, 50 mg/mL. Dissolve 20 mg of BCIP in 400 μL of dimethylformamide. Store at –20°C.

3. Methods

3.1. Labeling of Nucleic Acids with Photobiotin

The following protocol is suggested for the labeling of 1–25 μg of nucleic acid with photobiotin. The initial steps of the labeling reaction should be carried out in subdued light (e.g., working in dark room with door ajar to allow minimum light necessary to carry out the initial manipulations).

1. Dissolve the photobiotin acetate in water to a concentration of 1 μg/μL.

2. In a sterile Eppendorf tube or a siliconized glass microcapillary tube (5–50 μL), combine equal volumes (1–25 μL) of the photobiotin acetate solution and the DNA (or RNA) sample to be labeled. The nucleic acid solution should be 0.5–1.0 μg/μL and dissolved in either water or neutral 0.1

m*M* EDTA. Carefully heat seal the ends of the microcapillary tube if this method has been chosen.

3. Place the Eppendorf tube in crushed ice with the lid open
or lay the sealed microcapillary beneath the surface of
water in a shallow dish cooled on ice. This is to ensure that
the nucleic acid sample does not overheat during photolysis. *All subsequent steps can now be carried out in normal
light.* Irradiate the solution for 20 min at 10 cm beneath the
lamp source (*see* section 2). When 2 µg or more of photobiotin acetate is used, nitrogen bubbles normally appear in
the microcapillary tube.

4. Transfer the reaction solution to 50 µL of 0.1*M* Tris-HCl,
pH 9.0, 1.0 m*M* EDTA in an Eppendorf tube, and increase
the volume to 100 µL(if necessary) with water.

5. Add 100 µL of butan-2-ol to the above solution and mix
well. Spin for 1 min in an Eppendorf (or similar) centrifuge. Carefully remove and discard the top red butan-2-
ol phase.

6. Repeat the butan-2-ol extraction. The aqueous phase
should now be colorless and concentrated to 30–35 µL. If
not, make to 35 µL with water.

7. Add 5 µL of 3*M* sodium acetate solution, pH 5.2, and mix.
Add 2.5 vol ethanol (approximately 100 µL) and mix.

8. Precipitate the biotin-labeled nucleic acid by chilling in
solid dry ice for 15 min or at –20°C overnight.

9. Centrifuge the sample (Eppendorf centrifuge/4°C) for 15
min. Wash the pellet once with cold 70% (v/v) ethanol,
dry under vacuum, and dissolve in 0.1 m*M* EDTA.

10. Determine the exact probe concentration by measuring
the absorbance at 260 nm. Store the biotin-labeled probe
at –20°C.

3.2. Hybridization Protocol

The biotin-labeled DNA or RNA can be used as a hybridization probe in the same way as a radioisotope-labeled probe.
Reviews of standard dot-blot and Southern blotting techniques

as well as general hybridization methods may be found in references 2–4 and Vol. 2 of this series. Hybridization conditions in our laboratory are adapted from Thomas (5). Hybridization parameters and the stringency of posthybridization washings should be optimized for the particular system under study and the probe being used; e.g., refer to ref. *1* for DNA (probe)/RNA (target) hybridization protocol.

The following protocol has been found suitable for DNA/DNA hybridizations:

1. After dot or Southern blotting, bake the nitrocellulose filter for 2 h at 80°C under vacuum.
2. Place the filter in a heat-sealable polythene bag and add prehybridization buffer (0.08 mL/cm² nitrocellulose). Prehybridize overnight at 42°C.
3. Biotin-labeled probe preparation: Double-stranded DNA probes should be denatured by heating at 95°C for 5 min and snap cooled on ice *or* by brief treatment (10 min at room temperature) with an equal volume of 100 mM NaOH prior to addition to the hybridization buffer. Single-stranded probes (e.g., M13 DNA, RNA) require no pretreatment before addition to the hybridization buffer. The recommended biotin-labeled probe concentration in the hybridization buffer is 20–50 ng/mL.
4. For hybridization, add the probe to the hybridization buffer just before use. Remove the prehybridization buffer from the bag and add the hybridization buffer containing the biotin-labeled probe to the filter (0.08 mL/cm²nitrocellulose). Remove air bubbles and heat seal the bag. Hybridize in a shaking water bath (gentle agitation) for 20–24 h at 42°C.
5. Carry out the following posthybridization washes with agitation:
   a. Wash filters in 300 mL of 2x SSC, 0.1% SDS for 15 min at room temperature. Repeat two more times.
   b. Wash filters in 300 mL of 0.1x SSC, 0.1% SDS for 20 min at 65°C. Repeat two more times.

6. Gently blot the filter between two sheets of filter paper. The resulting slightly moistened filter is then placed in a heat-sealable polythene bag for colorimetric detection.

3.3. Colorimetric Detection of the Biotin-Labeled Probes

The following protocol (1), adapted from Leary et al. (6), describes the conditions for the detection of the biotin-labeled probe on a 50-cm^2 sheet of nitrocellulose. We recommend 0.08 mL solution/cm^2 filter.

1. Place the filter in a heat sealable polythene bag. Add 4 mL of 3% (w/v) BSA in Buffer 1. Seal the bag and incubate at 42°C for 20–30 min.
2. Remove the BSA solution from the bag and replace it with 4 mL of avidin-alkaline phosphatase (1 µg/mL) in Buffer 1. (The avidin-alkaline phosphatase should be prepared immediately before use from the 500x stock solution.) Incubate at room temperature for 10–15 min. Gently agitate the bag contents occasionally.
3. Remove the filter from the bag and carry out the following washes with agitation (each wash should be at least 300 mL).
    a. Wash thrice for at least 20 min each time in Buffer 1.
    b. Wash two times for at least 10 min each time in Buffer 2.
    Transfer the filter to a new polythene bag.
4. In a polypropylene or glass tube, prepare 4.0 mL of the substrate solution (0.3 mg/mL NBT and 0.2 mg/mL BCIP in Buffer 3) as follows: Add 16 µL of NBT solution (250x stock) to 4.0 mL of Buffer 3, and gently mix by inverting the tube; 16 µL of BCIP solution (250x stock) is then added to this solution, followed by gentle mixing. This substrate solution can be freshly prepared just prior to use, or it can be stored in aliquots at −20°C for 1–2 wk.
5. Add the substrate solution to the filter, seal the polythene bag, and allow the color reaction to proceed in subdued light or in the dark.

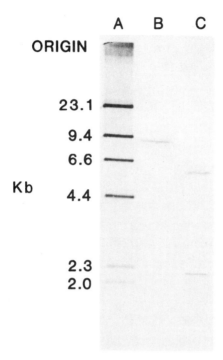

Fig. 2. Southern blot hybridization analysis of chicken genomal DNA with a photobiotin-labeled histone H5 DNA probe. Chicken genomic DNA (8 μg) was digested to completion with Hind III (lane B) or SacI (lane C), separated electrophoretically on a 1% agarose gel, and transferred to nitrocellulose. After prehybridization, the filter was hybridized with a photobiotin-labeled 2.5-kb DNA fragment containing the whole chicken histone H5 gene. Hybridization was done for 23 h at 42°C; probe concentration was 20 ng/mL and detection time was 2.5 h. Hind III cut λ DNA (20 ng) labeled with photobiotin was used for markers (lane A). Bands of 9.0 (in lane B), 6.2, and 2.1 kb (in lane C) were detected (from J.L. McInnes, S. Dalton, P.D. Vize, and A.J. Robins, unpublished data).

6. Terminate the color development in Buffer 4.
7. Filters can be stored dry in a sealed polythene bag and should always be protected from strong light.

An example of the use of a photobiotin-labeled DNA probe to detect the unique chicken histone H5 gene by Southern hybridization analysis is shown in Fig. 2.

4. Notes

4.1. Labeling of Nucleic Acids with Photobiotin

1. Photobiotin™ acetate (BRESA) is *light sensitive* and is supplied as a solid in a light-proof canister that should only be opened under subdued light. Photobiotin acetate is stable for at least a year when stored at –20°C in the dark. It is also likely to be stable for long periods when stored dry at 4°C.

2. Photobiotin acetate as a solution in water is stable for at least a year when stored –20°C in the dark. A solution of photobiotin acetate in a sealed Eppendorf tube is stable on the laboratory bench under normal lighting conditions for a minimum of several hours as determined by no change in its spectrum (Fig. 3). It is considered advisable to only handle photobiotin acetate under subdued light, however, and for as short a period as practicable.

3. Since photobiotin acetate will react with any organic material, it is *essential* that nucleic acids to be used as probes be *highly purified* and free from any contaminating DNA, RNA, proteins, or buffers such as Tris. Inorganic salts may also inhibit the biotinylation of nucleic acids. We employ routine nucleic acid purification procedures (3,7) in preparing double-stranded and single-stranded (M13) DNA probes, and, prior to the photoactivation step, the nucleic acid solution to be biotin-labeled is dissolved in water or 0.1 mM EDTA *only*. Additional purification procedures that may be of use include spermine precipitation and/or Sephacryl chromatography.

4. The use of molar ratios of photobiotin acetate to nucleic acid *higher* than those recommended may result in the precipitation of the nucleic acid.

5. If the biotinylation of large amounts of nucleic acid is required (e.g., >100 µg), the nucleic acid/photobiotin acetate solution can be irradiated either:
   a. As a series of microcapillary tubes (e.g., 100 µL volume in each),

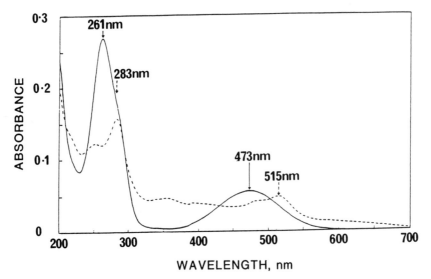

Fig. 3. Absorbance spectrum of photobiotin acetate before and after photolysis. An aqueous solution of photobiotin acetate (0.5 µg/ µL) was diluted with water to a volume of 1.2 mL for spectral analyis before (solid line) and after (broken line) photolysis; reproduced from ref. *1*.

    b. As drop(s) on a sterile plastic or glass dish, again ensuring that the irradiated solution remains cold.

6. Ultraviolet/visible spectral analysis (Fig. 3) can be used to check the quality of the photobiotin acetate sample and/ or to ensure that the photolysis reaction has gone to completion. Photolysis results in a darkening of the solution.

7. If 10–15 µg or more of nucleic acid is used in the biotinylation reaction, then the ethanol pellet after biotinylation should be clearly visible as an orange/brown color. A white nucleic acid pellet at this stage is indicative that the reaction has gone poorly or not at all. If less than 10 µg of nucleic acid is used, then the nucleic acid pellet is normally difficult to see.

8. Ethanol precipitation of the biotin-labeled nucleic acid may also be achieved quantitatively if the solution is cooled to 0°C on ice for 1 min (*8*).

9. The normal recovery rate of biotin-labeled nucleic acid achieved using the recommended protocol is approximately 70%.

4.2. Hybridization Protocol/Colorimetric Detection of Biotin-Labeled Probes

1. Most of our hybridization studies have involved the use of nitrocellulose filters. Lower backgrounds have been routinely achieved using nitrocellulose when compared to the use of nylon-based membranes (*see* Note 13 below).
2. Nitrocellulose filters should never be touched with fingers —it is preferable to wear gloves. Avoid excess pressure on and creasing of the filters—handle only at edges. Points of contact can often lead to increased background color.
3. The type of plastic bag used for the hybridization reaction is important. Low sensitivity of detection of target sequences may be caused by adsorption of biotin-labeled probe by the plastic. The type of plastic may also contribute to excessive background color during color development. We routinely use polythene (0.5 mm thickness) bags.
4. Several dilutions of the biotin-labeled nucleic acid to be used as probe (e.g., pg or ng range) may be spotted on to the nitrocellulose filter as a check for both the biotinylation and color development reactions.
5. Biotinylated DNA markers have proven very useful during Southern hybridization analysis. For example, a Hind III digest of λ DNA can be labeled with photobiotin and used as a marker during gel electrophoresis. After transfer to nitrocellulose by standard techniques, followed by the hybridization reaction, marker DNA bands can be located during the color development. With 20 ng of marker DNA per track, strong purple-colored bands appear in approximately 15–30 min.
6. Photobiotin-labeled single-stranded M13 DNA probes appear stable for at least 12 mo when stored at –20°C. There

has been no evidence of breakdown of photobiotin-labeled double-stranded DNA probes after 8 mo at –20°C. In comparing single-stranded and double-stranded photobiotin-labeled DNA probes over an 8-mo period, no major differences in levels of sensitivity have been observed.

7. Avidin–alkaline phosphatase (Sigma, A-2527) is routinely used in our laboratory for colorimetric detection. Other detection protocols (e.g., BRL: streptavidin, biotinylated polymer of alkaline phosphatase; and Enzo Biochem: streptavidin/biotinylated acid phosphatase complex) appear to be no more sensitive.

8. The color development should be carried out in subdued light or in the dark in order to decrease nonspecific background color. Maximum color development normally is obtained within 3–4 h. The optimal period of color development will vary, however, depending upon the amount of biotin-labeled probe annealed to the target nucleic acid, and may range from 5 min to 16 h (overnight). The hybridization signal will be most evident on only one side of the filter. Overnight incubations can result in background problems.

9. If colorimetric detection cannot be carried out immediately after hybridization, the nitrocellulose filters can be stored desiccated in sealed polythene bags for several weeks.

10. Nitrocellulose filters are not reused since the dyes NBT and BCIP appear to bind irreversibly to the nitrocellulose.

11. The color intensity of the hybridization signal on the nitrocellulose filter may decrease on drying. To restore the signal to its original strength, simply rehydrate the filter.

12. To minimize background coloration, we recommend thorough washing of the filters (gentle agitation) both at the posthybridization and the colorimetric detection steps. If backgrounds are found to be unacceptable, then they can usually be decreased by lowering the probe concentration during hybridization. If *persistent* background problems

arise with your particular protocol, try omitting the probe as a means of further investigating the problem.

13. We have carried out limited studies with two nylon-based membranes: Gene Screen (NEN) and Zeta Probe (BioRad Laboratories). Acceptable backgrounds have been achieved with these membranes provided a strong blocking procedure (9) is followed after hybridization [see section 3.3, step 1; incubate with 3% (w/v) BSA at 65°C for 60 min instead of the 42°C incubation for 20–30 min].

14. Nitrocellulose filters may be photographed under tungsten illumination on Kodak Technical Pan Film 2415 (100 ISO) using a green tricolor (Kodak No. 58) filter. Standard high-contrast processing is recommended.

References

1. Forster, A.C., McInnes, J.L., Skingle, D.C., and Symons, R.H. (1985) Non-radioactive hybridization probes prepared by the chemical labelling of DNA and RNA with a novel reagent, photobiotin. *Nucleic Acids Res.* **13**, 745–761.

2. Hames, B.D. and Higgins, S.J. (1985) *Nucleic Acid Hybridization: A Practical Approach* IRL Press, Oxford and Washington.

3. Maniatis, T., Fritsch, E.F., and Sambrook, J. (1982) *Molecular Cloning: A Laboratory Manual.* Cold Spring Harbor Laboratory, Cold Spring Harbor, New York.

4. Meinkoth, J. and Wahl, G. (1984) Hybridization of nucleic acids immobilized on solid supports. *Anal. Biochem.* **138**, 267–284.

5. Thomas, P.S. (1983) Hybridization of denatured RNA transferred or dotted to nitrocellulose paper. *Meth. Enzymol.* **100**, 255–266.

6. Leary, J.J., Brigati, D.J., and Ward, D.C. (1983) Rapid and sensitive colorimetric method for visualizing biotin-labeled DNA probes hybridized to DNA or RNA immobilized on nitrocellulose: Bio-blots. *Proc. Natl. Acad. Sci. USA* **80**, 4045–4049.

7. Barker, J.M., McInnes, J.L., Murphy, P.J., and Symons, R.H. (1985) Dotblot procedure with [^{32}P]DNA probes for sensitive detection of avocado sunblotch and other viroids in plants, *J. Virol. Meth.* **10**, 87–98.

8. Focus: BRL/Life Technologies Inc. (1985) Ethanol precipitation of DNA. **7**(4), 1–2.

9. *Focus*: BRL/Life Technologies Inc. (1985) DNA detection spot. **7**(2), 11.

Chapter 31

Enzyme-Labeled DNA Probes

J. Lesley Woodhead, Rachel Fallon, Hermia Figueiredo, and Alan D. B. Malcolm

1. Introduction

Many research workers and clinicians working in routine laboratories are now using the "tools" developed by molecular biologists over the past 10 years to produce DNA sequences that can be used as probes to detect specific genes. Thus it is possible to screen food for salmonella (1) or other organisms in a test much more quickly than conventional methods involving culturing the organisms. The presence of certain types of human papillomavirus in cervical smears can be used as an indicationof neoplasia (2). DNA probes can also be used to screen antenatally for diseases such as sickle cells anemia (3) or β^{o}-thalassemia (4).

Initially, the DNA probe was labeled using phosphorus (^{32}P)-deoxyribonucleotides and then detected by autoradiography. The use of ^{32}P has several drawbacks, however: it has a short half-life (14 d), is expensive, and presents a safety hazard. We present here a method involving direct linkage of the enzyme horseradish peroxidase (HRP) to the DNA, which is then used as a probe for specific DNA sequences.

The method used is that of Renz and Kurz (5). Two main steps are involved. First, the enzyme is covalently linked to a synthetic polymer, which carries many primary amino groups. This positively charged macromolecule can then be cross-linked to the single-stranded DNA, which is to be used as a probe. The macromolecule used is polyethyleneimine G-35, which has a molecular weight of about 1400, but any other polymeric cation could be used.

2. Materials

All reagents are of Analar quality, and deionized distilled water is used.

2.1. Construction of Enzyme-Labeled Probe

1. Horseradish peroxidase (grade I, Boehringer). Store at 4°C, desiccated.
2. p-Benzoquinone (Sigma). Store in the dark. A stock solution of 30 mg/mL is freshly made in ethanol.
3. Sephadex G-100 (Pharmacia).
4. Polyethyleneimine G-35 (obtained from BASF, Ludwigshafen, FDR).
5. Linear DNA probe 1 μg in 20 μL of 5 mM sodium phosphate, pH 6.8.
6. 5% (v/v) glutaraldehyde (Sigma).
7. Polyethyle glycol 8000 (Sigma).
8. 1.5M L-Glycine in 5 mM sodium phosphate, pH 6.8.
9. 90 mM Sodium phosphate, pH 6.0.
10. 0.15M NaCl.
11. Freshly made 1M sodium bicarbonate.

2.2. Hybridization of Enzyme-Labeled Probe

1. 3 mg/mL solution of tRNA (Boehringer).
2. Stock 10% (w/v) sodium dodecyl sulfate (SDS).
3. 40% (w/v) Solution of polyethylene glycol 8000 (Sigma).

4. Deionized formamide (Analar BDH) prepared as follows: Add 5 g of Amberlite monobed mixed resin (BDH) to 50 mL of formamide and stir for 4–6 h. Filter and store at –20°C in the dark.
5. 50x Denhardt's solution (6), which can be stored at –20°C. 1x Denhardt's contains 0.02% bovine serum albumin (Sigma, Fraction V), 0.02% Ficoll (Pharmacia), 0.02% polyvinylpyrrolidone (Pharmacia).
6. 20x SET (1x SET is 0.15M NaCl, 0.03M Tris-HCl, pH 8.0, 1 mM EDTA). Can be stored at room temperature.

2.3. Staining for HRP Using o-Dianisidine

1. o-Dianisidine (Sigma). This should be stored in the dark. *It is carcinogenic and should be treated with care.*
2. Ethanol.
3. 30% (w/v) Hydrogen peroxide (Sigma).
4. 1M Imidazole stock.
5. 1M Tris-HCl, pH 7.4

3. Methods

3.1. Preparation of Solution A

This solution contains polyethyleneimine conjugates and can be stored at 4°C without significant deterioration for several months.

1. Add 200 μL of 90 mM sodium phosphate to 20 mg of HRP in an Eppendorf tube.
2. To the above solution add 60 μL of a freshly prepared solution containing 30 mg/mL of p-benzoquinone in ethanol.
3. Leave the mixture rotating in the dark for 60 min at 37°C. Use a rotating wheel in an oven or hot room, and cover the tube with aluminium foil.
4. To remove unreacted benzoquinone, pass the solution through a 2-mL column packed with G-100. Run and

equilibrate the column with 0.15M NaCl. A Pasteur pipet can be used as the column with a small plug of siliconized glass wool in the bottom. Two peaks should be seen: a brown peroxidase peak, followed by a yellow p-benzoquinone peak.

5. Pool the brown HRP-containing fractions (approximately 1.5 mL), increase the pH by adding 180 μL of freshly made 1M sodium bicarbonate solution, and then add 2.7 mL of polyethyleneimine (PEI), which is supplied as a solution.

6. Allow crosslinking to take place by rotating the tube in the dark, at 37°C, as before, but for 20 h.

7. Decrease the pH by dialyzing against several changes of 5 mM sodium phosphate buffer, pH 6.8, over 24 h.

8. The resultant Solution A can be kept at 4°C as mentioned before.

3.2. Construction of the Probe

If the DNA to be used is in the form of a circular double-stranded plasmid, it must first be linearized with any suitable restriction enzyme so that it can be denatured, as follows:

1. Heat 1 μg of DNA in 20 μL of sodium phosphate, pH 6.8, at 100°C for 3 min, and then plunge into a bath of solid carbon dioxide and propanol. [*Note*: This step is even carried out when the probe is already single-stranded (e.g., M13 cloned DNA), since complementary sequences within the same strand may produce hairpin loop-type structure and cut down the efficiency of crosslinking.]

2. Add 20 μL of Solution A to the frozen pellet.

3. As soon as the mixture has thawed, add 6 μL of a 5% (v/v) glutaraldehyde solution. This will crosslink the polyethyleneimine–enzyme conjugates to the bases on the DNA, mainly via the amino groups.

4. Incubate at 37°C for 10 min.

5. To separate the DNA–PEI–HRP from the free PEI–HRP

conjugates, add 28 µL of a 40% solution (w/v) of poly-ethyleneglycol 8000.

6. The DNA–enzyme complexes will precipitate and can be separated by centrifugation in a microfuge for 6 min.

7. Dissolve the precipitate in 20 µL of 1.5*M* L-glycine in 5 m*M* sodium phosphate, pH 6.8. This solution can then be used for hybridization.

3.3. Hybridization Enzyme-Labeled Probe

This method can be used for detecting DNA that has been transferred to cellulose nitrate filters by Southern blotting (*see* Vol. 2 of this series) or has been applied the filter using a dot or slot blotter, such as can be bought from BioRad. Other membranes such as Zeta Probe or Biodyne have been found to give higher backgrounds with detection methods using enzymes.

1. Soak the filters in 4x SET, 10x Denhardt's, and 0.1% sodium dodecyl sulfate (SDS) for 1 h at 38°C.

2. Incubate the filters for another hour at 38°C in 20 mL of prehybridization buffer [50% (v/v) deionized formamide, 2x Denhardt's, 4x SET, 0.1% SDS, and 30 µg/mL yeast tRNA].

3. Hybridize the filters overnight at 38°C in 10 mL of prehy-bridization buffer, which also contains 6% polyethyle-neglycol and 500 ng of enzyme-labeled probe.

4. Wash the filters three times in 50 mL of 50% deionized formamide, 0.4% SDS, 0.5% SET, at 38°C for 60 min. Then wash at room temperature for 2 x 20 min in 2x SET.

5. Prior to staining (*see* following section) soak the filters in 0.1*M* sodium phosphate, pH 8.0.

3.4. Staining for HRP Using o-Dianisidine

1. Incubate the filters at room temperature in the dark in the substrate solution. This contains, per 10 mL, 6 mg of *o*-dianisidine in 2 mL of ethanol, 0.03% hydrogen peroxide, 10 m*M* imidazole in 10 m*M* Tris-HCl, pH 7.4.

2. Incubate with gentle shaking for 30 min to 2 h. The *o*-dianisidine is converted to a brown compound by hydrogen peroxide.

4. Notes

1. This staining method can detect about 27 amol of a plasmid DNA (~5 kb), which is about 100 pg. Using the luminescent detection system (*see* Chapter 32) as little as 13 amol (approximately 50 pg) of a single band of DNA can be detected.
2. *o*-Dianisidine is used as a sensitive substrate for HRP, despite being carcinogenic. Tetramethyl benzidine (a safer alternative), a sensitive substrate for HRP (*8*), does not detect this enzyme-labeled probe on filters and binds to DNA itself, which can produce spurious results.

References

1. Fitts, R., Diamond, M., and Hamilton, C. (1983) DNA-DNA hybridisation assay for detection of Salmonella spp. in foods. *Appl. Environ. Microbiol.* **46**, 1146–1151.
2. Wickenden, C., Malcolm, A.D.B., Steele, A., and Coleman, D.V. (1985) Screening for wart virus infection in normal and abnormal cervices by DNA hybridisation of cervical scrapes. *Lancet* **i**, 65–67.
3. Orkin, S.H., Little, P.F.R., Kazazian, H.H., and Bochm, C.D. (1982) Improved detection of the sickle mutation by DNA analysis. *N. Eng. J. Med.* **307**, 32–36.
5. Renz, M. and Kurz, C. (1984) A colorimetric method for DNA hybridisation. *Nucleic Acids Res.* **12**, 3435–3444.
6. Denhardt, D. (1966) A membrane filter technique for the detection of complementary DNA. *Biochem. Biophys. Res. Commun.* **23**, 641–645.

Chapter 32

Luminescent Detection
of Specific DNA Sequences

J. Lesley Woodhead, Hermia Figueiredo,
and Alan D. B. Malcolm

1. Introduction

The process whereby part of the energy evolved in a chemical reaction is emitted as light is known as luminescence. When luminol is oxidized by hydrogen peroxide in the presence of horseradish peroxidase (HRP), light is evolved (1). This is shown in Fig. 1. Luciferin not only enhances the reaction, but prolongs the emission over several minutes (2). The light may be measured in a luminometer or visualized as the blackening of photographic film, as will be described.

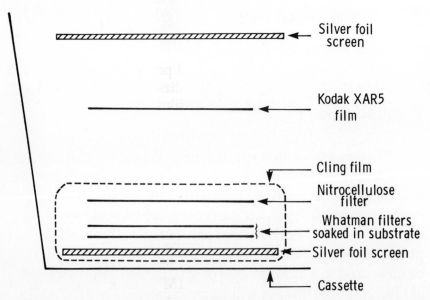

Fig. 1. Oxidation of luminol in the presence of hydrogen peroxide and HRP (the catalyst).

This method can be used for DNA sequence detection on blots that have been hybridized with biotinylated DNA (Chapter 33) or with HRP-labeled DNA (Chapter 31). The method has advantages over conventional methods using (^{32}P)-labeled DNA in that exposure time is measured in minutes rather than hours and it is safer to work with.

Fig. 2. Autoradiography cassette used for the detection of HRP using luminescent detection.

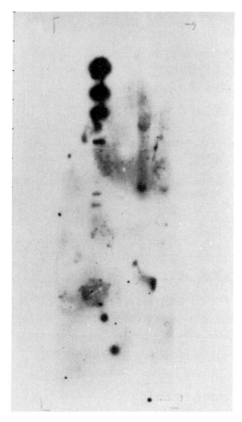

Fig. 3. The detection of biotinylated probe DNA with streptavidin–peroxidase. Peroxidase was visualized with luminol and luciferin after a 60-min exposure to X-ray film.

2. Materials

1. Autoradiography cassette.
2. Whatman 3MM paper.
3. Kodak XAR5 X-ray film to fit the cassette.
4. A stock solution containing 1.35 mM H_2O_2, 0.625 mM luminol, 18 µM Luciferin in 0.1M sodium phosphate, pH 8.0. Luminol and Luciferin can be obtained from Sigma, Poole, Dorset. The solutions should be freshly made.

3. Method

1. Dampen a pad of Whatman 3MM paper with the stock solution of luminol and Luciferin.
2. Place the pad in the autoradiograph cassette and put the nitrocellulose filter that has been treated with streptavidin-peroxidase (Chapter 33) on the pad, as shown in Fig. 2.
3. Expose the filter to a preflashed X-ray film for an optimal exposure of 30 min.

4. Notes

1. As shown in Fig. 3, it is possible to detect a single copy gene sequence on a human blot prepared from 20 µg of total human DNA digested with EcoRI. The 201-base pair MstII fragment of the β-globin gene is also highly homologous to β-globin; hence two bands are clearly visible.

References

1. Auses, J.P., Cook, S.L., and Maloy, J.T. (1975) Chemiluminescent enzyme method for glucose. *Anal. Chem.* **47**, 244–249.
2. Whitehead, T.P., Thorpe, G.H.G., Carber, T.J.N., Groneutt, C., and Kricka, L.J. (1983) Enhanced luminescent procedure for sensitive determination of peroxidase-labelled conjugate in immunoassay. *Nature* **305**, 158–159.

Chapter 33

Detection of DNA Sequences Using Biotinylated Probes

J. Lesley Woodhead, Hermina Figueiredo, and Alan D. B. Malcolm

1. Introduction

It was found (1) that biotinylated analogs of dUTP, which contain a biotin molecule covalently bound to the C-5 position of the pyrimidine via an allylamine linker arm, can be used as substrates for DNA polymerase. Figure 1 shows the structure of biotin-11-dUTP, a commonly used substrate for biotinylation by the method of nick translation (2). The substituted DNA molecule will have similar hybridization properties to the unsubstituted DNA. After hybridization to the test DNA, the biotin probe can be located by a variety of methods, including incubation with streptavidin–peroxidase conjugate, which will be described here.

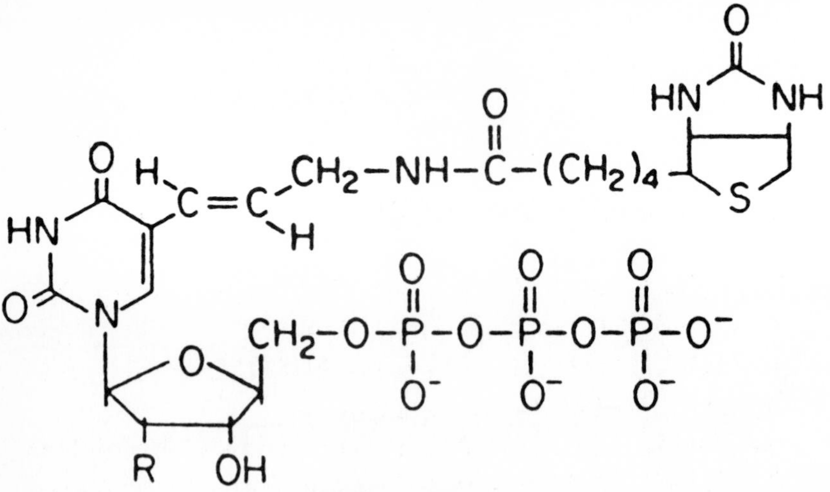

Fig. 1. Structure of biotin-11-UTP.

As already mentioned, the biotinylated dUTP can be incorporated into the DNA by nick translation. Two enzymes are used: DNAse I, which nicks the double-stranded probe, and DNA polymerase I, which repairs the nicked DNA. The enzymes must be present in correct ratios, and it is usual to buy them already made in the correct buffer as a "nick translation kit," sold by suppliers such as Amersham and BRL.

2. Materials

2.1. Biotinylation of DNA by Nick Translation

1. Nick translation kit from Amersham International. (Each kit contains DNAse I, DNA polymerase, and solutions of unlabeled deoxyribouncleotides, with full instructions.)
2. Biotin-11-dUTP (BRL).
3. Sephadex G-50 (Pharmacia).
4. Double-stranded DNA to be biotinylated.

5. 3x SET.
6. Cellulose nitrate strips, divided into 1 x 1 cm squares.
7. Materials for detection of biotinylated DNA; *see below*.

2.2. Hybridization of Biotinylated DNA

1. Nitrocellulose filters on which DNA has been immobilized using the method of Southern (3).
2. 10x SSPE (1x SSPE = 180 mM NaCl, 10 mM sodium phosphate, 1 mM EDTA, pH 7.4).
3. 50x Denhardt's solution (1x Denhardt's = 0.02% BSA, 0.02% polyvinylpyrrolidine 360, and 0.02% Ficoll 400).
4. 10% SDS.
5. 3 mg/mL sheared salmon sperm DNA.
6. 50% (w/v) dextran sulfate.
7. 20x SSC (1x SSC = 0.15M NaCl, 0.015M sodium citrate, pH 7.0).

2.3. Detection of Biotinylated DNA with Streptavidin–Peroxidase Conjugate

1. Streptavidin–peroxidase conjugate from Boots-Celltech (our sample was a kind gift from Dr. John Wright of the same).
2. Tween-20 (Sigma).
3. *o*-Dianisidine (Sigma).
4. Blocking buffer: 100 mM NaCl, 100 mM Tris-HCl, pH 7.5, 3 mM MgCl$_2$, 0.5% Tween 20.
5. Conjugate solution: streptavidin peroxidase (5 µg/mL) in 100 mM NaCl, 100 mM Tris-HCl, 3 mM MgCl$_2$, 0.05% Tween 20.
6. *o*-Dianisidine stain: This contains, per 10 mL, 6 mg *o*-dianisidine in 2 mL ethanol, 0.03% hydrogen peroxide, 100 mM Tris-HCl, pH 7.4, 10 mM imidazole (freshly made, kept dark).

3. Method

3.1. Biotinylation of DNA by Nick Translation

Note: Nick translation can only be carried out with double-stranded DNA. Single-stranded probes may be biotinylated use the oligolabeling method (4).

1. Add 1 µg of DNA to 50 µL of 50 mM Tris-HCl, 5 mM MgCl$_2$, pH 7.8, containing 2.5 U of polymerase I, 250 ng of DNAse I, 250 ng of BSA, and 1 nmol each of dATP, dCTP, dGTP, and biotin-11-dUTP.
2. Incubate at 15°C for 90 min.
3. Pack a 1-mL siliconized glass Pasteur pipet with Sephadex G-50 using 3x SET to equilibrate and run the column.
4. Load the nick-translated DNA onto the column and collect two drop (approximately 100 µL) fractions.
5. Assay the fractions for biotin, by spotting out 1-µL amounts onto the cellulose nitrate strips and using the antibody method described below.
6. Two active peaks should be found, the former containing biotinylated-DNA, and the latter, free biotinylated nucleotides. Pool the first peak for hybridization.

3.2. Hybridization of Biotinylated DNA

1. Incubate nitrocellulose filters on which DNA has been immobilized in 4x SSPE, 6x Denhardt's, 300 µg/mL sheared salmon sperm DNA at 65°C for 6 h.
2. Transfer filters to hybridization buffer 4x SSPE, 2x Denhardt's, 200 µg/mL salmon sperm DNA, 0.1% SDS, 10% dextran sulfate, and 200 ng/mL denatured biotinylated DNA probe. Hybridize for 16 h at 65°C.
3. Wash twice in 2x SSC, 0.1% SDS at room temperature for 5 min.
4. Wash twice in 0.2% SSC, 0.1% SDS, room temperature for 5 min.

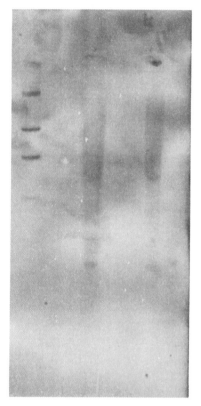

Fig. 2. Lane 1 contains 2 ng of Hind III digest of λ DNA hybridized to 0.5 µg of biotinylated λ DNA. Lanes 2 and 3 contain 20 and 15 µg of EcoRI-digested human DNA, respectively. These were hybridized with 1 µg of biotinylated JL200. The biotinylated probes were detected with streptavidin–peroxidase and stained with the chromogen o-dianisidine.

5. Wash twice stringently for 5 min in 0.16x SSC, 0.15% SDS at 50°C.
6. Rinse filters briefly in 2x SSC, 0.1% SDS at room temperature.

3.3. Detection of Biotinylated DNA with Strepatvidin–Peroxidase Conjugate

1. Incubate filters in blocking buffer at 22°C for 90 min (5).
2. Incubate in conjugate solution for 15 min at 22°C.

3. Wash three times in 100 mL of blocking buffer.
4. Incubate filters in the dark for 2 h with the *o*-dianisidine solution. Alternatively the filters can be stained with luminol using the method described in Chapter 32.

4. Notes

1. An example of human DNA blot can be seen in Fig. 2. It has been stained with *o*-dianisidine. A single copy sequence can be seen when 20 µg of DNA are loaded into a single track.

References

1. Langer, P.R., Waldrop, A.A., and Ward, D.C. (1981) Enzymatic synthesis of biotin-labeled polynucleotides: Novel nucleic acid affinity probes. *Proc. Natl. Acad. Sci. USA* **78**, 6633–6637.
2. Rigby, P.W.J., Dieckmann, M., Rhodes, C., and Berg, P. (1977) Labelling DNA to high specific activity *in vitro* by nick translation with DNA polymerase I. *J. Mol. Biol.* **113**, 237–251.
3. Southern, E.M. (1975) Detection specific sequences among DNA fragments separated by gel electrophoresis. *J. Mol. Biol.* **98**, 503–517.
4. Feinberg, A.P. and Vogelstein, B. (1984) A technique for radiolabelling DNA restriction fragments to high specific activity. *Anal. Biochem.* **137**, 266–267.
5. Chan, V.T.W., Fleming, K.A., and McGee, J.O.D. (1985) Detection of subpicogram quantities of specific DNA sequences on blot hybridization with biotinylated probes. *Nucleic Acids Res.* **13**, 8083–8091.

Chapter 34

Crosslinking of Single-Stranded DNA to Resins

J. Lesley Woodhead, Jane A. Langdale, and Alan D. B. Malcolm

1. Introduction

Specific fragments of single-stranded DNA may be covalently immobilized to solid supports for various reasons. The support can be used as an affinity column for purifying DNA from a test sample, to detect viral or bacterial DNAs, or for antenatal diagnosis of genetic diseases such as sickle cell anemia using a sandwich hybridization method developed in this laboratory (1).

Trial experiments have been carried out (2) comparing the linking of DNA restriction fragments to various resins including Sephadex G-50, G-200 cellulose, and Sephacryl S-200 and Sephacryl S-500. Two main methods were tested: that of Allfrey and Inoue (3), which uses carbodiimide to link the terminal phosphate of the DNA to the resin, and the diazotization method of Seed (4). The latter method, with minor modifications (2), was found to give higher values of covalent coupling with lower backgrounds, when compared to the carbodiimide method. DNA fragment size (0.2–3 kb) had no significant effect on efficiencies of immobilization by diazotization, whereas small fragments (<341 base pairs) were coupled most efficiently using carbodiimide. Of all the resins tested, Sephacryl S-500 was found to be the most suitable, giving higher levels of coupling compared to other resins. Thus we will describe here the Seed method (4) of linking DNA to Sephacryl S-500 as reported (2).

Details of practical procedures are divided into four sections: preparation of the aminophenylthioether derivative of the Sephacryl S-500; diazotization of the APTE resin; assay of DNA linked to resin by radiolabeling; and assay of linked DNA by micrococcal nuclease digestion.

2. Materials

All reagents used (unless stated otherwise) are of analar quality.

2.1. Preparation of Aminophenylthioether Derivative of Sephacryl S-500

It is essential that distilled deionized water be used for all solutions. The derivative will be stable in the dark at 4°C for up to 1 year in water.

1. 1M NaOH.
2. Sephacryl S-500.
3. 1,4-Butanediol diglycidylether (from Aldrich Chemical Co., Gillingham, Dorset).
4. Acetone.
5. 2-Aminothiophenol (from Aldrich Chemical Co., ibid).
6. 0.1M HCl.

2.2. Diazotization of APTE Resin

1. 1.8M HCl.
2. 10 mg/mL solution of sodium nitrite (freshly made).

3. Starch iodide paper.
4. A stock solution of 25 mM sodium phosphate, pH 6.0, and dimethylsulfoxide (DMSO) 20/80 (v/v).
5. DNA solution: prepared by dissolving dried down DNA pellet in 50 μL of sodium phosphate, pH 8.0, and adding 200 μL of DMSO. The amount of DNA added to this amount of solution can vary from 140 to 560 pmol of DNA and is suitable for linking to 1 g of Sephacryl.
6. 0.4M NaOH.
7. 10 mM Tris-HCl, pH 7.5, 1 mM EDTA.

2.3. Assay of DNA Linked to Resin by Radiolabeling

1. A trace amount; 100 ng of [^{32}P]kinase-labeled DNA.

2.4. Assay of DNA Linked to Resin by Micrococcal Nuclease Digestion

1. 0.1M Sodium borate, pH 8.8.
2. 0.1M Calcium chloride.
3. *Staphylococcus aureus* nuclease (micrococcal nuclease) (from Boehringer Mannheim, Lewes, East Sussex).

3. Methods

3.1. Preparation of 2-Aminophenylthioether Derivative of Sephacryl S-500

1. Wash 20-g aliquots of Sephacryl S-500 very well with distilled water and transfer to a 50-mL plastic screw top tube.
2. Add 20 mL of 1M NaOH and 1.7 mL of 1,4-butanediol diglycidyl ether. Rotate tube slowly end-over-end at room temperature overnight.
3. Filter the resin using a Buchner funnel, pressing out as much aqueous solution as possible by means of a spatula. Return moist resin to the plastic tube.

4. Add 23 mL of acetone and 2.86 mL of 2-aminothiophenol before rotating as before, overnight. Take care to seal the bottle since aminothiophenol is toxic.
5. In a fume hood, filter the resin in a Buchner with 200 mL each in turn of acetone, 0.1M HCl, water, 0.1M HCl, and water again.
6. Resuspend the resin in water. This APTE derivative is stable in the dark for up to a year at 4°C.

3.2. Diazotization of the APTE Resin and Linkage of DNA

1. Wash 1 g of the APTE-Sephacryl S-500 in a sintered glass funnel with 30 mL of distilled water and then resuspend in 3.33 mL of 1.8M HCl plus 1 mL of water.
2. Keep on ice for 30 min.
3. Add 33 µL aliquots of the 10 mg/mL sodium nitrite. After each addition, test for excess nitrite by using starch iodide paper, which will remain blue when sufficient aliquots have been added to the APTE-Sephacryl. Leave about 5 min between each addition of the nitrite to allow diazotization to occur. Typically 4 x 33 µL aliquots are necessary. If less than three are required, this indicates low diazotization, which will reduce the amount of DNA that can be bound to the Sephacryl subsequently. In this case start again.
4. Immediately transfer the resin to an ice-cold sintered glass filter funnel and wash with 50 mL of ice-cold water. Then wash with 5 mL of ice-cold 25 mM sodium phosphate, pH 6.0: DMSO (20/80 v/v). The Sephacryl will turn a darker yellow during this last step.
5. When the resin becomes dry, put about half of it into an Eppendorf tube and add 250 µL of the DNA solution.
6. Incubate for 2 d at room temperature, while rotating end-over-end as before.
7. Finally pour into a sintered glass funnel and wash with water and then with 0.4M NaOH at 4°C. Wash and store in 10 mM Tris-HCl, pH 7.5, 1 mM EDTA.

Once the DNA is linked, it remains relatively stable at 4°C. Over 50% of the DNA remains bound after 6 mo. The following are two methods for determining the amount of DNA that has been linked by this method.

3.3. Assay of DNA Linked to Resin by Radiolabeling

1. Incorporate a small amount of [^{32}P]kinase labeled DNA into the reaction described above and count by Cerenkov counting.
2. After incubating the DNA with the Sephacryl to immobilize it to the resin (see above), wash the resin well and count again.
3. Work out the percentage coupling by dividing the final cpm measured by the initial cpm added.

3.4. Assay of DNA Linked to Resin by Micrococcal Nuclease Digestion

1. Wash 50 mg of the resin with 0.1M sodium borate, pH 8.8, and transfer to an Eppendorf tube.
2. Suspend in 400 µL of sodium borate and 15 µL of 0.1M CaCl$_2$. Add Staphylococcus nuclease to a concentration of 300 U/mL and incubate at 37°C overnight.
3. Centrifuge and measure the absorbance of the supernate at 260 nm. Use a 1-cm path length cuvet. The digested DNA will have an optical density of 1 at a concentration of 38 µg/mL. This is compared to an optical density of 1 for a 50 µg/mL solution of native DNA.

4. Note

1. The radiolabeling method gives a higher estimate for coupling efficiency (about 80%) than does the nuclease method (typically 50%). The latter method, which is a measure of the amount of DNA accessible to nuclease, also

reflects the fraction of the DNA that is available for hybridization.

References

1. Langdale, J.A. and Malcolm, A.D.B. (1985) A rapid method of gene detection using DNA bound to Sephacryl. *Gene* **36**, 201–210.
2. Langdale, J.A. (1985) Gene detection using immobilized DNA probes. PhD Thesis, University of London.
3. Allfrey, V.G. and Inoue, A. (1978) Affinity chromatography of DNA binding proteins on DNA covalently attached to solid support. *Meth. Cell. Biol.* **17**, 253–270.
4. Seed, B. (1982) Diazotizable acrylamine cellulose papers for coupling and hybridisation of nucleic acids. *Nucleic Acids Res.* **10**, 1799–1810.

Chapter 35

Extraction of RNA from Plants

Adrian Slater

1. Introduction

Current research into the structure and function of plant genes involves the application of many elaborate techniques for gene cloning and analysis. The isolation of pure, intact plant mRNA is required at many stages in this process, e.g., for generation and screening of cDNA clones, for characterization and mapping of cloned genes, and for the study of gene expression. The isolation of RNA from plants, however, often presents more problems than many of the sophisticated procedures used subsequently. Not only is there the problem of RNA degradation by endogenous and exogenous nucleases (common to RNA isolation from any organism), there is also the particular problem of extracting RNA from plant material that may be rich in starch, pectins, phenolics, and various other secondary products. Many of these plant products are not removed during conventional phenol extraction procedures, such as the one described previously in this series (1). Clearly the problems encountered in obtaining clean, undegraded RNA from any particular plant tissue will vary according to the level of nuclease activity and the range of possible contaminants, and it might be expected that different solutions would be required in each case. However, the method presented here has been used successfully to isolate RNA from leaf, root, shoot, fruit, and callus tissue from a number of different plant species.

This method was originally developed to isolate translatable mRNA from tomato fruit (2) and is a combination of a number of different procedures. The first stage involves preparation of a cleared homogenate, based on a method for isolation of tomato fruit polyribosomes (3). Following the removal of protein by phenol and chloroform extraction, much of the remaining carbohydrate is removed, along with DNA and small RNA molecules, by washing with 3M sodium acetate (4). The final stage of the procedure involves passing the cytoplasmic RNA down a cellulose column, prior to oligo-dT cellulose chromatography, to remove contaminants that would otherwise copurify with poly (A)$^+$ RNA on the oligo-dT cellulose (5).

Tomato fruit RNA isolated by this procedure has been used successfully for in vitro translation (2), cDNA cloning (6,7), hybrid-release translation (6,8), and RNA blot hybridization (6).

2. Materials

2.1. Equipment and Chemicals

1. Pestle and mortar: Clean the surface of the pestle and mortar with Decon 90, distilled water, then ethanol, and allow to dry. Cover with foil and precool at −70°C to avoid cracking when cooled rapidly with liquid nitrogen.
2. Polyvinylpolypyrollidone (Sigma): This is a crosslinked insoluble form of polyvinylpyrollidone (PVP).
3. Miracloth (Calbiochem) (for filtration of gelatinous grindates): Cut into convenient shapes for filtration (approximately 20-cm squares), moisten with distilled water, and autoclave.
4. Sigmacell 50 (Sigma): Suspend in 0.1M NaOH and autoclave at 15 psi for 20 min. Pour a small column in a sili-

conized Pasteur pipet plugged with a small wad of siliconized glass wool and wash extensively with sterile distilled water, then with cellulose binding buffer, to remove the brown supernatant. [Cellulose preparations may contain lignins, which can bind poly(A)$^+$ RNA (9), unless removed by alkali treatment] Glassware for cellulose and oligo-dT cellulose should be siliconized with Repelcote (Hopkins and Williams), washed with distilled water, and baked at 180°C for 2 h.

2.2. Solutions

1. Homogenization buffer: 0.2M Tris-HCl, pH 8.5, 0.2M sucrose, 30 mM magnesium acetate, 60 mM potassium chloride. Autoclave at 10 psi to avoid caramelization of the sucrose, and store at room temperature. Immediately before use, add 2-mercaptoethanol to a final concentration of 0.31% (v/v).
2. 10% (w/v) Sodium dodecyl sulfate (SDS): Prepare in sterile distilled water and store at 60°C.
3. 3M Sodium acetate, pH 6.0: Autoclave and store at room temperature.
4. 0.15M Sodium acetate, pH 6.0, 0.5% (w/v) SDS: This can be prepared from solutions 2 and 3 above. Autoclave and store at room temperature.
5. Water saturated phenol. Add 25 mL of sterile distilled water to 100 g of phenol AR and dissolve by stirring. It is often more convenient and safer to add the appropriate volume of water to an unopened bottle of phenol. The phenol solution should be stored at 4°C in the dark bottle. The solution should remain colorless for a few months, but should be discarded when any coloration appears.
6. Cellulose binding buffer: 10 mM Tris-HCl, pH 7.5, 0.5M NaCl, 1 mM EDTA, 0.5% (w/v) SDS. Autoclave and store at room temperature.

Unless otherwise stated, normal precautions to prevent exogenous nuclease contamination should be observed, i.e., all solutions and appropriate plastic ware should be autoclaved at 15 psi, and glassware should be sterilized by dry heat at 180°C for at least 2 h. Disposable gloves should be worn throughout to prevent contamination with "finger" nucleases (10). A more detailed description of precautions necessary to avoid ribonuclease contamination can be found elsewhere (e.g., ref. 11).

3. Method

3.1. Homogenization and Phenol Extraction

1. Freeze the plant tissue by immersion in liquid nitrogen. Large pieces of tissue should be cut into small (approximately 5-mm) pieces with a clean scalpel or scissors and frozen immediately (see Note 2 in section 4). The tissue can be stored under liquid nitrogen for months (glass scintillation vials are convenient containers) or at –70°C for some weeks.

2. Grind a known weight of plant material to a powder under liquid nitrogen in a mortar and pestle precooled with liquid nitrogen. Keep the material cold by frequent addition of nitrogen during the grinding (see Note 3 in section 4).

3. Add polyvinylpolypyrollidone to the powder (17.5 mg/g lant material) (see Note 4 in section 4) and then homogenization buffer (1.75 mL/g plant material). Continue grinding until the buffer thaws.

4. Filter the homogenate by squeezing through two layers of Miracloth into a beaker, on ice (see Note 5 in section 4).

5. Transfer the filtrate to centrifuge tubes or buckets and centrifuge at 20,000g for 10 min at 4°C in a precooled rotor (see Note 6 in section 4).

6. Pour off the supernatant into a flask (large volume) or 50-mL glass centrifuge tube (volumes less than 10 mL), add 0.05 vol of 10% (w/v) SDS, and mix.

7. To the cloudy solution, add an equal volume of water-saturated phenol and mix extensively. Then add the same volume of chloroform AR and mix again. Allow the solution to warm up to room temperature during these steps (*see* Note 7 in section 4).

8. Centrifuge at 25,000g for 10 min at 20°C. Remove the lower organic phase with a Pasteur pipet, leaving the aqueous phase and interphase.

9. Re-extract the aqueous phase with an equal volume of chloroform, mixing thoroughly to disperse the interphase. Centrifuge and remove the organic phase as in step 8.

10. Repeat step 9 until the interphase is reduced to a thin disc of insoluble protein. Two chloroform extractions are usually sufficient to achieve this.

11. Remove the upper aqueous phase with a Pasteur pipet, taking care not to disturb the interphase. Add 0.05 vol of 3M sodium acetate, pH 6.0, and 2.5 vol of absolute ethanol. Mix well and cool to –20°C. The RNA can be stored under ethanol at –20°C or used once complete precipitation has occured.

3.2. Sodium Acetate Wash

1. Centrifuge the ethanol precipitated RNA at 12,000g for 10 min at 4°C in glass Corex tubes (Corning). Pour off the supernatant and drain the tubes by inverting on tissue paper for a few minutes.

2. Resuspend the pellet in cold 3M sodium acetate, pH 6.0. Use a volume of sodium acetate approximately 10 times the volume of the pellet. Vigorous vortex mixing or repeated pipeting may be required to resuspend the pellet completely (*see* Note 8 in section 4). Leave on ice for 20 min.

3. Centrifuge at 12,000g for 30 min at 4°C.

4. Repeat the wash with 3M sodium acetate until clean RNA, indicated by a clear, opalescent pellet, is obtained. This pellet may be much smaller than the original ethanol precipitate.

5.   Resuspend the pellet in 0.15M sodium acetate, pH 6.0, 0.5% (w/v) SDS, and add 2.5 vol of ethanol. Store at –20°C (*see* note 9 in section 4).

3.3. Cellulose Column Chromatography

1.   Centrifuge the ethanol precipitated RNA at 10,000g for 10 min at 4°C. Pour off the supernatant and drain by inverting the tube on tissue paper. Dry the pellet by placing the tube in a vacuum desiccator. Cover the tube with parafilm and prick holes in the cover before drying, to avoid nuclease contamination or losing the pellet during evacuation and release of the vacuum.
2.   Dissolve the pellet in cellulose binding buffer to obtain an RNA concentration of about 1 mg/mL.
3.   Pass the solution down a cellulose column prepared in a Pasteur pipet with a bed volume of about 1 mL, described in section 2. RNA preparations high in certain carbohydrate contaminants are viscous at this stage and tend to clog up the column. In this case, apply pressure with a Pasteur pipet teat. Much of the viscous material appears to remain trapped at the top of the column (*see* note 10 in section 4).
4.   Wash the column with an equal volume of binding buffer and pool with the nonbound RNA sample.
5.   The RNA in binding buffer can be precipitated by addition of 2.5 vol of ethanol, and stored at –20°C, or directly applied to an oligo-dT cellulose column for isolation of poly (A)$^+$ RNA (*12,* and *see* Volume 2).

4. Notes

1.   Steps 1–5 of the homogenization are the most critical with regard to degradation of RNA by released plant nucleases. It is necessary to work rapidly at this stage and ensure that the temperature of the homogenate does not rise above

4°C. Plant cell nucleases are inhibited at high pH (13), but it is conceivable that this method will not be applicable to particularly nuclease-rich plant tissues. Ribonuclease inhibitors [e.g., vanadyl ribonucleoside complex (14) or human placental ribonuclease inhibitor (15)] could be worth trying in this instance. Alternative methods for plant RNA extraction are suggested at the end of this section.

2. Immediate freezing of tissue after detachment from the plant and cutting into smaller pieces is essential in order to avoid artifacts caused by the rapid appearance of wound-induced mRNAs (16).

3. For large amounts of tissue or particularly tough plant material, it may be more convenient to use an electric coffee grinder than a mortar and pestle. The bowl of the grinder can be chilled briefly with a small amount of liquid nitrogen and will grind frozen plant material rapidly, so there is little chance of the tissue heating up. A small domestic coffee grinder may also be cheaper than a conventional mortar and pestle, particularly considering the short life-span of the latter when used regularly with liquid nitrogen.
   If a large amount of material is used in a mortar and pestle, it may be helpful to transfer the frozen powder to a second mortar cooled only to 0°C, before adding the homogenization buffer. Otherwise it can take a long time for a large amount of material to thaw out in the original liquid nitrogen-chilled mortar.

4. Polyvinylpolypyrollidone removes phenolics from plant cell extracts (17). Its use does not appear to be essential for extraction of RNA from all plant tissues.

5. Miracloth is particularly effective for removing plant cell debris, especially cell wall material. Several layers of muslin, however, may make a reasonable substitute in the filtration step.

6. The centrifugation step removes intact nuclei, plastids, and mitochondria. The supernatant will therefore contain

predominantly cytoplasmic RNA, much of which will be associated with polyribosomes. It is possible to isolate intact polyribosomes at this stage and separate them by centrifugation through sucrose gradients (3).

7. The rationale for this particular variation of phenol, then chloroform, extraction is described by Palmiter (4).

8. High molecular weight RNA is insoluble in 3M Na-acetate, whereas small RNA, DNA, and many polysaccharides are soluble in this solution (4). The volume of 3M Na-acetate is fairly critical, since it is possible to lose higher molecular weight RNA if the suspension is too dilute. The resuspended pellet should give an opalescent, rather viscous, solution. The initial pellet will probably be opaque (either white or colored) and may be compacted and difficult to resuspend. In this case, the pellet should be completely dissolved in a small volume of 0.15M Na-acetate, followed by addition of 4M Na-acetate (both pH 6.0). This should be left on ice for 30 min for the large RNA to come out of solution.

9. At this stage it is possible to translate the cytoplasmic RNA in an in vitro translation system (2). The sodium acetate-washed pellet should be washed twice with 0.1M potassium acetate, pH 7.0, in 80% (v/v) ethanol, and twice with 95% ethanol, as described elsewhere in this volume (8).

10. Cellulose was originally used to remove carbohydrate contaminants from RNA preparations, prior to isolation of poly (A)$^+$ RNA on oligo-dT cellulose (5). It appears that starch and other polysaccharides bind to cellulose under high salt conditions and elute in low-salt buffers, thus co-purifying with the poly (A)$^+$ RNA. Thus, tomato fruit poly (A)$^+$ RNA, made by passing cytoplasmic RNA directly down oligo-dT cellulose contains a greater proportion of carbohydrate (material absorbing at 230 nm) and translates less efficiently than the original cytoplasmic RNA.

A related benefit of the cellulose column step is that viscous contaminating material present in many plant

RNA preparations at this stage would rapidly clog up an oligo-dT cellulose column. Much of this contaminating material is trapped on top of the cellulose column, and the RNA preparation can be forced through the cellulose column under pressure, using a Pasteur pipet teat. It may be also advantageous to remove this material even if it is not necessary to purify poly (A)$^+$ RNA (e.g., for RNA blot hybridization).

11. A number of alternative methods are available for extraction of RNA, many of which are applicable to plant material. A method similar to that described in this series for extraction of total RNA by the detergent and phenol method (1) has been applied successfully to tomato leaves (16), for example. It may be necessary to use chaotropic agents to isolate RNA from nuclease-rich plant material. Guanidine salts have been used to extract RNA from a number of nuclease-rich sources (e.g., ref. 18), including plants. Another chaotropic agent that has been applied particularly to plant RNA isolation is sodium perchlorate, in situations in which conventional phenol extraction has been found to be unsatisfactory (e.g., ref. 19).

References

1. Slater, R.J. (1985) The extraction of total RNA by the detergent and phenol method, in *Methods in Molecular Biology* Vol. 2 (Walker, J., ed.) Humana, Clifton, New Jersey.

2. Grierson, D., Slater, A., Spiers, J., and Tucker, G.A. (1985) The appearance of polygalacturonase mRNA in tomatoes: One of a series of changes in gene expression during development and ripening. *Planta* **163**, 263–271.

3. Speirs, J., Brady, C.J., Grierson, D., and Lee, E. (1984) Changes in ribosome organisation and messenger RNA abundance in ripening tomato fruits. *Aust. J. Plant Physiol.* **11**, 225–233.

4. Palmiter, R.D. (1974) Magnesium precipitation of ribonucleoprotein complexes. Expedient techniques for the isolation of undegraded polysomes and messenger ribonucleic acid. *Biochemistry* **13**, 3606–3615.

5.   Mozer, T.J. (1980) Partial purification and characterisation of the mRNA for α-amylase from barley aleurone layers. *Plant Physiol.* **65**, 834–837.

6.   Slater, A., Maunders, M.J., Edwards, K., Schuch, W., and Grierson, D. (1985) Isolation and characterisation of cDNA clones for tomato polygalactouronase and other ripening-related proteins. *Plant Mol. Biol.* **5**, 137–147.

7.   Maunders, M.J., Slater, A., and Grierson, D. (1985) Generation and Use of cDNA Clones for Studying Gene Expression, in *Molecular Biology and Biotechnology* (Walker, J.M. and Gingold, E.M., eds.) Royal Society of Chemistry, London.

8.   Slater, A. (1987) Hybrid-Release Translation, in *Methods in Molecular Biology*, Vol. 4 (Walker, J., ed.) Humana, Clifton, New Jersey.

9.   Kitos, P.A., Saxon, G., and Ames, H. (1972) The isolation of polyadenylate with unreacted cellulose. *Biochem. Biophys. Res. Comm.* **47**, 1426–1437.

10. Holley, R.W., Apgar, J., and Merrill, S.H. (1961) Evidence for the liberation of a nuclease from human fingers. *J. Biol. Chem.* **236**, PC42.

11. Maniatis, T., Fritsch, E.F., and Sambrook, J. (1982) *Molecular Cloning, A Laboratory Manual* Cold Spring Harbor Laboratories, Cold Spring Harbor, New York.

12. Slater, R.J. (1985) Purification of Poly (A)-Containing RNA by Affinity Chromatography, *Methods in Molecular Biology* Vol. 2 (Walker, J., ed.) Humana, Clifton, New Jersey.

13. Davies, E., Larkins, B.A., and Knight, R.H. (1972) Polyribosomes from peas. An improved method for their isolation in the absence of ribonuclease inhibitors. *Plant Physiol.* **50**, 581–584.

14. Gray, J.C. (1974) The inhibition of ribonuclease activity and the isolation of polysomes from leaves of the French bean, *Phaseolus vulgaris. Arch. Biochem. Biophys.* **163**, 343–348.

15. Blackburn, P., Wilson, G., and Moore, S. (1977) Ribonuclease inhibitor from human placenta. *J. Biol. Chem.* **252**, 5904–5910.

16. Smith, C.J.S., Slater, A., and Grierson, D. (1986) Rapid appearance of an mRNA correlated with ethylene synthesis encoding a protein of molecular weight 35,000. *Planta* **168**, 94–100.

17. Loomis, W.D. (1974) Overcoming problems of phenolics and quinones in the isolation of plant enzymes and organelles. *Meth. Enzymol.* **31**, 528–544.

18. McGookin, R. (1985) RNA Extraction by the Guanidine Thiocyanate Procedure, in *Methods in Molecular Biology*, Vol. 2 (Walker, J., ed.) Humana, Clifton, New Jersey.

19. Nelson, C.E. and Ryan, C.A. (1980) *In vitro* synthesis of pre-proteins of vacuolar compartmented proteinase inhibitors that accumulate in leaves of wounded tomato plants. *Proc. Natl. Acad. Sci. USA* **77**, 1975–1979.

Chapter 36

Isolation of Plant Nuclei

Richard D. Henfrey and Robert J. Slater

1. Introduction

The isolation of plant nuclei is a useful first step in many experiments concerned with the mechanism and control of gene expression in plants. For example, isolated nuclei can be used for the isolation of nuclear components such as chromosomal proteins (1), for the study of the processing of primary transcripts (2,3), for the assay and characterization of RNA polymerase activities (4–7), or for the measurement of transcription rates of specific genes (8) (see Chapter 37). A method is described here for the isolation of a crude preparation of intact plant nuclei, with an additional protocol for nuclei purification on a discontinuous gradient of Percoll (9) (see Note 1 in section 4). Centrifugation of crude nuclei preparations through Percoll gradients removes much of the contaminating cytoplasmic material such as starch grains (10), and the Percoll step appears to reduce the ribonuclease activity associated with nuclei (10). It is therefore recommended for transcription experiments. The crude nuclei preparation alone may be adequate for some work, for example, when attempting RNA polymerase assays for the first time, when very small amounts of tissue are involved, for preliminary experiments on the characterization of enzyme activities, or for the isolation of nuclear components that may be lost during purification.

The method described is suitable for most plant tissues, although the mortar and pestle homogenization step described is not particularly suitable for most plant tissue cultures, since cell breakage is inefficient. As a guideline, 10–12 g of 3-d-old, etiolated wheat shoots yield approximately 6×10^6 nuclei following Percoll purification.

2. Materials

1. All solutions are stored at 4°C. All glassware should be cleaned with a detergent solution, rinsed with distilled water, and baked at 150°C for 4 h. Eppendorf tubes should be autoclaved at 15 psi for 15 min. These procedures are to remove contaminating RNAse, DNAse, and protease activities.
2. Gloves should be worn at all times to prevent contamination from RNAse, DNAse, and protease activities present on the hands.
3. Nuclear isolation buffer (NIB): 0.44 M Sucrose, 2.5% (w/v) Ficoll (mol wt 400,000), 5.0% (w/v) Dextran 40, 25 mM Tris-HCl (pH 7.6), 10 mM MgCl$_2$, 10 mM 2-mercaptoethanol, and 5% (v/v) Triton X-100, (1 L) (see Note 2 in section 4). This is made up in sterile distilled water since it cannot be autoclaved.

    It is convenient to prepare a 1 M stock solution of Tris-HCl and a 500 mM stock solution of MgCl$_2$. These can be autoclaved separately.
4. Nuclear resuspension buffer (NRB): 50 mM Tris-HCl (pH 7.8), 10 mM MgCl$_2$, 10 mM 2-mercaptoethanol, and 20% (v/v) glycerol (200 mL).
5. 40, 60, and 80% (v/v) Percoll (Pharmacia) solutions that are 0.44 M sucrose, 25 mM Tris-HCl (pH 7.5), and 10 mM MgCl$_2$ (50 mL).

    Note: It is advisable, to make up the Percoll solutions in a laminar flow cabinet since the Percoll is liable to become microbially contaminated when exposed to the air.

6. Miracloth (Cambridge Bioscience).
7. 85% (w/v) Aqueous sucrose solution.

3. Method

3.1. Isolation of Crude Nuclei

1. The harvested tissue is placed into a chilled mortar and pestle. One volume (i.e., 1 mL/g of fresh weight tissue) of nuclear isolation buffer (NIB) is added and the tissue ground until it forms a homogeneous suspension. The homogenate is then diluted with approximately 2 vol of NIB and filtered through two layers of Miracloth (presoaked in NIB). The residue remaining on the Miracloth is washed with at least 5 vol of NIB (*see* Note 3 in section 4).
2. The combined filtrates are divided among 50 mL polypropylene centrifuge tubes and centrifuged at 350*g* for 9 min, and the supernatant is discarded.
3. The pellets are then resuspended in NIB (20 mL), combined, and centrifuged again at 350*g* for 9 min, and the supernatant is discarded.
4. The "crude" nuclear pellet can then be resuspended in 1–5 mL of nuclear resuspension buffer (NRB) (approximately 1 mL/10 g of fresh weight starting material) and used immediately or stored (*see* step 5 in section 3.2). Alternatively, if the nuclei are to be further purified on a discontinuous gradient of Percoll (*9,10*), the final "crude" nuclear pellet is resuspended at a concentration of 1 mL/10 g of starting material in NIB instead of NRB (*see* Note 4 in section 4).

3.2. Percoll Purification of Nuclei

1. Prepare Percoll gradients (1 tube/10 g of starting material) on ice, using a laminar flow cabinet, in 15-mL glass centrifuge tubes. Three milliliters of 85% sucrose is added to each tube, followed by 3 mL layers of 80, 60, and 40%

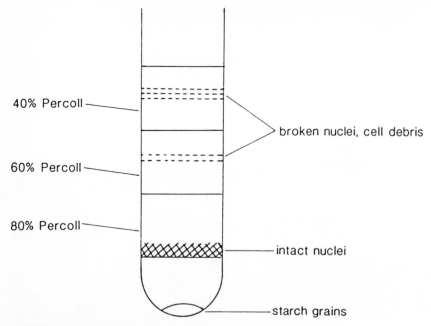

Fig. 1. Diagram depicting separation of a nuclear suspension on a discontinuous gradient of Percoll.

Percoll solutions, respectively. The nuclei suspension is then layered on top of the 40% Percoll layer (1 mL of nuclear suspension per tube).

2. Centrifuge the gradients at 4000g for 30 min at 4°C in a swing-out rotor. Most of the nuclei band is in the 80% Percoll layer, just above the sucrose layer (*see* Fig. 1).

3. Remove the nuclei by carefully placing the tip of a Pasteur pipet above them, and slowly draw them off, taking as little of the sucrose as possible. The nuclei are pooled and placed into a 15-mL glass centrifuge tube on ice.

4. To remove the Percoll, add 10 mL of NIB to each tube, mix the contents by gentle inversion of the tubes, then centrifuge at 350g for 5 min. Pour off the supernatant, and repeat once using NIB, then once using NRB.

5. The final nuclear pellet is resuspended in 1-5 mL of NRB (1 mL/10 g tissue) and can be stored at –70°C for several months without loss of RNA polymerase activity.

4. Notes

1. Percoll is a medium for density gradient centrifugation of cells, viruses, and subcellular particles. Percoll is composed of colloidal silica coated in polyvinylpyrrolidone (PVP), which renders the material completely nontoxic and ideal for use with biological materials. Centrifugation of Percoll results in spontaneous formation of a density gradient because of the heterogeneity of particle sizes in the medium. Percoll can be used for formation of gradients by high-speed centrifugation. The sample can be premixed with the medium and then separated on the gradient *in situ*. Percoll is supplied as a sterile solution.

2. For some experiments it may be desirable to omit the Triton X-100 from the nuclear isolation buffer. This is because the Triton X-100 strips any cytoplasmic material from the nuclear envelope. Although omission of the detergent will result in a less well purified product, it may preserve some nuclear components that may be lost in the presence of Triton X-100.

3. Filtering the homogenate is best performed by placing two pieces of Miracloth at 90° to each other (since the pores are elliptical in shape) and stretching them across a plastic ring. The Miracloth can be held in place with an elastic band, and then placed inside a funnel. The homogenate is then poured onto the Miracloth, which is presoaked for 5 min in nuclear isolation buffer, and the homogenate washed by pouring isolation buffer over its surface. The filtrate is collected in a 250-mL precooled conical flask.

4. The volumes and the number of tubes used here are ideal for 10–20 g of tissue. If more tissue is used, this can create problems in overloading the homogenization and Miracloth filtration stages, and overloading on the Percoll gradients. If more nuclei are required, it is best to carry out several preparations in parallel or use larger apparatus.

References

1. Mayes, E.L.V. and Walker, J.M. (1984) Putative high mobility group (HMG) non-histone chromosomal proteins from wheat germ. *Int. J. Peptide Protein Res.* **23**, 516–520.

2. Manley, J.L., Sharp, P.A., and Gefter, M.L. (1979) RNA synthesis in isolated nuclei: Identification and comparison of adenovirus 2 encoded transcripts synthesized *in vitro* and *in vivo*. *J. Mol. Biol.* **135**, 171–197.

3. Manley, J.L., Sharp, P.A., and Gefter, M.L (1979) RNA synthesis in isolated nuclei: In vitro initiation of adenovirus 2 major late mRNA precursor. *Proc. Natl. Acad. Sci. USA* **76**(1), 160–164.

4. Calza, R.E., Oelke, S.M., and Lurquin, P.F. (1982) Transcription in nuclei isolated from carrot protoplasts: Effects of exogenous DNA. *Feb. Lett.* **143**, 109–114.

5. Bouman, H., Mennes, A.M., and Libbenga, K.R. (1979) Transcription in nuclei isolated from tobacco tissues. *Feb. Lett.* **101**, 369–372.

6. Slater, R.J., Venis, M.A., and Grierson, D. (1979) Characterisation of ribonucleic acid synthesis by nuclei isolated from Zea Mays. *Planta* **144**, 89–93.

7. Mennes, A.M., Boumann, H., Van Der Burg, M.P.M., and Libbenga, K.R. (1978) RNA synthesis in isolated tobacco callus nuclei and the influence of phytohormones. *Plant Sci. Lett.* **13**, 329–339.

8. Beach, L.R., Spencer, D., Randall, P.J., and Higgins, T.J.V. (1985) Transcriptional and post-transcriptional regulation of storage protein gene expression in sulphur-deficient pea seeds. *Nucleic Acids Res.* **13**(3), 999–1013.

9. Pertoft, H., Laurent, T.C., Laas, T., and Kagedal, L. (1978) Density gradients prepared from colloidal silica particles coated with polyvinyl-pyrrolidone (Percoll). *Anal. Biochem.* **28**, 271–282.

10. Luthe, D.S. and Quatrano, R.S. (1980) Transcription in isolated wheat nuclei. I. Isolation of nuclei and elimination of endogenous ribonuclease activity. II. Characterisation of RNA synthesized *in vitro*. *Plant Physiol.* **65**, 305–313.

Chapter 37

In Vitro Transcription
in Plant Nuclei

Richard D. Henfrey and Robert J. Slater

1. Introduction

Isolated plant nuclei can be used for fundamental studies on the transcription apparatus. Total RNA polymerase activity can be measured using plant nuclei, and by using different α-amanitin concentrations in the enzyme assay, the individual RNA polymerase I, II, and III activities can be measured (1–5). The assay procedure involves the incubation of nuclei in the presence of the four substrates for RNA synthesis: ATP, GTP, CTP, and UTP. If one of these precursors is supplied as a radiolabeled molecule, transcription can be detected as incorporation of radioactivity into acid-insoluble material. Following incubation, transcription products are precipitated with TCA, collected and washed on glass fiber filter discs, and counted by liquid scintillation counting.

Isolated plant nuclei can also be used as a means to study the control of expression of specific genes. Control of gene expression is often associated with changes in cellular mRNA content. These changes could occur by two possible mechanisms: a change in the rate of transcription and/or a change in the stability of transcripts. In vitro transcription followed by isolation of radiolabeled RNA, and analysis by hybridization against unlabeled probes bound to a membrane filter, can be used to investigate changes in transcription rate. This is often referred to as the "run-off transcription assay" and can be used in experiments attempting to correlate transcription rates with changes in hormone concentration or other developmental processes (6–9). The protocol for run-off transcription experiments is essentially the same as that for RNA polymerase assays. The differences are that in the run-off transcription assay the TCA precipitation step is replaced by an RNA extraction procedure, and a relatively large quantity of a ^{32}P-labeled nucleotide is substituted for ^{3}H-UTP. This is because the intended outcome of the experiment is nucleic acid hybridization followed by autoradiography.

2. Materials

All glassware should be cleaned with a detergent solution, rinsed with distilled water, and baked at 150°C for 4 h. Eppendorf tubes should be autoclaved at 15 psi for 15 min. These procedures are for inactivating contaminating RNAse, DNAse, and protease activities.

Gloves should be worn at all times to prevent contamination from RNAse, DNAse, and protease activities present on the hands.

2.1. RNA Polymerase Assay

1. 1.5-mL Eppendorf tubes.
2. TM buffer: 50 mM Tris-HCl, pH 7.8, 10 mM 2-Mercaptoethanol. Autoclave and store at 4°C.
3. Nucleoside triphosphate solution: 10 mM of each of the nucleoside triphosphates ATP, GTP, and CTP together, made up in TM buffer. Make 5 mL of this solution, aliquot it into 0.5-mL Eppendorf tubes, and store at –20°C (*see* Note 1 in section 4).

4. UTP solution: 10 m*M* UTP in TM buffer. Store at –20°C.
5. [³H]-UTP solution: This is purchased at high specific radioactivity in 50% aqueous ethanol and is stored at –20°C. An aliquot in an Eppendorf tube (typically 100 μCi) is evaporated to dryness and made up to 0.5 mL with 100 μ*M* UTP (5 μL UTP solution and 495 μL TM buffer). This is sufficient for 100 assays, adding 5 μL (1 μCi) per assay (*see* Note 2 in section 4). Store at –20°C.
6. Make up two stock aqueous solutions of α-amanitin; one, 2 mg/mL, and the other, 10 μg/mL α-amanitin.
7. Stock ammonium sulfate solution (0.5–5*M*, depending on requirements) (*see* note 3 in section 4). Autoclave and store at 4°C.
8. 10% (w/v) TCA containing 0.01*M* sodium pyrophosphate. Store at 4°C.
9. 5% (w/v) TCA containing 0.01*M* sodium pyrophosphate. Store at 4°C.
10. Absolute alcohol.

2.2. Incubation for Analysis of Transcripts

1. 1.5-mL Eppendorf tubes.
2. TM buffer: 50 m*M* Tris-HCl, pH 7.8, 10 m*M* 2-mercaptoethanol. Autoclave and store at 4°C.
3. Nucleoside triphosphate solution: 10 m*M* each of the nucleoside triphosphates ATP, CTP, and UTP together, made up in TM buffer. Make 5 mL of this solution, aliquot it into 0.5-mL Eppendorf tubes, and store at –20°C.
4. GTP solution: 2.5 m*M* GTP in TM buffer. Store at –20°C.
5. α-³²P-GTP solution: This is purchased at high specific radioactivity in 50% aqueous ethanol and is stored at –20°C. 10–200 μCi is evaporated to dryness and is made up to 10 μL with 250 μ*M* GTP (1 μL GTP solution and 9 μL TM buffer). This is sufficient to bring the final concentration of [³²P]-GTP plus unlabeled GTP to 50 μ*M*, when 10 μL of [³²P-GTP solution is added in a reaction volume of 50 μL (*see* Note 4 in section 4).

6.   Stock ammonium sulfate solution (0.5–5*M*, depending on requirements) (*see* Note 3 in section 4). Autoclave and store at 4°C.

2.3. Extraction of Nucleic Acids

1.   Phenol mixture: 500 g of phenol crystals, 70 mL of *m*-cresol, 0.5 g of 8-hydroxyquinoline, 150 mL of water. The phenol and *m*-cresol should be colorless. If not, they must be redistilled. The solution is intended to be water-saturated. Store in a dark bottle at 4°C for up to 2 mo. The solution darkens in color with age because of oxidation. Discard the solution if the color darkens beyond light brown. The *m*-cresol is an optional component that acts as an antifreeze and an additional deproteinizing agent.
2.   Detergent solution: 1 g of Sodium tri-isopropylnaphthalene sulfonate (TPNS), 6 g of sodium 4-amino salicylate, 5 mL of phenol mixture. Make to 100 mL in 50 m*M* Tris-HCl (pH 8.5). Mix the TPNS with the phenol mixture before adding the other components. Store as for phenol mixture.
3.   Deproteinizing solution: Phenol mixture and chloroform :isoamyl alcohol (24:1) mixed 1:1 by volume. Store as for phenol mixture.
4.   Absolute alcohol.
5.   Sodium acetate buffer, 0.15*M* (pH 6.0 with acetic acid) containing 5 g/L sodium dodecyl sulfate (SDS). Store at room temperature.

2.4. Removal of DNA

1.   Tris-magnesium acetate buffer: 50 m*M* Tris-HCl (pH 7.4) containing 2 m*M* magnesium acetate. Autoclave and store at –20°C.
2.   DNAse solution: 0.5 mg/mL RNAse-free DNAse I (*see* Chapter 3 in Vol. 2 of this series) in Tris-magnesium acetate buffer. Store at –20°C in batches to avoid repeated freeze-thawing.

3. Method

Nuclei are isolated by the Percoll method described in Chapter 36; 25 µL of nuclei preparation are required per assay. Every reaction should be carried out at least in duplicate.

3.1. RNA Polymerase Assay

1. For each assay, add 2.5 µL of nucleoside triphosphate solution (ATP, GTP, CTP), 5 µL of [^{3}H]-UTP, 2–5 µL of $(NH_4)_2SO_4$ (*see* Note 3 in section 4), and sufficient TM buffer to bring the volume to 25 µL in an Eppendorf tube. For control tubes omit the nucleoside triphosphate solution and increase the TM buffer *pro rata* (*see* Note 7 in section 4). The inhibitor of RNA polymerase II and III, α-amanitin, can be added if desired (*see* Note 6 in section 4).
2. Start the reaction by the addition of 25 µL of nuclei preparation (*see* Note 8 in section 4). This will contribute 10% (v/v) glycerol, 5 mM 2-mercaptoethanol, and 5 mM MgCl$_2$ to the assay (*see* Note 5 in section 4).
3. Incubate the tubes at 25°C for 30 min. (*See* also Note 9 in section 4).
4. Stop the reaction by the addition of 10 µL of UTP solution (this lowers the background count slightly) and 1 mL of ice-cold 10% TCA. Mix the tube's contents thoroughly.
5. Leave the tubes on ice for 1–4 h. The time is not critical.
6. Collect the precipitates on GF/C filter discs prewetted with 5% TCA solution. Wash the filters with 3 x 5 mL of ice-cold 5% TCA solution and 1 x 5 mL of ethanol. Dry the filters and estimate the radioactivity in toluene/PPO/POP-OP scintillation cocktail with a scintillation counter. It is very important that the filters are completely dry before placing them in scintillation cocktail, otherwise this could cause quenching of the scintillation process. If the filter is completely dry, which normally takes approximately 1 h after washing with ethanol, it will appear transparent

when placed in scintillant. If the filter is not dry, it will appear opaque and must be removed and allowed to dry completely.

3.2. Incubation for Analysis of Transcripts

1. For each experiment add to an Eppendorf tube: 2.5 μL of nucleoside triphosphate solution (ATP, CTP, UTP), 10–200 μCi [α-^{32}P]-GTP (*see* Note 2 in section 4), 2–10 μL of $(NH_4)_2SO_4$ (*see* Note 3 in section 4) and TM buffer to bring the volume to 25 μL.

2. Add 25 μL of nuclei preparation, and incubate the tube at 25°C for 5–60 min (*see* Note 9 in section 4).

3. To stop the reaction and lyse the nuclei, add a minimum of 50 μL of detergent solution at 4°C. For easier handling of the aqueous phase in subsequent manipulatins, however, it is better to increase the volume of the aqueous phase by adding 200 μL of detergent solution.

4. Add an equal volume of deproteinizing solution. Agitate the mixture to maintain an emulsion for 10 min at room temperature.

5. Spin the tubes in a microcentrifuge for 5 min. This separates the tube contents into three phases: an upper aqueous phase, a denatured protein interphase, and a lower phenol phase.

6. Carefully remove the upper aqueous phase with a pipet and retain in a second Eppendorf tube.

7. Re-extract the remaining phenol and protein phases by adding an additional 150 μL of detergent solution to the original Eppendorf tube. Shake, centrifuge, and remove the aqueous phase as before, and pool it with the aqueous phase from step 6.

8. Add an equal volume of deproteinizing mixture (400 μL) to the combined aqueous phases and shake for a further 5–10 min. Spin the tubes in a microcentrifuge for 5 min, then carefully remove the aqueous phase.

9. Unicorporated GTP is removed by the processes outlined in steps 9–12. To the aqueous phase, after RNA extraction procedures, add 2.5 vol of ethanol, mix, and then leave at –70°C for 30 min (or at least 2 h at –20°C).
10. Spin for 5 min in a microcentrifuge in a cold room, and decant the supernatant.
11. Resuspend the pellet in 200 μL of 0.2% SDS, add 200 μL of 2*M* ammonium acetate, then precipitate the nucleic acid as described in step 9.
12. Repeat step 11.
13. Spin for 5 min in a microcentrifuge in a cold room, and decant the supernatant. Drain any remaining alcohol from the precipitate.
14. If desired, DNA can be removed by the method described in steps 15–18. If not, the nucleic acids can be resuspended in the desired volume of hybridization fluid.

3.3. Removal of DNA

1. Collect the nucleic acid precipitate by centrifugation and wash at least twice with 70% alcohol at room temperature to remove SDS.
2. Dissolve the nucleic acid preparation in 0.4 mL of Tris-magnesium acetate buffer at 0–4°C.
3. Add 0.1 mL of DNAse solution, and incubate in an ice bath for 30 min.
4. Deproteinize with phenol/chloroform solution and precipitate the RNA with alcohol as previously described, i. e. steps 4–9.

4. Notes

1. For RNA synthesis, nuclei are incubated with all four ribonucleoside triphosphates, one of which is radiolabeled. The concentrations must be at least 50 μ*M* for CTP, GTP, and UTP, and 100 μ*M* for ATP for maximal activity. Routinely, 500 μ*M* of each of the three unlabeled triphos-

phates is used, and 10 µM of the labeled triphosphate (20 µM if the label is in ATP).

2. The choice of labeled nucleotide depends upon whether the RNA transcripts are to be isolated and analyzed, or whether only a measurement of RNA polymerase activity is required. If the latter is desired, the ^{3}H-nucleotides are the most suitable. These are relatively stable (chemical decomposition approximately 2%/mo at –20°C) and less expensive than ^{32}P-nucleotides. If acid-insoluble precipitates of ^{3}H-labeled nucleic acids are collected on glass fiber filters, the efficiency of scintillation counting is between 20 and 40% depending on the instrument, allowing for successful in vitro transcription experiments with 1.0 µCi of labeled nucleotide per reaction.

    If more detailed analysis of the RNA synthesized in vitro is required, i.e., the run-off transcription assay, then ^{32}P is the radionuclide of choice. The high energy of β radiation from ^{32}P means that autoradiography can be used as the detection method. If the gene of interest is expressed at a low level (contributing, say, less than 5% of the total mRNA), then a high level of ^{32}P-labeled nucleotide (100–200 µCi) may be required to obtain sufficient incorporation for hybridization experiments.

3. RNA polymerase activity is stimulated by the inclusion of monovalent cations supplied in the form of salts such as $(NH_4)_2SO_4$. With isolated plant nuclei, RNA polymerases I and III (i. e., α-amanitin-resistant RNA polymerase activity—*see* Note 6 for information on α-amanitin) are most active at 0.05–0.10M $(NH_4)_2SO_4$, whereas RNA polymerase II (i. e., α-amanitin-sensitive activity) is optimized at 0.25–0.5M ammonium sulfate (1–6).

    In measuring total RNA polymerase activity, it is recommended to use a range of ammonium sulfate concentrations for the plant nuclei under investigation, to determine the optimum concentration. To optimize for RNA polymerases I and III, and RNA polymerase II, nuclei should be incubated over a range of $(NH_4)_2SO_4$ concentrations in the absence and presence of 0.5 µg/mL α-amanitin (*see* Note 6).

It is convenient to make up a concentrated solution of $(NH_4)_2SO_4$ of between 0.5 and 5*M*, according to the range of ammonium sulfate concentrations being investigated, and to add 2–5 µL of this solution to the reaction to give the required final concentration in the assay.

4. Any of the four [α-^{32}P]-NTPs can be used as the radiolabeled nucleotide precursor. Some of nuclei, however, transfer radiolabel from [α-^{32}P] UTP to the 3' end of some RNAs. Reactions are also known in which [α-^{32}P]-ATP (polyadenylation or tRNA maturation) or [α-^{32}P]-CTP (tRNA maturation) are added to the end of intact RNAs. In addition, [α-^{32}P]-GTP is the substrate in the capping reaction. Therefore, although similar rates are obtained regardless of which radiolabeled triphosphate is used, [α-^{32}P]-GTP is the best choice for most experiments since there is less incorporation of this nucleotide into RNA by reactions that are independent of transcription (9).

5. To synthesize RNA, the RNA polymerase enzymes require divalent (e.g., Mg^{2+} or Mn^{2+}) cations (1). The divalent cation is also required in the isolation of intact nuclei. The optimum divalent cation concentrations are generally broad and in the range of 5–10 m*M* Mg^{2+} or 1–2 m*M* Mn^{2+} (1). This is conveniently supplied via the nuclear resuspension buffer, but the divalent cation optima for the tissue used should be determined.

6. The fungal toxin α-amanitin interacts directly with RNA polymerase II and inhibits the propagation of RNA chains in in-vitro transcription reactions (1). All higher plant RNA polymerase I enzymes examined are refractory to α-amanitin inhibition (1). The relative activities of the different RNA polymerases are determined by assaying nuclei in the absence of α-amanitin (total enzyme activity), or with 0.5 µg/mL (RNA polymerase I and III-active) or 100

μg/mL (only RNA polymerase I-active) α-amanitin (6).
Stock α-amanitin solutions of 2 mg/mL and 10 μg/mL are
convenient: by adding 2.5 μL of one of these solutions to
the reaction, the desired concentration of 100 and 0.5 μg/
mL, respectively, is achieved.

7. There will be some nonspecific binding of radiolabeled
nucleotide to the filters, and there may be microbial
contamination or other spurious reactions giving rise to
false-positive results. Suitable controls are: zero time incubations, reactions carried out in the absence of the three
unlabeled nucleotides (although some incorporation may
be expected here), addition of ribonuclease, and reactions
carried out in the presence of known inhibitors of RNA
synthesis, e. g., α-amanitin.

8. To prevent shearing of the nuclei by passage through the
tips of automatic pipets, it is recommended to cut the first
2 mm off the end of the tips. The nuclei resuspension is
viscous, so care must be taken to ensure complete expul-
sion of the contents. It is also advisable to wipe the end of
the pipet tip before expelling the nuclei.

9. The assumption made in using the run-off transcription
technique to measure transcription rates is that the transcription rate of all genes in vitro reflects the activity in
vivo (9). This may not be the case for small transcripts,
which may initiate inefficiently, whereas large transcripts
are being elongated after in vivo initiation. In this situation, the large transcripts will be a larger proportion of
the RNA labeled in vitro than they would be after a pulse
label in vivo. To compensate for this, the transcription can
be carried out for a short period of time (5 min) when most
transcription should represent this elongation compared
with longer times (30 or 60 min) when variations in the
efficiency of initiation of different genes should be more
apparent. The relative rates of transcription of the gene are
measured for each time period. If the relative rates of
transcription are independent of the incubation period,

then the measured rate is probably an accurate reflection of the rate of transcription in vivo (9).

References

1. Guilfoyle, T. J. (1983) DNA-Dependent RNA Polymerases of Plants and Lower Eukaryotes, in *Enzymes of Nucleic Acid Synthesis and Modification* Vol. 2 (Jacob, S. T., ed.) CRC, Florida.
2. Guilfoyle, T. J., Lin, C. Y., Chen, Y. M., and Key, J. L. (1975) Enhancement of soybean RNA polymerase I by auxin. *Proc. Natl. Acad. Sci. USA* **72**, 69–72.
3. Chen, Y. M., Lin, C. Y., Chang, H., Guilfoyle, T. J., and Key, J. L. (1983) Isolation and properties of nuclei from control and auxin treated soybean hypocotyl. *Plant Physiol.* **56**, 78.
4. Guilfoyle, T. J. and Key, J. L. (1977) Purification and Characterisation of Soybean DNA-Dependent RNA Polymerases and the Modulation of Their Activities During Development, in *Nucleic Acids and Protein Synthesis in Plants* (Bogorad, L. and Weil, J., eds.) Plenum, New York.
5. Guilfoyle, T. J. (1980) Transcription of the cauliflower mosaic virus genome in isolated nuclei from turnip leaves. *Virology* **107**, 71–80.
6. Slater, R. J. (1987) *In Vitro* Transcription, in *Techniques in Molecular Biology* Vol. 2 (Walker, J. M., ed.) Croom Helm, London, Sydney, 203–227.
7. Beach, L. R., Spencer, D., Randall, P. J., and Higgins, T. J. V. (1985) Transcriptional and post-transcriptional regulation of storage protein gene expression in sulphur-deficient pea seeds. *Nucleic Acids Res.* **13**(3), 999–1013.
8. Mosinger, E., Batschauer, A., Schafer, E., and Apel, K. (1985) Phytochrome control of *in vitro* transcription of specific genes in isolated nuclei from barley (*Hordeum vulgare*). *Eur. J. Biochem.* **147**, 137–142.
9. Marzluff, W. F. and Huang, R. C. C. (1984) Transcription in Isolated Nuclei, in *Transcription and Translation: A Practical Approach* (Hames, B. D. and Higgins, S. J., eds.) IRL, Oxford, Washington DC, 89–129.

Chapter 38

Large-Scale Isolation of Ti Plasmid DNA

Kevan M. A. Gartland

1. Introduction

Use of the *Agrobacterium tumefaciens* Ti plasmids to introduce foreign genes into many plant species is now firmly established (*1*). Ti plasmids, which are oftern single copy molecules, typically of 180–220 kilobases (kb), are divided into three main groups: nopaline, octopine, and agropine types. This classification (*2*) is based on the synthesis of these unusual amino acids in plant cells transformed by these plasmids. Their synthesis is governed by the expression of T-DNA encoded genes that become integrated into the host plant DNA in the course of infection. A feature thought to be important in this transfer and integration is the 25-base pair (bp) imperfect direct repeat sequences (*3*) separated by the 22 kb of nopaline T-DNA, or the TL- (13 kb) and TR- (8 kb) of the bipartite octopine type T-DNA (*4.5; see* Fig.1). Other important features of the Ti plasmids include; the large virulence region, essential for transfer of the T-DNA; the ori region, responsible for the Ti plasmid replication functions; the opine catabolism region, which is associated with transfer of Ti plasmids from one bacterium to another (*6*).

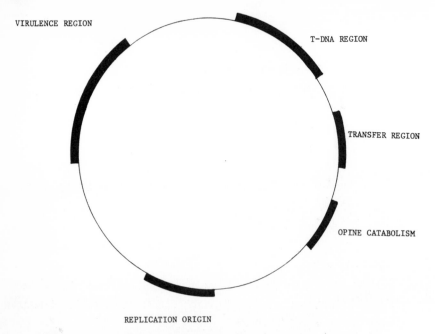

Fig. 1. Functional map of the *Agrobacterium tumefaciens*
Ti plasmid.

One important technique in the characterization of Ti plasmids is their large-scale isolation. The procedure described
here for the large-scale isolation of Ti plasmid DNA is modified
from the methods of Zaenen et al., Van Larebeke et al., and
Currier and Nester (7–9). In particular, an overt shearing step
has been omitted, and polyethylene glycol precipitation of
DNA has been included. Yields of 15–25 µg Ti plasmid DNA
per liter of bacterial culture may be obtained.

2. Materials

1. Peptone medium: 4 g/L bacto-peptone, 1 g/L yeast extract, 5 mM magnesium sulfate, pH 6.8.
2. TE buffer: 10 mM Tris-HCl (pH 8.0), 1 mM sodium EDTA.
3. Pronase B (Boehringer): 5 mg/mL in TE buffer.
4. Sodium dodecyl sulfate solution: 10% (w/v) in TE buffer.
5. 3M Aqueous sodium hydroxide.

6.   pH Electrode.
7.   2*M* Aqueous Tris-HCl (pH 7.0).
8.   5*M* Aqueous sodium chloride.
9.   50% (w/v) Polyethylene glycol (PEG), molecular weight 8000 (Sigma).
10. Cesium chloride (molecular biological grade).
11. Ethidium bromide: 5 mg/mL in TE buffer.
12. Refrigerated ultracentrifuge, vertical rotor, and soft-walled tubes capable of being centrifuged at 40,000 rpm.
13. 19 G needle (Sabre).
14. 2 mL Syringe (Gillette).
15. Isobutanol saturated with TE buffer.
16. Absolute ethanol, stored at –70°C.
17. 30 mL Corex tubes, with rubber adaptors (Corning). These may be siliconized, autoclaved, and rinsed 10 times with distilled water before use.
18. Microcentrifuge tubes.
19. Wide-bore micropipet tips.
20. UV illuminator (300 nm).
21. UV light-proof spectacles.
22. 70% (v/v) Ethanol.
23. Vacuum pump and desiccator.

3. Method

1.   Grow *Agrobacterium tumefaciens* in 1 L of peptone medium, at 30°C., with shaking, until late log phase is reached. This typically takes 16–20 h.
2.   Harvest the cells by centrifugation (10 min, 10,000 rpm, 4°C).
3.   Resuspend the pellets in 500 mL of TE buffer, and repeat the centrifugation step.
4.   Resuspend the cells in 80 mL of TE buffer.
5.   Add 10 mL of Pronase B, previously incubated for 90 min at 37°C.
6.   Add 10 mL of Sodium dodecyl sulfate solution. Mix gent-

ly, and incubate at 37°C for 1 h or until the bacterial cell lysate becomes viscous and less opaque.

7. Slowly add sodium hydroxide solution, until a pH of 12.3 is reached. Gentle (100 rpm) stirring may be required.

8. Stir for a further 30 min at 25°C.

9. Add 2M Tris-HCl until pH 8.5 is reached. The lysate should now have reduced viscosity (Note 1 in section 4).

10. Add 0.2 vol of 5M sodium chloride. Mix by gentle inversion, and leave on ice for 4–6 h.

11. Centrifuge to pellet the precipitate (10,000 rpm, 20 min, 4°C). Retain the supernatant (Note 2 in section 4).

12. Add 0.2 vol of 50% (w/v) polyethylene glycol 8000. Mix gently and leave on ice for a minimum of 6 h or overnight.

13. Centrifuge to collect the DNA precipitate (10,000 rpm, 15 min, 4°C). Discard the supernatant.

14. Air dry the DNA pellet for a minimum of 30 min.

15. Carefully resuspend the pellet in 2.5 mL of TE buffer, using wide-bore micropipet tips (Note 3 in section 4).

16. Add 5.0 mL of TE buffer, 8.6 g of caesium chloride, and 1.0 mL of ethidium bromide solution. Mix gently by inversion and transfer to ultracentrifuge tubes.

17. Ultracentrifuge at 40,000 rpm for up to 40 h at 20°C.

18. Pierce the top of the ultracentrifuge tube. Puncture the tube 2–3 mm below the two DNA bands visible in the middle of the tube under UV illumination. Carefully withdraw the lower plasmid-containing band using a 19-G needle and a 2-mL syringe, without disturbing the upper band of chromosomal DNA.

19. Remove ethidium bromide from the plasmid DNA by multiple partitioning with TE saturated isobutanol, each involving shaking the mixture, leaving the two phases to separate out, and discarding the upper, red phase. This should be repeated until no remaining trace of red color can be seen. Carry out two further partitions.

20. Add 2 vol of absolute ethanol, mix gently, and store overnight in Corex tubes at –20°C (Note 4 in section 4). Plasmid

DNA may be harvested by centrifugation (10,000 rpm, 15 min, 4°C).

21. Discard the supernatant without disturbing the plasmid DNA, and wash the pellet thoroughly with 70% ethanol. Brief recentrifugation may be required.

22. Evacuate the plasmid DNA pellet to dryness, typically 5–10 min, and carefully resuspend in 400 µL of TE buffer in a microcentrifuge tube, using wide-bore tips. Plasmid DNA may now be restricted or subjected to agarose gel electophoresis.

4. Notes

1. In step 9, should the viscosity of the lysate not be reduced, the denaturation and neutralization process may be repeated, commencing with step 4.

2. The sodium dodecyl sulfate–sodium chloride precipitate produced at step 11 contains cell debris, proteins, and chromosomal DNA.

3. Wide-bore micropipet tips should be used at step 15 is in order to avoid shearing of the very large plasmid DNA molecules. Alternatively, they can be produced by removing the apical 8 mm of ordinary micropipet tips, using a pair of scissors heated briefly in a Bunsen flame.

4. Alternatively, the plasmid DNA may be dialyzed for 24 h against several changes of TE buffer at 4°C in order to remove cesium chloride. In this case, 0.07 vol of 5*M* sodium chloride should be added before ethanol is added, and the DNA left to precipitate overnight.

References

1. An, G., Watson, B.D., Stachel, S., Gordon, M.P., and Nester, E.W. (1985) New cloning vehicles for transformation of higher plants. *EMBO J.* **4**, 277–284.

2. Davey, M.R., Gartland, K.M.A., and Mulligan, B.J. (1986) Transformation of the Genomic Expression of Plant Cells, in *Society for Experimental Biology, Plasticity in Plants Meeting* Durham.

3. Shaw, C.H., Watson, M.D., Carter, G.H., and Shaw, C.H. (1984) The right hand copy of the nopaline Ti plasmid 25 bp repeat is required for tumour formation. *Nucleic Acids Res.* **12**, 6031–6041.
4. Thomashow, M.F., Nutter, R., Montoya, A.L., Gordon, M.P., and Nester, E.W. (1980) Integration and organisation of Ti plasmid sequences in crown gall tumours. *Cell* **19**, 729–739.
5. Gielen, J., De Beuckeleer, M., Seurick, J., Deboeck, F., De Greve, H., Lemmers, M., van Montagu, M., and Schell, J. (1984) The complete nucleotide sequence of the T-DNA of the *Agrobacterium tumefaciens* plasmid pTiACH5. *EMBO J.* **3**, 83d5–846.
6. Kahl, G. and Schell, J., (1982) *Molecular Biology of Plant Tumours* Academic, London.
7. Zaenen, I., van Larebeke, N., Teuchy, H., van Montagu, M., and Schell, J. (1974) Supercoiled circular DNA in crown gall inducing *Agrobacterium* strains. *J. Mol. Biol.* **86**, 109–127.
8. Van Larebeke, N., Genetello, C., hernalsteens, J.P., Depicker, A., Zaenen, Il, Messens, E., van Montagu, M., and Schell, J. (1977) Transfer of Ti plasmids between *Agrobacterium* strains by mobilization with the conjugative plasmid RP4. *Mol. Gen. Genet.* **152**, 119–124.
9. Currier, T.C. and Nester, E.W. (1976) Isolation of covalently closed circular DNA of high molecular weight from bacteria. *Anal. Biochem.* **76**, 431–441.

Chapter 39

Transformation and Regeneration of Dicotyledonous Plants

Kevan M. A. Gartland and Nigel W. Scott

1. Introduction

The formation of crown galls on many dicotyledonous plants, as a result of *Agrobacterium tumefaciens* infection, has been well documented (*1*). The transfer of a small T-DNA region of the Ti plasmid to the nuclear genome of the plant host (*2*) and expression of the T-DNA growth hormone genes are responsible for the unorganized habit characteristic of crown gall tumors (*3*). Originally, it was thought that only dicotyledonous species could be infected, but a number of reports have recently suggested that monocotyledonous species may also be transformed by this means (*4–7*).

Molecular biologists have adapted this system for the transfer of foreign genes to plant cells, provided that 25 base pairs (bp) at the right-hand boundary of the T-DNA and the plasmid encoded vir region are present (*8*). A number of foreign genes, e.g., antibiotic resistances, have been shown to express within plant cells, if flanked in phase by promoter and polyadenylation signals known to function in plants (*9,10*).

The method described here utilizes such *Agrobacterium tumefaciens* vectors (*see* Chapter 38), which have also been deleted for the oncogenic growth hormone genes. When used with a tissue culture regime including suitable antibiotic selection and a high cytokinin:auxin ratio, they promote the direct regeneration from leaf discs of transformed shoots, rather than the unorganized tumors produced by wild type strains. Evidence for the transformed nature of these shoots may rapidly be obtained via nopaline testing, as well as continued resistance at antibiotic levels toxic to untransformed material. The techniques described here are perhaps the easiest way of studying the effects of new genetic material on plant development.

2. Materials

Note: Solvents for plant growth hormones are shown in parentheses. These may be stored at 4°C for several months. Many of these hormones are suspected teratogens and should be weighed out in a fume hood.

1. Plant material: The method described here is applicable to most dicotyledonous plant species. Leaves from sterile shoot cultures should be used (Note 1 in section 4), but this method can be carried out with surface-sterilized leaves from 6–8 wk-old plants.

2. Agrobacterium tumefaciens cultures: Disarmed Ti plasmid vectors have been described (9) in which genes 1, 2, and 4, responsible for the tumorous characteristic of wild-type Ti plasmids, have been deleted. Many of these vectors retain the entire nopaline synthase gene within the 25 bp T-DNA boundaries, together with a selectable marker gene, e.g., aminoglycoside phosphotransferase II from Tn 5, surrounded by nopaline synthase promoter and polyadenylation sequences known to function in plants. The T-DNA region may be on the same replicon as the vir region essential for integration; or it may be in the binary state, with the T-DNA on a separate, wide host range repli-

con. These vectors may be grown with shaking at 30°C overnight, by inoculating 5 mL of peptone medium with a single bacterial colony, in the presence of appropriate antibiotic selection.

3. Peptone medium: 4.0 g/L of bacto-peptone, 1.0 g/L of yeast extract, 2 m*M* magnesium sulfate, pH 6.8.

4. MSK3 Shoot-inducing medium: 4.6 g/L of Murashige and Skoog (*11*) powder (Flow Labs), 30 g/L of sucrose, 1.0 mg/L of 6-benzylaminopurine (dilute hydrochloric acid), 0.15 mg/L of 1-napthylacetic acid (ethanol), pH 5.8. This may be solidified by the addition of 9.0 g/L of agar.

5. MSK2 Rooting medium: 4.6 g/L of Murashige and Skoog powder, 30 g/L of sucrose, 0.1 mg/L of 1-napthylacetic acid (ethanol), pH 5.8.   This may be solidified by the addition of 9.0 g/L of agar.

6. Cefotaxime (Roussel Laboratories): This antibiotic, sometimes known as Claforan, may be added in aqueous solution after filter sterilization to a final concentration of 500 mg/L to MSK3 medium, in order to inhibit *Agrobacterium tumefaciens* growth.

7. Kanamycin sulfate: Aqueous filter-sterilized kanamycin may be added to MSK media to give a final concentration of 250 mg/L in order to select for putative kanamycin-resistant shoots.

8. Feeder suspension culture cells: Rapidly growing suspension cells of, e.g., *Catharantus roseus, Nicotiana tabacum,* or *Petunia hybrida,* are used to promote leaf disc growth immediately after *Agrobacterium tumefaciens* infection. Typically, 4–6-d-old suspensions, maintained by weekly subculturing in UM liquid medium, are used.

9. UM medium (12): 4.6 g/L of Murashige and Skoog powder, supplemented with 30 g/L of sucrose, 2.0 mg/L of 2,4-dichlorophenoxyacetic acid (ethanol), 0.25 mg/L of kinetin (dilute hydrochloric acid), 9.9 mg/L of thiamine, 9.5 mg/L of pyridoxine, 4.5 mg/L of nicotinic acid, 2.0 g/L of casein hydrolysate, pH 5.8.

10. 9- and 7.5-cm Diameter filter papers: 9-cm filter papers may be trimmed to approximately 8.5 cm diameter before sterilization, to aid insertion into 9-cm Petri dishes.

11. No. 5 cork borer.

12. 9-cm Diameter Petri dishes.

13. Schleicher and Schuell MN 214 chromatography paper.

14. Horizontal flat-bed electrophoresis tank.

15. 5 µL Drummond microcaps.

16. Hair dryer capable of blowing hot and cold air.

17. Nopaline test electrophoresis buffer: 5% (v/v) formic acid, 15% (v/v) acetic acid. This buffer is corrosive and should be made up in a fume cupboard.

18. Nonabsorbent cotton wool.

19. Nopaline standard: This is available from Sigma, should be dissolved in water at a concentration of 0.2 mg/mL, and stored at –20°C until required.

20. L-Arginine HCl incubation solution: 4.6 g/L of Murashige and Skoog powder, 30 g/L of sucrose, 100 mM L-arginine HCl, pH 5.8.

21. Staining solution A: 100 mL absolute ethanol containing 20 mg of phenanthrenequinone. This may be stored at –20°C in 15-mL aliquots. Phenanthrenequinone is a suspected carcinogen and should be used only in a fume cupboard with appropriate care.

22. Staining solution B: 60 mL of absolute ethanol, 40 mL of distilled water, 10% (w/v) sodium hydroxide. This may be stored at –20°C.

23. UV (300 nm wavelength) lamp.

24. UV light-proof spectacles.

25. White ceramic tiles for dissection of leaves. These may be wrapped in a double layer of aluminium foil prior to autoclaving.

26. Microcentrifuge tubes.

27. Plastic stirring rods.

28. Domestos (Lever Brothers) or another proprietary bleach.

3. Methods

Note: Government legislation may apply to the use of *Agrobacterium tumefaciens* strains containing recombinant DNA and the storage of regenerated transformed plants.

3.1. Leaf Disc Transformation Procedure

Described here is a modification of the method of Horsch et al. (*13*). Manipulations should be done in an appropriate biological safety cabinet. Techniques should, wherever possible, be carried out aseptically.

1. Punch discs of approximately 0.5 cm diameter using a no. 5 cork borer from sterile axenically grown leaves or from leaves previously surface-sterilized by 25 min of incubation in 8% (v/v) Domestos and six washes with sterile distilled water on a white ceramic tile (Note 1 in section 4).
2. Pour 5 mL of an overnight culture of the *Agrobacterium tumefaciens* strain being used into a 9-cm Petri dish. Add 20 mL of MSK3 liquid medium.
3. Add leaf discs to the bacteria and mix occasionally while incubating at 25°C for 90 min.
4. Meanwhile, pour 4 mL of feeder suspension cells onto the surface of 9-cm Petri dishes containing MSK3 solid medium (Note 2 in section 4).
5. Place a 9-cm, then a 7-cm, diameter sterile filter paper on top of the feeder layer.
6. After the 90-min incubation period, carefully remove the leaf discs and place them with the lower epidermis uppermost onto the 7-cm filter paper. Four or five discs may be placed onto each Petri dish.
7. Incubate at 25°C under artificial daylight conditions for 48 h.
8. Transfer the leaf discs to Petri dishes containing MSK3 solid medium, supplemented after sterilization with 500 mg/

L of cefotaxime and 250 mg/L of kanamycin (Notes 3 and 4 in section 4). Incubate as above for 35 d.

9. Shoots may appear after 15–21 d. These may usually be excised after 35 d and placed in tissue culture vessels containing either MSK3 solid medium, to allow the shoots to grow larger, or MSK2 solid medium to promote rooting. In either case, 500 mg/L of cefotaxime and 250 mg/L of kanamycin should be added after sterilization.

10. After rooting, shoots may be transferred to pots, ready for transfer to a greenhouse.

3.2. Nopaline Test Procedure

This technique, modified from the method of Aerts et al. (14), should be carried out in a fume cupboard, and gloves should be worn at all times.

1. Incubate 1-cm square pieces of leaf tissue from putative transformants, and uninfected leaf material, in 5 mL of arginine incubation medium at 25°C for 48 h.

2. Rinse the leaf pieces individually with distilled water and blot dry with filter paper.

3. Grind the leaf pieces in microcentrifuge tubes with a plastic stirring rod (Note 5 in section 4).

4. Centrifuge at 12000 rpm for 10 min at room temperature.

5. Apply 5-µL aliquots of the supernatants using microcaps at intervals of at least 2 cm along a line drawn 3.5 cm from one end of a piece of MN 214 paper on a clean surface. The spots should be of no more than 0.5 cm diameter and may be dried using a hot air blower. Include standards lanes or 5 µL aliquots of 0.2 mg/mL of nopaline; and a mixture of 5 µL of nopaline, and 5 µL of 0.2 mg/mL L-arginine HCl.

6. Carefully wet the electropherogram with opine test electrophoresis buffer using nonabsorbent cotton wool. Wet the paper to within 1 cm on either side of the sample spots.

7. Place the paper in an electrophoresis tank containing buffer so that it forms a bridge between the two chambers, with the sample spots nearest the positive electrode.

8. Electrophorese for 60 min at 400 V.
9. Suspend the electropherogram with the sample spots uppermost and dry with a hot air blower for 20 min.
10. Mix 15 mL each of staining solutions A and B.
11. Pass the electropherogram through the staining mixture, sample spots (+ve) end first (Note 6 in section 4).
12. Suspend the electropherogram with the sample spots uppermost, and dry with a cold air blower for 20 min.
13. View the electropherogram under UV illumination, using protective spectacles. Transformed shoots are indicated by the presence of a spot parallel to the nopaline standard (Note 7 in section 4). Sixty to eighty percent of shoots regenerated in this way are transformed.

4. Notes

1. Sterile shoot cultures of many plant species may be initiated by germinating surface-sterilized seeds on 3.6 g/L of Murashige and Skoog powder, 30 g/L of sucrose, pH 5.8, solidified with 9 g/L of agar in the dark. Seedlings are then transferred to an appropriate tissue culture vessel and subcultured by decapitation and rerooting of apical segments at monthly intervals. Surface sterilization times may vary slightly from species to species. For seeds, this is dependent on the seed size and morphology.
2. Alternative hormone combinations for shoot-promoting media include 1.0 mg/L of 6-benzylaminopurine, 2.0 mg/L of indole acetic acid (ethanol); 0.05 mg/L of 1-napthylacetic acid, 0.5 mg/L of 6-benzyl amino purine; and 1.0 mg/L of zeatin. A high ratio of cytokinins/auxins favors shoot formation. Should none of these hormone regimes promote shooting with a particular plant species, transformed callus tissue will be formed instead.
3. 1 g/L of filter-sterilized aqueous carbenicillin may be used instead of cefotaxime to prevent bacterial overgrowth in MSK3 medium. Carbenicillin should not, however, be

used with *Agrobacterium tumefaciens* strains carrying ampicillin-resistance genes.

4. The precise level of kanamycin used for selection of putative transformants is affected by the chimeric gene construct and the plant species used. In many species, however, particularly of the Solanceae, 250 mg/L has been appropriate. Constructs containing other plant-selectable marker genes, e.g. chloramphenicol resistance, necessitate the use of antibiotic concentrations able to inhibit growth of the leaf discs.

5. The addition of a very small amount of acid-washed sand may aid disruption of the leaf tissue.

6. It is very important in staining opine electropherograms to pass the sample spot end through the staining solution first to prevent smearing of the large arginine spot into the nopaline spot.

7. Nopaline may be partially converted into pyronopaline if stored at 4°C for long periods. This is characterized by two spots, rather than one, in the nopaline standard lane of an electropherogram. It has been reported (*15*) that, in some cases, uninfected tissue may give spots close to the expected position of nopaline. For this reason, comparison of infected and uninfected tissue must be made on each electropherogram.

References

1. Smith, E.F. (1916) The staining of Bacterium tumefaciens in tissue. *Plant Physiol.* **2**, 127–128.

2. Yadav, N.S., Postle, K., Saiki, R.K., Thomashow, M.F., and Chilton, M-D. (1980) T-DNA of a crown gall teratoma is covalently joined to host plant DNA. *Nature* **287**, 458–461.

3. Davey, M.R., Gartland, K.M.A., and Mulligan, B.J. (1986) Transformation of the Genomic Expression of Plant Cells, in *Society of Experimental Biology, Plasticity in Plants Meeting*, Durham.

4. Hernasteens, J.P., This-Thoong, L., Schell, J., and van Montagu, M. (1984) An *Agrobacterium* transformed cell sulture from the monocot *Asparagus officinalis*. *EMBO J.* **3**, 3039–3041.

5. Graves, A.C. F. and Goldman, S.L. (1986) The transformation of *Zea mays* seedlings with *Agrobacterium tumefaciens*. *Plant Mol. Biol.* **7**, 43–50.
6. Hooykaas-Van Slogteren, G.M.S., Hooykaas, P.J.J., and Schilperoort, R.A. (1984) Expression of Ti plasmid genes in monocotyledonous plants infected with Agrobacterium tumefaciens. *Nature* **311**, 763–764.
7. Grimsley, N., Hohn, T., Davies, J.W., and Hohn, B. (1987) *Agrobacterium*-mediated delivery of infectious maize streak virus into maize plants. *Nature* **3256**, 177–179.
8. Shaw, C.H., Watson, M.D., and Carter, G.H., (1984) The right hand copy of the nopaline Ti plasmid 25 bp repeat is required for tumour formation. *Nucleic Acids Res.* **12**, 6031–6041.
9. Herrera-Estrella, L., Depicker, A., van Montagu, M., and Schell, J. (1983) Expression of chimaeric genes transferred into plant cells using a Ti-plasmid derived vector. *Nature* **303**, 209–213.
10. Bevan, M.W. and Flavell, R.B. (1983) A chimaeric antibiotic resistance gene as a selectable marker for plant cell transformation. *Nature* **304**, 184–187.
11. Murashige, T. and Skoog, F. (1962) A revised medium for rapid growth and bioassays with tobacco tissue cultures. *Physiolog. Planta* **15**, 473–497.
12. Uchimiya, H. and Murashige, T. (1974) Evaluation of parameters in the isolation of viable protoplasts from cultured tobacco cells. *Plant Physiol.* **54**, 936–944.
13. Horsch, R.B., Fry, J.E., Hoffman, N.L., Eichholtz, D., Rogers, S.G., and Fraley, R.T. (1985) A simple and general method for transferring genes into plants. *Science* **227**, 1229–1231.
14. Aerts, M., Jacobs, M., Hernalsteens, J.P., van Montagu, M., and Schell, J. (1979) Induction and in vitro culture of *Arabidopsis thaliana* crown gall tumours. *Plant Sci. Lett.* **17**, 43–50.
15. Christou, P., Platt, S.G., and Ackerman, M.C. (1986) Opine synthesis in wild type plant tissue. *Plant physiol.* **82**, 218–221.

Chapter 40

Plant Protoplast Fusion

Neil Fish, Keith Lindsey, and Michael G. K. Jones

1. Introduction

It is now possible to regenerate plants from protoplasts of a wide range of species. As a result, genetic manipulation by protoplast fusion in vitro is now a realistic proposition. This area is exciting because protoplasts from different origins can be fused together to form new genome combinations that cannot be obtained by conventional means. Thus, protoplast fusion can be used to introduce novel germplasm into breeding programs. Practical examples include the introduction of resistance to potato leaf roll virus from the wild species *Solanum brevidens* by fusion with dihaploid potato (*Solanum tuberosum*) protoplasts (1) and the production of novel hybrids between tomato (*Lycopersicon esculentum*) and the sexually incompatible wild species (*Solanum rickii*) (2).

Using protoplast fusion, it is also possible to manipulate cytoplasm-encoded characters (i.e., chloroplasts and mitochondria) (3,4) and to study somatic cell genetics (e.g., mutant complementation) (5–7).

Plant protoplasts can be fused routinely using two main approaches. In the first, chemical fusogens are used to make the membranes of protoplasts adhere and then to fuse to form heterokaryons. In the second, electrofusion, electric fields are used initially to bring protoplasts into close contact and then to induce fusion.

1.1. Chemical Fusion

The first chemical fusogen to be used successfully was sodium nitrate (8,9), but fusion frequencies were low and in many cases the fusogen exhibited toxic effects. Keller and Melchers (10,11) used a combination of high Ca^{2+} and high pH. This method is suitable only for fusing mesophyll protoplasts and probably acts both by altering surface electrical properties of membranes while stabilizing them so fusion can occur.

Over the last decade, polyethlylene glycol (PEG) has been the most widely used chemical fusogen (12). PEG causes protoplasts to agglutinate and to become closely associated, and fusion usually occurs when the PEG is eluted. Higher fusion frequencies can be obtained by elution with a high Ca^{2+}/high pH solution, which presumably disturbs membrane charges and promotes the fusion of contacting membranes (13). PEG has been used successfully to obtain many somatic hybrids (14,15), and is often used in combination with other additives to increase the fusion frequency (e.g., DMSO; 16).

Although chemical fusogens have been routinely applied, they have several disadvantages. The process is not easily controlled, and fusion frequencies are variable. They also stress protoplasts, and so their use is restricted to more "robust" protoplast populations. There are also many different published techniques, most derived empirically; they are therefore difficult to compare directly (17). These problems have led to great interest in the more recently developed technique of electrofusion.

1.2. Electrofusion

Following a report that plant protoplasts held in close contact by electrodes could be fused by applying a short direct

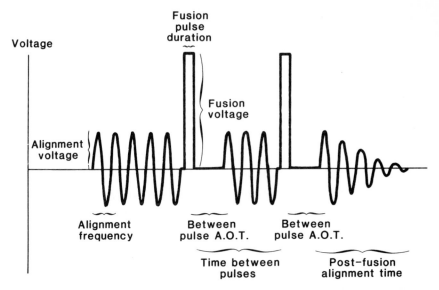

Fig. 1. Electrical output of electrofusion apparatus. (AOT = alignment off time).

current electric pulse (*18*), Zimmermann and coworkers developed the concept of electrofusion (*19*) for plant protoplasts.

When protoplasts are suspended in media of low conductivity, they become polarized by application of high-frequency (0.5–1.5 MHz) alternating electric current. This polarization causes the protoplasts to migrate to regions of higher field intensity. When two cells meet during this migration, they are strongly attracted to each other because they are dipoles. When a series of protoplasts align, they form relatively stable chains parallel to the field lines. Chain lengths depend on protoplast density and the field strength and duration of its application. Once in contact, protoplasts can be subjected to one or a series of short direct current pulses (10–300 μs, 1–3 KV/cm), and these induce the transient formation of pores that lead to fusion of contacting membranes. The collecting field is briefly reapplied to allow fusion to commence, then switched off to prevent possible damage caused by the alternating field passing through, rather than around, the protoplasts. The sequence of an electrofusion cycle is shown in Fig. 1.

The electrode systems used for animal cell fusion tend to be small, being made of platinum wires stretched on a slide, separated by 100–200 μm. This system has generally been found to be awkward to use for plant protoplasts, and wider electrode spacings (1–5 mm) have been used with flat electrodes (20–22). The use of wider electrode spacing has some consequences. These include the need to introduce protoplasts at higher densities; the majority of protoplast chains do not contact the electrodes; not all protoplasts become included in chains; electrical loads on the equipment are increased. The great advantage is that large numbers of protoplasts can conveniently be introduced and treated in one electrofusion cycle.

In the original work of Zimmermann and coworkers, carefully controlled square pulses were applied for fusion. This system is used in commercial electrofusion apparatus, and is described later. An alternative method was developed by Watts and King (20), in which the fusion pulse is supplied by the discharge of a capacitor. The waveform is a spike. This approach has the advantage of being simple and cheap to make, and has been used in electrofusion successfully (23). Use of the Zimmermann system has allowed detailed analysis of electrical properties of protoplasts in electrofusion, by the generation of pulse duration–fusion response curves (21,22,24,25). This information can be used to optimize fusion conditions for 1/1 (binary) fusions and in some cases to direct fusion between different protoplast types. This approach has also given viable hybrids (26–28). In general, electrofusion can give higher fusion yields than by using chemical fusogens, with a much greater degree of control over the products that are formed.

2. Materials

1. Polyethylene glycol of molecular weights: 1540, 6000, 8000 (Koch-light Laboratories Ltd., Colnbrook, Bucks, England).

2. High Ca^{2+} high pH fusion method: Solution A: Mannitol (72 g/L), $CaCl_2 \cdot 2H_2O$ (7.35 g/L), glycine (3.75 g/L). Adjust to pH 10–10.5 with 10N KOH. Solution B: Mannitol (72 g/L), $CaCl_2 \cdot 2H_2O$ (7.35 g/L). Make up both solutions fresh before use and filter sterilize.

3. Two-step PEG/high Ca^{2+}/high pH fusion method: Solution C: Glucose (72 g/L), $CaCl_2 \cdot 2H_2O$ (0.52 g/L), KH_2PO_4 (0.095 g/L). Solution D: PEG (1540) (500 g/L), $CaCl_2 \cdot 2H_2O$ (1.54 g/L), KH_2PO_4 (0.095 g/L), pH 5.5. Solution E. $CaCl_2 \cdot 2H_2O$ (7.36 g/L), glycine (3.75 g/L), glucose (54 g/L). Adjust to pH 10.5 with NaOH. Filter sterilize all solutions. Solutions C and D can be stored at –20°C. Make solution E fresh before use.

4. Two-step PEG/calcium nitrate fusion method: Solution F: NaCl (21.3 g/L), pH 6.0. Solution G: $Ca(NO_3)_2$ (23.6 g/L), PEG (6000) (250 g/L), mannitol (82 g/L). Adjust to pH 9.0 with KOH. Solution H: $Ca(NO_3)_2$ (64 g/L), pH 6.0. Filter sterilize all solutions and store at –20°C.

5. Single-step PEG/high Ca^{2+}/high pH fusion method: Solution I: KNO_3 (190 mg/L), $CaCl_2 \cdot 2H_2O$ (44 mg/L), $MgSO_4 \cdot 7H_2O$ (37 mg/L), KH_2PO_4 (17 mg/L), pH 5.6. Sterilize by autoclaving. Solution J: PEG (6000) (200 g/L), $CaCl_2 \cdot 2H_2O$ (35 g/L), glycine (4 g/L). Adjust to pH 10.5 with 10N KOH.

6. PEG/DMSO fusion method: Solution K (W5 solution): $CaCl_2 \cdot 2H_2O$ (18.37 g/L), NaCl (9.04 g/L), KCl (0.37 g/L), glucose (0.9 g/L). Solution L: Prepare solutions a and b:
   a. 1 g of PEG (8000) in 7 mL of H_2O.
   b. 2 g of NaOH in 35 mL of H_2O. Add glycine to give pH 10.0, then adjust volume to 50 mL.
   Sterilize by autoclaving. Add 1 mL of glycine-NaOH buffer b and 1 mL of dimethylsulfoxide to the PEG solution a and mix throughly to give solution L. Use immediately. Solution M (W5 solution): 2(N-morpholino) ethane sulfonic acid (MES) (9.76 g/L).

2.2. Electrofusion

1. Commercial apparatus: Zimmermann Apparatus GCA/ Precision Scientific Group, 3737 W. Cortland Street, Chicago, Illinois 60647, USA. Electro cell fusion system CFA-400, Krüss GmbH, Borsteler Chaussee 85-99a, 2000 Hamburg 61, FRG.

2. Workshop-made electrofusion apparatus: A fusion generator was built that incorporated an oscillator with a range of up to 1 MHz and 1–20 V peak–peak sinewave amplitude (for protoplast alignment), and a pulse generator with calibrated stepwise control of pulse width (10 μs–1 ms) and amplitude (40–200 V) (for protoplast fusion, *see* refs. *21,22*).

   An electrofusion apparatus was built in which protoplasts are aligned in an alternating field (500 kHz, 20 V RMS/cm) and fused by a pulse (800 V/cm, time constant 1 ms) discharged from 8 μF capacitors through a 160-ohm resistor in parallel with the fusion electrodes (*see* ref. *20*).

3. Electrodes: Analytical, parallel wires (0.1 mm diameter), 20 mm long, separation 1 mm (*21*); Preparative, lamellar electrodes (CFA404) Krüss GmbH; Parallel strip electrodes (*22*), parallel strips of brass glued to the surface of a glass microscope slide, seperated by 1 mm; "Transferable" electrodes (*20*), a series of thin brass or stainless-steel plates are grouped as parallel sheets, separated by 1 mm and dipped into suspensions of protoplasts in the square chambers of multiwell Sterilin culture dishes.

4. Protoplasts: good preparation, for example, from leaves, suspension cultured cells, or hypocotyls.

5. Electrofusion wash, 9% mannitol + 1 m*M*CaCl$_2$.

3. Methods

3.1. Chemical Fusions

When using the techniques outlined in this chapter, it is essential to have protoplast preparations free from cell debris to obtain good results. Protoplast fusion techniques are difficult to master, and it takes practice to obtain good reproducible results. Many manipulations must be carried out in a laminar flow bench. It is difficult to maintain sterile cultures, so a good aseptic technique is important. When estimating the success of a fusion experiment, it is useful to measure the fusion frequency. In this section on chemical fusion it is defined as:

$$\frac{\text{Number of heterokaryons}}{\text{Number of surviving protoplasts}} \times 100 = \%$$

A heterokaryon is a hybrid cell between the two fusion partners. Protoplast densities should be estimated using a hemocytometer and adjusted for the method used.

3.1.1. Fusion of Protoplasts Using High Calcium and High pH

This is a method suitable for fusing mesophyll protoplasts that are easily damaged by polyethylene glycol (*10*).

1. Mix protoplasts from each fusion partner at a 1/1 ratio in a glass centrifuge tube. Centrifuge the protoplast mixture at 100g for 5 min and remove the supernatant.
2. Resuspend the protoplasts in high calcium/pH fusion solution A to give a final density of 1 x 106 protoplasts/mL.
3. Pellet the protoplasts at 50g for 3 min.
4. Seal the centrifuge tube with Nescofilm, place it in a 37°C water bath, and incubate for 30–45 min. Take care not to disturb the protoplast pellet.
5. Remove the fusion solution A and replace with an equal volume of the postfusion wash B.
6. Incubate at room temperature for a further 30 min.
7. Remove B, gently add culture medium, and transfer into Petri dishes.

3.1.2. Protoplast Fusion Using a Two-Step Polyethylene Glycol and High Calcium, High pH Fusion Technique

This was the first published technique using PEG and high Ca^{2+}/pH to fuse plant protoplasts together (*13*). There have been many subsequent variations on this technique, some of which can be found in the references at the end of this chapter. This technique is suitable for fusing suspension and robust mesophyll protoplasts.

1.   Protoplasts from each fusion partner are mixed at a 1:1 ratio in a glass centrifuge tube. The protoplast mixture is centrifuged at 100g and the pellet resuspended in the fusion wash C to give a final density of 1 x 10^6/mL.
2.   Pipet 150 µL of the protoplast mixture onto a sterile cover-slip placed inside a 6-cm Petri dish. Allow the protoplasts to settle out.
3.   Add 450 µL of the PEG solution D carefully to the proto-plast droplet and incubate at room temperature for 40–50 min. This will cause the protoplasts to aggregate.
4.   Carefully add 500 µL of the high Ca^{2+}/pH wash E, and after 10 min add a further 500 µL. This washing will induce protoplast fusion.
5.   After 5 min, add 1 mL of protoplast culture medium.
6.   The protoplasts will remain adhering to the coverslip. Gently wash the protoplasts five times at 5-min intervals with a total of 10 mL of a protoplast culture media.
7.   Add culture media to produce the final protoplast plating density.

3.1.3. Fusion of Protoplasts Using a Two-Step Polyethylene Glycol, Calcium Nitrate Fusion Technique

This PEG fusion technique is suitable for fusing suspension and less robust mesophyll protoplasts (29).

1.   Protoplasts from each fusion partner are mixed in a 1:1 ratio in a glass centrifuge tube. The protoplast mixture is centrifuged at 100g for 5 min and the pellet resuspended in fusion wash E to give a final density of 1 x 10^6/mL.
2.   Take a 6-cm diameter Petri dish and place two 1-mL drops of the PEG solution G in the middle, 0.5 cm apart.

3. Transfer 0.5 mL of the protoplast suspension between the two drops of PEG.
4. After 10 min of incubation at room temperature, add 3 mL of the CaNO$_3$ solution H to the protoplast droplet, to induce fusion.
5. After a further 5 min transfer the protoplasts to a glass centrifuge tube and spin at 100g for 5 min.
6. Remove the supernatant and add culture media.

3.1.4. Protoplast Fusion Using a One-Step Polyethylene Glycol, High Calcium, High pH Fusion Technique

This technique (Foulger, D., Fish, N., Bright, S.W.J., and Jones, M.G.K., unpublished results) has been used in our laboratory to produce somatic hybrids between *S. tuberosum* and *S. brevidens*. It is only suitable for fusing suspension protoplasts or mesophyll with suspension protoplasts.

1. Protoplasts from each fusion partner are mixed at a 1:1 ratio in a glass centrifuge tube. The protoplast mixture is centrifuged at 100g for 5 min and resuspended in a wash medial at a final density of 1 x 10^6/mL.
2. Place 0.5 mL of the protoplast suspension in the middle of a 6-cm Petri dish.
3. Allow the protoplasts to settle for 5 min, and gently agitate the dish to accumulate them in the middle of the drop.
4. Add 1 mL of the PEG solution J around the protoplast drop. Incubate at room temperature for 10–15 min.
5. Add 3 mL of wash I, remove, and add 3 mL of fresh wash.
6. Repeat the 3-mL washes of culture, and add culture media to give the correct plating density.

3.1.5 Fusion of Protoplasts Using a Polyethylene Glycol, Dimethylsulfoxide Technique

This PEG technique gives higher fusion frequencies, but imposes great stresses on protoplasts, and is only suitable for suspension protoplast or suspension/mesophyll fusions (*16*).

1. Protoplasts from each fusion partner are mixed at a 1:1 ratio in a glass centrifuge tube. The protoplast mixture is centrifuged at 100g and the pellet resuspended in W5 solution K to give a density of 4.5 x 10^6/mL.
2. Arrange four drops of the fusion solution L in a rectangular pattern with 1–2-mm gaps between them in a 6-cm Petri dish.
3. Add a drop of the protoplast mixture between the gaps of the PEG droplets.
4. The droplets will coalesce with the protoplasts floating to the top. After 5 min tilt the dish to mix all the solutions thoroughly.
5. After 10 min dilute the suspension with 6 mL of the post-fusion wash M and incubate at room temperature for 2 h.
6. Gently pipet the protoplast suspension to break up the large protoplast aggregates.
7. Pipet into a glass centrifuge tube and centrifuge at 50g for 2 min. Resuspend the protoplasts in culture medium.

3.2. Electrofusion

3.2.1. Construction of Pulse Duration–Fusion Response Curves

These experiments can be carried out rapidly, at an analytical level, to optimize electrofusion parameters for any protoplasts of interest.

1. Carefully pipet protoplasts at a suitable density in electrofusion wash between analytical electrodes on a microscope slide placed on a microscope stage, so that about 500 protoplasts are present when a coverslip is gently laid over the electrodes.
2. Connect electrodes to Zimmermann or Krüss electrofusion apparatus.
3. Apply the alternating field at 1 MHz (alter the frequency if protoplasts rotate until they do not do so) and 15–30 V amplitude. Protoplasts should align rapidly. (If they do

not, the salt concentration may be too high—wash and resuspend the protoplasts in fresh electrofusion wash).

4. Having set the pulse characteristics (e.g., start with one pulse of 10 μs at 1500 V/cm, apply the fusion pulse when the average chain length is 5–10 protoplasts).

5. Maintain the alternating field (continuous), but reduce the amplitude to about 30% of the former value. This holds the chains in place, so that the fusion characteristics can be scored.

6. Over the following 10–15 min, the number of protoplasts aligned in chains that fuse can be scored—including the number involved in fusion events. The record should include (a) experiment description; (b) electrical parameters: alternating field (frequency, amplitude), pulse conditions (amplitude, duration, number, separation), electrode information (type, separation); (c) data on protoplast fusion: for a series of fields between the electrodes:

Not aligned	Aligned fusion products	
	0 2 3 4 5 6 7 8	(These are the numbers of
e.g., a	b c d e	protoplasts involved in any
		fusion event, category 0 =
		unfused, i.e., single.
		2 = 2 protoplasts fused
		together, and so on)

These data yield the following results:

Percentage aligned	Percentage of aligned protoplasts fused
$\dfrac{b + 2c + 3d + 4e}{a + b + 2c + 3d + 4e} \times 100$	$\dfrac{2c + 3d + 4e}{b + 2c + 3d + 4e} \times 100$
Percentage of 1/1 fusion products	Percentage of total protoplasts involved in fusion events
$\dfrac{c}{c + d + e} \times 100$	$\dfrac{2c + 3d + 4e}{a + b + 2c + 3d + 4e} \times 100$

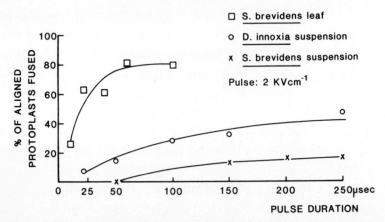

Fig 2. Pulse duration–fusion response curves for mesophyll and suspension protoplasts after electrofusion in mannitol. Leaf protoplasts of *S. brevidens* fuse more readily than suspension protoplasts of *Datura innoxa* and *S. brevidens*.

7. Similar data from pulses of 10, 20, 50, and 100 µs at pulses of 1.5 and 2.0 kV/cm yield pulse duration–fusion response curves of the type shown in Fig. 2. From these, the parameters giving optimum fusion are chosen—usually the value for the shortest pulse duration that gives maximum percentage fusion.

8. These parameters may be modified slightly (e.g., reduction in chain length and pulse length) to obtain the highest frequency of 1:1 fusions (21,22). The presence of 1 mM CaCl$_2$ in the fusion medium increases the fusion frequency of suspension culture protoplasts, which are commonly less fusogenic (25).

3.2.2 Large-Scale Electrofusions

1. Protoplasts: As above, density 5 x 10^5 to 2 x 10^6/mL in electrofusion wash.

2. Electrodes: Lamellar or parallel brass or stainless steel strips stuck to microscope slide.

3. Introduce 0.5 mL of protoplast preparation into the electrode chamber, and insert electrodes (lamellar type).

4. Follow the procedure outlined above, using optimin conditions determined in small-scale analytical experiments.
5. Ten to fifteen minutes after electrofusion, dilute the protoplasts to their final culture density with appropriate volumes of culture medium.

4. Notes

4.1. Selection of Heterokaryons

After fusion it may be necessary to select the heterokaryons from the resulting mixed population of protoplasts. Several strategies are possible, and some examples are given below.

1. Fusion: By using harsh fusion conditions, it may be possible to select against one of the fusion partners (16).
2. Auxotrophic mutant complementation: Nitrate reductase-deficient mutants, for example, with cofactor or enzyme lesions, are only able to grow when an exogenous ammonium supply is present in the medium. By growing the callus obtained after fusion on a minimal medium containing nitrate alone, parental nitrate reductase mutants will be unable to grow (5,7).
3. Mechanical isolation: Fluorescent labeling of each fusion partner allows hybrid cells to be identified easily, mechanically isolated and cultured separately.
4. Drug or herbicide resistance: If a fusion partner with drug or herbicide resistance is used, growth of fusion products in the presence of the drug or herbicide can be used to inhibit growth of the nonresistant fusion partner, so enriching the culture with hybrid material (14).
5. Mass selection: After fusion, all the resulting calluses can be grown, and selection of hybrids is made on the regenerated plants. This approach is more labor-intensive, but is often the only approach if no suitable selection system (e.g., for crop species) is available. We have used mass

selection to obtain somatic hybrids between *S. tuberosum*
and *S. brevidens*. However, with the introduction of
electrofusion, it is now possible to obtain higher fusion frequencies. This reduces the need for strict selection at an
early stage, since a high proportion of the plants obtained
after fusion will be hybrids.

6.   Fluorescence-activated cell sorter (FACS): The use of
     different fluorescent stains, or one stain and endogenous
     red fluorescence of chlorophyll in mesophyll protoplasts,
     has allowed successful use of FACS machines to discrimi-
     nate between heterokaryons. Electrofusion is well suited
     to this approach. However, the expense and need to
     modify FACS machines for plant protoplast work means
     that their availability will be limited.

4.2. High Ca²⁺/High pH Fusion Method

1.   When adding the fusion solution, the protoplasts will start
     to agglutinate.
2.   On adding the postfusion wash, the protoplast aggregates
     will start to loosen.
3.   Using this technique it is possible to obtain fusion frequen-
     cies of 1–2%.
4.   A modification of this technique has been used in our
     laboratory to produce somatic hybrids between *S.
     brevidens* and a dihaploid *S. tuberosum*. Protoplasts from
     these species are fragile, so fusion is achieved by incubat-
     ing the protoplasts for 10–15 min in solution A followed by
     30 min in culture media.

4.3. Two-Step PEG/High Ca²⁺/High pH Fusion Method

1.   When the PEG is added, the protoplasts will stick to the
     coverslip. When the high Ca^{2+}/pH solution is added, the
     tightly bound protoplasts will begin to fuse. After fusion
     the protoplasts will remain stuck down for 1–2 d. As they
     synthesize cell walls they will be released.

2. Using this technique it is possible to obtain fusion frequencies of 5–10%.

4.4. Two-Step PEG/Ca(NO₃)₂ Fusion Method

4.4. Two-Step PEG/Ca(NO$_3$)$_2$ Fusion Method

1. By suspending the protoplasts in NaCl, they will not stick to the Petri dish.
2. Using this technique it is possible to obtain fusion frequencies of 3–7%.

4.5. Single-Step PEG/High Ca^{2+}/High pH Fusion Method

1. The protoplasts will adhere to the bottom of the Petri dish. By repeated gentle washings it is possible to remove most of the cell debris after fusion.
2. This fusion technique will damage mesophyll protoplasts. Thus it is possible to select against them at the fusion stage.
3. Suspension/mesophyll heterokaryons are easily identifiable.
4. Using this technique it is possible to obtain fusion frequencies of 5–10%.

4.6. PEG/DMSO Fusion Method

1. Using this technique it is possible to obtain fusion frequencies of up to 17%.

4.7. Electrofusion Method

1. For mixtures of protoplasts that exhibit similar fusion characteristics (e.g., two populations of leaf protoplasts), mix in a ratio of 1:1 and use the electrofusion parameters as described above.
2. For mixtures of protoplasts with different fusion characteristics, (e.g., leaf and suspension culture protoplasts), use the settings giving optimum fusion for the more responsive population. With ratios of 1:5 to 1:10 (responsive/less responsive), there is then preferential heterokaryon formation (21).

3. For large-scale electrofusion, the use of a transferable lamellar electrode to fit, for example, the 20 x 20 mm square wells of repli dishes (Sterilin), allows large numbers of protoplasts to be fused very rapidly (20).

References

1. Austin, S.A., Baer, M.A., and Helgeson, J.P. (1985) Transfer of resistance to potato leaf roll virus from *Solanum brevidens* into *Solanum tuberosum* by somatic fusion. *Plant Sci.* **39**, 75–82.
2. O'Connell, M.A. and Hanson, M.R. (1986) Regeneration of somatic hybrid plants found between *Lycopersicon esculentum and Solanum rickii. Theor. Appl. Genet.* **72**, 59–65.
3. Aviv, D., Bleichman, S., Avzee-Cronen, P., and Galun, E. (1984) Intersectional cytoplasmic hybrids in *Nicotiana*. Identification of plastomes and characteristics in *N. sylvestris* + *N. rustica* hybrids housing *N. sylvestris* nuclear genomes. *Theo. Appl. Genet.* **67**, 499–504.
4. Izhav, S. and Power, J./B. (1978) Somatic hybridization in *Petunia*: A male sterile cytoplasmic hybrid. *Plant Sci. Lett.* **14**, 49–55.
5. Marton, L, Sidorov, V., Biasini, G., and Maliga, P. (1982) Complementation in somatic hybrids indicates four types of nitrate reductase lines in *Nicotiana plumbaginifolia. Mol. Gen. Genet.* **187**, 1–3.
6. Glimelius, K., Eriksson, T., Grafe, R., and Muller, A.J. (1978). Somatic hybridisation of nitrate reductase-deficient mutants of *Nicotiana tabacum* by protoplast fusion. *Physilo. Plant.* **44**, 273–277.
7. Grafe, R. and Muller, A.J. (1983) Complementation analysis of nitrate reductase deficient mutants of Nicotiana tabacum by somatic hybridisation. *Theor. Appl. Genet.* **66**, 127–130.
8. Power, J.B., Cummins, S.E., and Cocking, E.C. (1970) Fusion of isolated plant protoplasts. *Nature* **225**, 1016–1018.
9. Carlson, P.S., Smith, H.H., and Dearing, R.D. (1972) Parasexual interspecific plant hybridisation. *Proc. Natl. Acad. Sci. USA* **69**, 2292–2294.
10. Keller, W.A. and Melchers, G. (1973) The effect of high pH and calcium on tobacco leaf protoplast fusion. *Z. Naturforsch.* **28c**, 737–741.
11. Melchers, G. and Labib, G (1974) Somatic hybridisation of plants by fusion of protoplasts. 1. Selection of light resistant hybrids of 'haploid' light sensitive varieties of tobacco. *Mol. Gen. Genet.* **135**, 277–294.
12. Kao, K.N. and Michayluk, M.R. (1974) A method for high-frequency intergeneric fusion of plant protoplasts. *Planta* **115**, 355–367.
13. Kao, K.M., Constabel, F., Michayluk, M.R., and Gamborg, O.L. (1974) Plant protoplast fusion and growth of intergeneric hybrid cells. *Planta* **120**, 215–227.

14. Power, J.B., Frearson, E.M., Hayward, C., George, D., Evans, P.K., Bennly, S.F., and Cocking, E.C. (1976) Somatic hybridisation of *Petunia hybrida* and *P. parodii. Nature* **263**, 500–502.
15. Melchers, G., Sacristan, M.D., and Holden, A.A. (1978) Somatic hybrid plants of potato and tomato regenerated from fused protoplasts. *Carlsberg Res. Comm.* **43**, 203–218.
16. Menczel, L. and Wolfe, K. (1984) High frequency of fusion induced in free suspended protoplast mixtures induced by polyethylene glycol and dimethylsulfoxide at high pH. *Plant Cell Rep.* **3**, 196–198.
17. Negrutiu, I., De Brouwer, D., Watts, J.W., Sidorov, I.V., Dicks, R., and Jacobs, M. (1986) Fusion of plant protoplasts: A study using auxotrophic mutants of Nicotiana plumbaginifolia, Viviani. *Theor. Appl. Genet.* **72**, 279–286.
18. Sende, M., Takeda, J., ABE, S., and Nakamura, T. (1979) Induction of cell fusion of plant protoplasts by electrical stimulation. *Plant Cell Physiol.* **220**, 1441–1443.
19. Zimmermann, U. and Scheurich, P. (1981) High frequency fusion of plant protoplasts by electric fields. *Planta* **151**, 26–32.
20. Watts, J.W. and King, J.M. (1984) A simple method for large-scale electrofusion and culture of plant protoplasts. *Bio. Sci. Rep.* **4**, 335–342.
21. Tempelaar, M.J. and Jones, M.G.K. (1985a) Fusion characteristics of plant protoplasts in electric fields. *Planta* **165**, 205–216.
22. Tempelaar, M.J. and Jones, M.G.K. (1985b) Directed electrofusion between protoplasts with different responses in a mass fusion system. *Plant Cell Rep.* **4**, 92–95.
23. Watts, J.W., Doonan, J.H., Cove, D.J., and King, J.M. (1985) Production of somatic hybrids of moss by electrofusion. *Mol. Gen. Genet.* **199**, 349–351.
24. Tempelaar, M.J. and Jones, M.G.K. (1985c) Analytical and Preparative Electrofusion of Plant Protoplasts, in *Oxford Surveys of Plant Molecular and Cell Biology* (Miflin, B.J., ed.).
25. Tempelaar, M.J., Duyst, A., de Vlas, S.J., Symonds, C., and Jones, M.G.K. (1987) Modulation of the electrofusion response in plant protoplasts. *Plant Sci.* **48**, 99–105.
26. Bates, G.W. and Hasenkampf, C.A. (1985) Culture of plant somatic hybrids following electrical fusion. *Theor. Appl. Genet.* **70**, 227–233.
27. Kohn, H., Schieder, R., and Schieder, O. (1985) Somatic hybrids in tobacco mediated by electrofusion. *Plant Sci.* **38**, 121–128.
28. de Vries, S.E., Jacobsen, E., Jones, M.G.K., Lonnen, A.E.H.M., Tempelaar, M. J., Wybrandi, J., and Feenstra, W. (1987) Somatic hybridization of amino acid analogue-resistant cell lines of potato (*Solanum tuberosum L.*) by electrofusion. *Theo. Appl. Genet.* **73**, 451–458.

29. Hein, T., Przewozny, T., and Schieder, O. (1983) Culture and selection of somatic hybrids, using an auxotrophic cell line. *Theo. Appl. Genet.* **64,** 119–122.

Chapter 41

Plant Tissue Culture

Michael G. K. Jones, Neil Fish,
and Keith Lindsey

1. Introduction

Plant tissues normally grow in an organized fashion in which specific cell types differentiate from nonspecialised meristematic cells. Plant developmental processes can be modified by culture in vitro in a suitable nutrient medium and with the application of plant growth regulators. The interactions of the main growth regulators can be complex, but at the simplest level auxins can cause cell enlargement and division, cytokinins cause cell division, gibberellins cause elongation, and abscisic acid inhibits growth (1). When meristems are cultured, it is possible to maintain organization and multiply such meristems, and this is the basis of micropropagation. If a suitable combination of growth regulators is chosen, however, cultured tissues can be grown in a disorganized and undifferentiated way to form callus. Single isolated wall-less cells, protoplasts, can also be released from tissues following enzymatic degradation of cell walls, and these can be cultured to form callus. In both cases the aim is to regenerate intact plants from disorganized callus, usually by manipulating growth regulator

concentrations so that the cytokinin/auxin ratio is increased
(2). Plant regeneration may proceed by embryogenesis—in
which a bipolar structure resembling a zygotic embryo germi-
nates to produce a shoot and a root—or by organogenesis—in
which one organ (usually the shoot) develops first, and excised
shoots are rooted in a separate stage. Plantlets produced in this
way may be potted initially under high humidity, then grown
on to mature plants.

Plants regenerated from cells in which the meristematic
organization has been maintained are normally identical to the
parental plants.

Practical applications using this approach are widely employed, and include micropropagation, germplasm storage,
embryo rescue, virus elimination, haploid production, and
ploidy manipulation. Plants regenerated after a callus phase
may well show some differences from parental tissues, however—a phenomenon known as "somaclonal variation" (3).
This aspect must be borne in mind in any experiments that
include a callus phase.

Clearly the field of plant tissue culture is an immense subject. In this chapter aspects involving propagation of meristematic tissues will not be considered further. This is because
the main aim of molecular work described in this book is to
alter the genotype of plants in specific ways by the introduction
of foreign or modified DNA. At some stage it is therefore necessary to introduce such DNA into a single cell and to regenerate whole plants in which all the cells are modified in the same
way. The obvious starting point here is single protoplasts. In
most cases, however, shoots regenerated from excised tissue
pieces (explants) are not chimeras, and so the technically more
demanding procedures of protoplast isolation and culture may
not be required.

The choice of methods described in this chapter is necessarily very limited. Those chosen relate to the regeneration of
model and crop species that may be used in association with
genetic manipulation—in particular, protoplast fusion, tran-

sient expression, and stable integration of introduced DNA. The methods are subdivided into some standard tissue culture procedures, regeneration from explants, and isolation, culture, and regeneration of protoplasts. Since there is much published work on model species (e.g., tobacco), the methods described here have a bias toward major crop species.

2. Materials

2.1. Hardware

1. Filter sterilization units (e.g., Nunc, Sartorius).
2. Osmometer (for protoplast work).

2.2. Transfer and Culture

1. Laminar flow cabinet (for all manipulations requiring sterility).
2. Incubator or culture room (controlled temperature, light, humidity).
3. Orbital shaker (for cell suspension cultures).
4. Low-speed bench centrifuge, such as the Hettich Universal (Tuttlingen, West Germany). 38-, 50-, and 280-μm mesh stainless-steel sieves (e.g., Endecotts Ltd., Lombard Road, London SW19 3UP, UK).

2.3. Media

Many media formulations have been devised to suport the growth of cultured plant cells. The most widely used formulation, MS, was developed by Murashige and Skoog (Table 1). Other media include B5 (4), in which the ammonium nitrate is replaced by a lower level of ammonium sulfate, N6 (5), used particularly for cereal anther culture, and E1, which is useful for legume culture (modified SL; 6, see ref. 7 for more details).

Table 1
Murashige and Skoog's Medium[a]

	Stock solution		Final, mg/L	Volume of stock/L, mL
Macronutrient				
KNO_3			1900	
NH_4NO_3			1650	
$CaCl_2 \cdot 2H_2O$			440	
$MgSO_4 \cdot 7H_2O$			370	
KH_2PO_4			170	
Iron	200 x			5
Na_2EDTA	672	in 100 mL	37.25	
$FeSO_4 \cdot 7H_2O$	556		27.85	
Micronutrients	100 x			10
$MnSO_4 \cdot 4H_2O$	2230		22.3	
$ZnSO_4 \cdot 4H_2O$	860		8.6	
H_3BO_3	620		6.2	
KI	83	in 11 mL	0.83	
$Na_2MoO_4 \cdot 2H_2O$	25		0.25	
$CuSO_4 \cdot 5H_2O$	2.5		0.025	
$CoCl_2 \cdot 6H_2O$	2.5		0.025	
Vitamins	1000 x			1
Glycine	200		2.0	
Nicotinic acid	50	in 100 mL	0.5	
Pyridoxine HCl	50		0.5	
Thiamine HCl	10		0.1	
Sucrose			Various amounts for different protocols: *see* section 3	
Myo-inositol			100	
Agar			8000	
Auxins/cytokinins, as required; *see* section 3				
pH 5.7				

[a]All weights are given in milligrams. Stock solutions can be stored in aliquots in plastic bottles or bags at −20°C and thawed out for use.

2.3.1. Media production

The production of standard media, such as MS and B5, has been simplified by the availability of powdered media that may require the addition of water, sucrose, growth regulators, vitamins, and agar. Suppliers include Flow Laboratories Inc. (7655 Old Springhouse Road, McLean, Virginia, 22102, USA or PO Box 17, Second Avenue, Industrial Estate, Irvine KA12 8NB, Scotland, UK, and Gibco Laboratories (3175 Staley Road, Grand Island, NY, 14072, USA, or Gibco-Biocult Ltd., PO Box 35, 3 Washington Road, Sandyford Industrial Estate, Paisley, Scotland PA3 3EP, UK).

2.3.2. Procedure for Powdered Media

1. Add about 750 mL of glass double-distilled, deionized water to a beaker.
2. Weigh out the appropriate weight of powder (e.g., 4.71 g/ L for MS medium).
3. Add the powder to the water, stir with a magnetic stirrer.
4. Add the required supplements and proceed as from step 5 below. Media can also be made conveniently from a series of stock solutions as shown in Table 1.

2.3.3. Procedure for Media from Stock Solutions (see Table 1)

1. Add about 700 mL of glass distilled water to a 1-L beaker.
2. Add each macronutrient and dissolve (using a magnetic stirrer).
3. Add the appropriate volume of each stock solution.
4. Add sucrose and myo-inositol and appropriate growth regulators (*see* section 3), with stirring.
5. Adjust pH to 5.7 with 1*N* KOH or 1*N* HCl.
6. Add agar if required (*see* section 3).
7. Make up to 1 L with glass-distilled water.
8. Keep the agar suspended by shaking or by dissolving it first by heating to 70°C, and dispense the medium into

suitable Erlenmeyer flasks or medicine bottles to no more than three-quarters full.

9. Cap the containers with a double layer of aluminum foil or loose fitting lid (do not make an airtight seal), and autoclave for 20 min at 121°C (15 lb/in² pressure, 103 kPa).

10. Transfer the containers to a laminar flow bench, and when suitably cool (30–40°C) swirl the flasks to mix the agar evenly throughout the medium and dispense into sterile containers.

Glassware and instruments can be sterilized by dry heat, as well as autoclaving, using an oven set to 180°C for 30 min.

2.4. Ultrafiltration

Media containing heat labile components (e.g., indole acetic acid, gibberellic acid, zeatin, glutamine) or complex mixture of sugars (e.g., KM8P) (*8*) should be filter sterilized by suction of the medium through a 0.2-μm diameter pore size filter. Suitable units are available (e.g., from Nalgene Labware, Nalge Co., Rochester, NY, 14602, USA), Whatman Ltd. (Springfield Mill, Maidstone, Kent ME14 2LE, UK). If agar media are required, they can be filter sterilized at twice the final concentration, and mixed with an equal volume of autoclaved double-strength agar that has been allowed to cool to about 60°C.

2.5. Growth Regulators

1. Water-soluble: ABA (abscisic acid), GA_3 (gibberellic acid).

2. Dil. HCl-soluble: BAP (benzyl amino purine), kinetin, zeatin.

3. Ethanol soluble: 2,4-D (2,4-dichlorophenoxy acetic acid), IAA (indole-3-yl-acetic acid), NAA (α-naphthalene acetic acid).

4. Procedure: Dissolve the growth regulator in a small volume of 0.5*N* HCl, 1*M* KOH or absolute ethanol, heat slightly if required to dissolve, and gradually dilute to 100 mL with distilled water.

5.   Stocks: 2,4-D, IAA, NAA, and gibberellic acid, 200 mg/L; kinetin and zeatin 100 mg/L; BAP, 50 mg/L.

2.6. Media for Regeneration from Explants of Solanum tuberosum

1.   Medium A: MS powder (4.7 g/L), sucrose (30 g/L), BAP (2.25 mg/L), GA_3 (5 mg/L), NAA (0.186 mg/L), Agar (6 g/L), pH 5.8.
2.   Medium B: MS powder (4.7 g/L), sucrose (30 g/L), BAP 2.25 mg/L), GA_3 (5 mg/L), agar (6 g/L), pH 5.8.

2.6.1. Enzymes for Protoplasts

1.   Macerozyme R10, Cellulase R10, Cellulase Onozuka RS (Kinki Yakult Mfg. Co. Ltd., 8-21 Shingikancho, Nishinomiyua 622 Japan).
2.   Meicelase (Meiji Seika Kaisha Ltd., International Division, 4-16 Kyobashi 2 chome, Chuo-ku, Tokyo 104, Japan).
3.   Pectolyase Y-23 (Seishin Pharmaceutical Co. Ltd., 4-13 Koamicho, Nihobashi, Tokyo, Japan).
4.   Rhozyme HP-150 (Rohm & Haas Co., Independence Mall West, Philadelphia, PA 19105, USA).
5.   Cellulysin, Macerase (Calbiochem-Behring Corporation, PO Box 12087, San Diego, CA 92112, USA).
6.   Driselase, Hemicellulase, Pectinase, Cellulase (Sigma Chemical Co. PO Box 14508, St. Louis, MO 63178, USA).
7.   The above enzymes are also available from R.W. Unwin & Co. Ltd., Prospect Place, Welwyn, Herts. AL6 9EW, UK. Pollock & Poole Ltd., Ladbroke Close, Woodley, Reading RG5 4DH, UK.

Enzyme mixtures should be prefiltered through Whatman filter paper, then filter sterilized. They can be stored at –20°C in suitable aliquots.

2.7. Media for the Isolation of Mesophyll Protoplasts of Solanum tuberosum

1.   Conditioning medium: KNO_3 (190 mg/L), $CaCl_2 \cdot H_2O$ (44 mg/L), $MgSO_4 \cdot 7H_2O$ (37 mg/L), KH_2PO_4 (17 mg/L), Na_2-

EDTA (3.7 mg/L), $FeSO_4 \cdot 7H_2O$ (2.8 mg/L), H_3BO_3 (0.6 mg/L), $MnCl_2 \cdot 4H_2O$ (2 mg/L), $ZnSO_4 \cdot 7H_2O$ (0.9 mg/L), KI (0.08 mg/L), $Na_2MoO_4 \cdot H_2O$ (0.03 mg/L), $CuSO_4 \cdot 5H_2O$ (0.003 mg/L), $CoSO_4 \cdot 7H_2O$ (0.003 mg/L), thiamine HCl (0.05 mg/L), glycine (0.2 mg/L), nicotinic acid (0.5 mg/L), pyridoxine HCl (0.05 mg/L), folic acid (0.05 mg/L), biotin (0.005 mg/L), myo-inositol (10 mg/L), NAA (2.0 mg/L), BAP (0.5 mg/L), pH 5.6.

2. Preplasmolysis solution: KNO_3 (190 mg/L), $CaCl_2 \cdot 2H_2O$ (44 mg/L), $MgSO_4 \cdot 7H_2O$ (37 mg/L), KH_2PO_4 (17 mg/L), mannitol (82 g/L), pH 5.6.

3. Protoplast wash solution: KNO3 (190 mg/L), $CaCl_2 \cdot 2H_2O$ (44 mg/L), $MgSO_4 \cdot 7H_2O$ (37 mg/L), KH_2PO_4 (17 mg/L), mannitol (93.5 g/L), pH 5.6.

4. Enzyme solution: KNO_3 (190 mg/L), $CaCl_2 \cdot 2H_2O$ (44 g/L), $MgSO_4 \cdot 7H_2O$ (37 mg/L), KH_2PO_4 (17 mg/L), meicelase P (15 g/L), pectolyase Y23 (1 g/L), mannitol 89.6 g/L), pH 5.6. Adjust osmolarity to 570 mOsm, filter sterilize store at –20°C until required.

2.8. Solutions for the Isolation of Wheat Suspension Protoplasts

1. Source material: Wheat suspension culture (e.g., C82d), diluted fivefold at weekly intervals in MS containing 3% sucrose, 1–5 mg/L of 2,4-D, maintained in Erlenmeyer flasks on a gyratory shaker (100 rpm) (22).

2. Culture medium: MS containing 3% (w/v) sucrose, 1 mg/L 2,4-D (adjusted to 750±10 mOsm with mannitol, pH 5.8).

3. Enzyme mixture: 2% (w/v) cellulase R10, 0.3% (w/v) pectolyase Y23. Adjust osmolality to 750 ± 10 mOsm, pH 5.6. The osmolality is determined in an osmometer. The osmolality of the solution is reduced by dilution with culture medium or water, and is increased by adding mannitol (0.166 g/L of mannitol in water = 1 mOsm).

3. Methods

3.1. Surface Sterilization

1. Immerse the tissue in 10% (v/v) commercial bleach (sodium hypochlorite, containing 3 or 4 drops of teepo to act as a wetting agent) for 15 min in a sealed container. Gently invert, but do not shake. Decant the bleach and wash the material with six changes of sterile distilled water.
2. After sterilization, localized necrosis of tissue may be observed, especially at cut surfaces. This should not prove a problem and the tissue should quickly recover. If the tissue is killed, however, reduce the incubation time or the bleach concentration [e.g., to 5% (v/v)]. If cultures become contaminated, it will be necessary to increase the incubation time or the bleach concentration. Contaminated cultures may also be caused by faulty aseptic techniques.
3. Some tissues can contain systemic bacteria and fungi. This presents more of a problem but can sometimes be solved by including the antibiotic cefotaxime at 50 mg/L.
4. For stem tissues, cut ends can be sealed by dipping into molten paraffin wax to prevent saturation of the tissue with sterilant. A brief immersion in absolute ethanol (15 s–1 min) can be included before hypochlorite treatment for compact tissues, e.g., stems, tubers.

3.2. Callus Initiation

3.2.1. Choice of Explant

Evidence is accumulating that the origin of the explanted tissue may have a profound effect on both the metabolic and the developmental behavior of the explant, and callus tissue produced from it, when maintained in a particular culture environment (*see* refs. *9,10*). The nature of the explant appears to be important in at least three respects: (1) the region of the plant from which it is obtained; (2) the developmental age of the tissue of explantation; and (3) the genotype of the plant itself. For example, it is now possible to regenerate intact plants from tissue culture of all the major cereal crops (*see* below and ref. *11*), but in order to do so, it is necessary to initiate embryogenic callus from a limited number of tissues (usually from immature embryos, young inflorescences, or young leaves).

3.2.2. Initiation of Callus and Suspension cultures from Carrot

A description of the induction of callus cultures from carrot (*Daucus carota*) tap root tissue will serve to illustrate the general principle of this technique.

1. Use only disease-free and undamaged tap roots. Wash them free of all soil under running tap water, and surface-sterilize them in absolute ethanol and sodium hypochlorite solution as indicated above.
2. Using flame-sterilized forceps and scalpel, slice the carrot into 2–3-mm thick discs in a sterile 9-cm Petri dish, then excise blocks of tissue, about 0.5 x 0.5 cm squares from the region of the disc containing the meristematic cambial tissue (which is bound on both sides by vascular tissue, Fig. 1).
3. Place the explanted tissue onto plates of MS medium supplemented with 1 mg/L (5×10^{-6}M) 2,4-D, 3 g/L sucrose (carrot medium), and 0.8–1% (w/v) agar, and seal with Nescofilm to prevent drying out and contamination of the cultures. Maintain the cultures at 25°C in continuous cool white fluorescent illumination at a photon flux density of approximately 20 µmol/m²/s.
4. After approximately 7 d, swelling of the tissue is observed, followed by a crystalline appearance caused by cell division at the cut surfaces. At 4–6 wk after explantation, isolate callus from the explant and transfer it to fresh medium at intervals of 3–4 wk. Since the disorganized proliferation of cells in culture is a known source of genetic

Explanted region

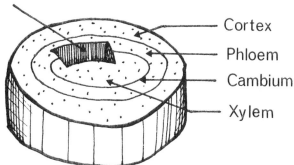

Cortex

Phloem

Cambium

Xylem

Fig. 1. Diagrammatic representation of a slice of carrot taproot showing the site of tissue explantation for callus initiation. The explant should contain the cambial tissue, sited between the xylem and phloem.

instability (*12*), it may be preferable, when attempting to introduce foreign DNA, to avoid a callus stage if possible. Therefore, the transformation, e.g., by *Agrobacterium*, of leaf discs, which can be induced to form shoots or roots with a minimum of callus formation, is a useful approach (*see* below and ref. *13*). To initiate suspension cultures from carrot callus, the following procedure is employed.

a. Using flame-sterilized forceps, transfer approximately 1 g of fresh weight of rapidly dividing and friable callus to 60 mL of carrot medium in 250-mL Erlenmeyer flasks. The flasks should be capped with a double layer of sterile aluminum foil. Flame the neck of the flask before and after each manipulation.

b. Maintain the flasks on orbital shakers at 95–150 rpm with a displacement of 1.5 cm. This allows dispersion of the callus and aeration of the cells. The culture conditions of light and temperature should be the same as for callus culture.

c. Suspension cultures grow more rapidly than callus, and subculture should be performed at intervals of 1–2 wk by dilution of 10 mL of the culture into 50 mL of fresh medium, by transfer with a sterile wide-bore glass pipet.

3.2.3. Induction of Embryogenesis in Cultured Carrot Cells

For the purpose of genetic engineering, the most valuable asset of plants is their relative plasticity of development, and, unlike animal cells, it is possible to exploit this by inducing the redifferentiation of intact plants from undifferentiated cultured cells. This can be performed by the induction of roots and shoots directly (*see below*), or via an embryogenic reorganization. We will now describe the simple procedure for inducing embryoids (bipolar structures resembling natural embryos) in cultured carrot cells.

It is often observed that embryogenesis occurs toward or into the stationary phase of culture of carrots and other species, when the supply of nutrients and growth regulators has become exhausted (*10,14*), and the process can be induced prematurely by experimentally limiting the availability of auxins and cytokinins.

1. Use rapidly growing cultures.
2. For suspended cell cultures, remove old medium by sieving through sterile stainless-steel of 280 μm pore size. Large aggregates can be removed from the suspension by sieving similarly through stainless steel mesh (*15*) or through glass beads (*16*) to produce a homogenouly sized culture.
3. Transfer the suspended cells or callus tissue to MS basal medium (MS medium as in Table 1, but with sucrose, auxins, cyokinins, and agar omitted), plus 30 g/L of sucrose.
4. Embryos should appear within 1 wk and reach maturity within 3 wk. Development into plantlets may be encouraged by the inclusion of $10^{-7}M$ zeatin in the medium (*15*).

3.3. Establishment of Shoot Cultures

3.3.1. Shoot Cultures of Brassica napus from Seeds

1. Surface-sterilize seed.
2. Place the seeds on MS with 2% (w/v) sucrose, medium solidified with 0.75–1.0% (w/v) agar in a 9-cm diameter

sterile Petri dish. Incubate for 24 h at 25°C in the dark to allow the seeds to germinate.

3. Transfer the seeds into the light in an incubator or growth room (e.g., 25°C, 16-h d, 160–180 µmol/m²/s from white daylight fluorescent tubes).
4. After the seeds have germinated, subculture them onto new medium. After 1–2 wk cut the stem into 1–2 cm segments containing the apical or axillary meristems. Transfer the stem segments to fresh medium in a larger container (e.g., 9 cm diameter glass jar). To maintain healthy shoot cultures, it is necessary to subculture the material every 3–4 wk.

3.3.2. Solanum tuberosum *Shoot Cultures from Tuber Sprouts*

1. Allow tubers to sprout in the dark at room temperature.
2. Excise the sprouts and surface sterilize them.
3. Cut each sprout into 1–2-cm segments containing at least one bud.
4. Transfer each sprout segment to MS medium solidified with agar containing 2% (w/v) sucrose and up to 0.5 mg/L BAP and incubate in the light as above.
5. After a period of 3–4 wk, individual stem nodes are excised and subcultured onto fresh medium containing 0.05 mg/L BAP.

3.4. Regeneration from Explants of Solanum tuberosum

This technique is suitable for the regeneration from explants of a wide range of tetraploid and dihaploid *S. tuberosum* lines (17,18).

1. Surface-sterilize mature, fully expanded leaves from growth chamber-grown plants.
2. Cut 1-cm squares from the leaf with a scalpel, or discs with a cork borer, (excluding the mid-rib), and place them lower (abaxial) side down on medium A.

3. Incubate the explants from 14 d at 25°C in 70 μmol/m²/s from warm fluorescent lights in a 16-h d.
4. After 14 d, callus develops from the cut edge of the explant. Regeneration is achieved by transferring the explant to medium B for a further 40–80 d.
5. On medium B, many shoot primordia will form to develop into shoots. These shoots can be excised and established as shoot cultures.

3.5. Regeneration from Immature Embryos of Wheat

1. Check the size of the immature embryos present on plants 10–15 d after anthesis (pollination). These should be <2 mm in length. The size varies with growth conditions and position of the developing grain in the spike.
2. Remove inflorescences of the correct age, take out developing grains, dehusk, and surface-sterilize for 10–15 min in 7% calcium hypochlorite (this is a saturated solution— filter off undissolved hypochlorite) plus a drop of detergent (e.g., Tween). Rinse 6 times in sterile distilled water.
3. With a dissecting microscope in a laminar flow bench, carefully remove the remaining sheets of cells that overlay the embryo, and dissect out. (Developing embryos contact the endosperm via the scutellum and can be detached easily.)
4. With a graticule eyepiece or a piece of scaled paper under the Petri dish, check that the length of the excised embryo is 1–2 mm.
5. Place the embryo scutellum side up on the culture medium (embryo axis contacting the medium—immature embyos tend to germinate if placed the other way up).
6. Culture in low light at 25°C.
7. Two types of callus will develop—one is watery and friable; the other, which develops mainly from the scutellum, is compact. The latter becomes nodular, and after 2–3 wk embryogenic structures can be observed, which go on to develop green shoots.

8. Excise the shoots plus a small piece of basal callus, transfer to MS medium containing 30 g/L sucrose, and allow to root.
9. After a few weeks, carefully wash away agar from the roots and transfer to vermiculite, watered with nutrient solution (e.g., MS salts) in a humid incubator.
10. Once established, transfer the plantlet to soil and grow to maturity.

3.6. Isolation of Mesophyll Protoplasts of Solanum tuberosum

1. From sterile shoot cultures, subculture the apical meristems onto fresh medium. Take the remaining leaves and weigh out 1 g of tissue and place in 50 mL of conditioning medium at 4°C overnight.
2. Slice preconditioned leaves (5x mm strips, e.g., with a multiblade scalpel), and incubate for 30 min in 25 mL of preplasmolysis solution.
3. Transfer preplasmolysed leaves to 20 mL of enzyme solution, and incubate for 3–4 h at 25°C in the dark on an orbital shaker (40 rpm).
4. Carefully pipet the protoplast suspension through 50- and 38-µm sieves to remove undigested tissue aggregates and debris and then centrifuge at 75g for 5 min to pellet the protoplasts.
5. Resuspend the protoplast pellet in 2 mL of wash medium, then layer over a 30% percoll solution (3 mL percoll + 7 mL dilution solution) and centrifuge at 100g for 5 min. Cell debris will be pelleted, and protoplasts will float as a band on the surface of the percoll.
6. After centrifugation, carefully remove the green bands at the percoll–wash interface with a Pasteur pipet and resuspend in 10 mL of wash medium. Centrifuge at 75g for 5 min.
7. Remove the supernatant and resuspend the pellet in 10 mL of wash medium. Centrifuge at 75g for 5 min.

8.   Resuspend the pellet in 5 mL of wash medium and count the protoplast density using a hemocytometer.

3.7. Wheat suspension culture protoplasts

3.7.1. Procedure

1.   Transfer suspended cells, 5–8 d after subculture, into centrifuge tubes, allow the cell aggregates to settle out, measure the packed cell volume, and remove the culture medium.

2.   Suspend the cells in enzyme mixture (20 mL of enzyme solution to 3–4 mL of packed cell volume), transfer to 9-cm Petri dishes, and incubate for 3–6 h on an orbital shaker (30 rpm) at 25°C in the dark.

3.   Gently filter the enzyme mixture plus protoplasts through 50- and 38-μm diameter stainless-steel sieves to filter out undigested cells and debris.

4.   Transfer to centrifuge tubes, pellet protoplasts by centrifugation at 100g for 10 min.

5.   Remove the supernatant, resuspend in a small volume of culture medium (or mannitol alone at 750 ± 10 mOsm, pH 5.8), and layer over a cushion of 25% sucrose. Centrifuge for 30 min at 700g.

6.   Remove protoplasts from the top of the sucrose cushion and wash twice more in culture medium by pelleting (100g, 10 min), and resuspend.

7.   Pass the protoplasts through a 38-μm mesh sieve (to separate any aggregates), count the density using a hemocytometer, and adjust ot 1–5 x 10^5/mL in culture medium. Incubate in Petri dishes at 25°C in the dark.

4. Notes

4.1. Shoot Cultures

To obtain rapidly growing shoot cultures, it is important to subculture only from the most vigourous shoots. Contaminated cultures should be discarded. If cultures are sealed and airtight, then ethylene can accumulate, resulting in stunted plants with small leaves. This can be avoided by keeping screw caps loose or by making ventilation slits in any sealing plastic film.

4.2. Regeneration from Solanum tuberosum

1. Although this method works for many different genotypes of *S. tuberosum*, some genotypes respond better than others.
2. When using medium B, it is necessary to ventilate the culture vessel to allow for rapid shoot growth.

4.3. Regeneration from Immature Embryos of Wheat

1. This procedure is suitable for the culture of immature embryos of a range of species of the Gramineae (*11*).
2. For wheat, most cultivars are responsive. This is not the case for all cereals (*11*).

4.4. Isolation of Mesophyll Protoplasts

1. For reliable yields of viable protoplasts, only healthy leaves from fast-growing shoot cultures should be used as source material.
2. Protoplasts from some cultivars are prone to damage during isolation. This may be reduced by adding additional $CaCl_2 \cdot 2H_2O$ (0.88–1.47 g/L) to the enzyme and wash media, and by reducing the centrifugation speed (*19,20*). The centrifugation step can only be optimized empirically for each experimental system.
3. Once viable protoplasts of *S. tuberosum* are obtained, it is possible to regenerate plants using published protocols (*21,22*).

4.5. Isolation of Wheat Suspension Culture Protoplasts

1.   This procedure is suitable for the isolation and culture of a range of suspension protoplast systems.
2.   The sucrose cushion, and final 38-μm mesh sieving, may be omitted if the initial digest gives clean protoplasts that do not clump together.
3.   The protoplasts obtained are suitable for direct gene transfer experiments, fusion, or longer-term culture. The frequency of divisions may depend on the genotype of the culture used.

References

1.   George, E.F. and Sherrington, P.D. (1984) *Plant Propagation by Tissue Culture Exegetics*, Everseley, pp. 709.
2.   Jones, M.G.K. and Karp, A. (1985) Plant Tissue Culture Technology and Crop Improvement, in *Advances in Biotechnological Processes* vol. 5, A.R. Liss, New York.
3.   Larkin, P.J. and Scowcroft, W.R. (1981) Somaclonal variation—a novel source of genetic variability from cell cultures for plant improvement. *Theor. Appl. Genet.* 60, 197–214.
4.   Gamborg, O.L., Miller, R.A., and Ojima, K. (1968) Nutrient requirements of suspension cultures of soybean root cells. *Exp. Cell. Res.* 50, 151–158.
5.   Chu, C.C. (1978) The N6 Medium and Its Applications to Anther Culture of Cereal Crops, in *Proceedings of the Symposium on Plant Tissue Culture* Science, Peking.
6.   Phillips, G.C. and Collins, G.B. (1980) Somatic embryogenesis from cell suspension cultures of red clover. *Crop Sci.* 20, 323–326.
7.   Gamborg, O.L. (1984) Plant Cell Cultures: Nutrition and Media, in *Cell Culture and Somatic Cell Genetics of Plants*. (Vasil, I.K., ed.) Academic, Orlando.
8.   Kao, K.N. and Michayluk, M.R. (1975) Nutritional requirements for growth of Vicia hajastana cells and protoplasts at very low population densities in liquid media. *Planta* 126, 105–110.
9.   Meins, F. (1986) Determination and Morphogenetic Competence in Plant Tissue Culture, in *Plant Cell Culture Technology* (Yeoman, M.M., ed.) Blackwell, Oxford.
10. Lindsey, K. and Yeoman, M.M. (1985) Dynamics of Plant Cell Cultures, in *Cell Culture and Somatic Cell Genetics of Plants* vol. 2 (Vasil, I.D., ed.) Academic, Orlando.

11. Bright, S.W.J. and Jones, M.G.K., eds. (1985) *Cereal Tissue and Cell Culture* Nijhoff/Dr. W. Junk, Dordrecht.
12. Scowcroft, W.R. and Ryan, S.A. (1986) Tissue Culture and Plant Breeding, in *Plant Cell Culture Technology* (Yeoman, M.M., ed.) Blackwell, Oxford.
13. Comai, L., Facciotti, D., Hiatt, W.R., Thompson, G., Rose, R.E., and Stalker, D.M. (1985) Expression in plants of a mutant are A gene from *Salmonellya typhimuriumn* congers resistance to glyphosate. *Nature* **317**, 741–744.
14. Lindsey, K. and Yeoman, M.M. (1983) The relationship between growth rate, differentiation and alkaloid accumulation in cell cultures. *J. Exp. Bot.* **34**, 1055–1065.
15. Ammirato, P.V. (1984) Induction, Maintenance and Manipulation of Development in Embryogenic Cell Suspension Cultures, in *Cell Culture and Somatic Cell Genetics of Plants* vol. 1 (Vasil, I.K., ed.) Academic, Orlando.
16. Warren, G.S. and Fowler, M.W. (1978) Cell number and cell doubling times during development of carrot embryoids in suspension culture. *Experientia* **34**, 356.
17. Karp, A., Risiott, R., Jones, M.G.K., and Bright, S.W.J. (1984) Chromosome doubling in monohaploid and dihaploid potatoes by regeneration from cultured leaf explants. *Plant Cell Tiss. Org. Cult.* **3**, 363–373.
18. Wheeler, V.A., Evans, N.E., Foulger, D., Webb, K.J., Karp, A., Franklin, J., and Bright, S.W.J. (1985) Shoot formation from explant cultures of fourteen potato cultivars and studies of the cytology and morphology of regenerated plants. *Ann. Bot.* **55**, 309–320.
19. Foulger, D. and Jones, M.G.K. (1986) Improved efficiency of genotype dependent regeneration from protoplasts of important potato cultivars. *Plant Cell Rep.* **5**, 72–76.
20. Fish, N. and Karp, A. (1986) Improvements in regeneration from protoplasts of potato and studies on chromosome stability. 1. The effect of initial culture media. *Theor. Appl. Genet.* **72**, 405–412.
21. Creissen, G.P. and Karp, A. (1985) Karyotypic changes in potato plants regenerated from potato. *Plant Cell Tiss. Org. Cult.* **4**, 171–182.
22. Maddock, S.E. (1986) Suspension and protoplast culture of hexaploid wheat (*Triticum aestivum*, L.) *Plant Cell Rep* **6**, 23–26.

Chapter 42

Direct Gene Transfer into Plant Protoplasts

Keith Lindsey, Michael G. K. Jones, and Neil Fish

1. Introduction

The techniques of plant molecular biology have advanced rapidly in the last 5 yr, and, for a number of plant species, including some crops, it is now possible to bypass traditional plant breeding techniques and introduce specific genes directly. The earliest reports of the transformation of plants with foreign genes exploited the fact that the soil bacterium *Agrobacterium tumefaciens*, which induces the formation of crown galls, transfers part of a plasmid (the tumor-inducing, Ti, plasmid) into its host (e.g., ref. *1*). The isolation and modification of this plasmid by the insertion of one or more structural genes together with regulatory elements has provided the means of genetically manipulating intact plants, cultured plant tissues, and protoplasts.

Unfortunately, this approach has been largely unsuccessful for the introduction and expression of foreign genes in monocotyledonous plants, which include the economically valuable cereal crops (but *see* refs. 2,3); these species appear not to be susceptible to the bacterium, and fail to develop galls if inoculated. Recently, however, it has been demonstrated (4) that specific DNA fragments can be introduced into *Nicotiana tabacum* protoplasts in the absence of either *A. tumefaciens* or its Ti plasmid, and transformed plants can be regenerated subsequently. This evidence has opened up the whole area of plant genetic engineering, particularly since the technique has been successfully applied to members of the Gramineae, including *Triticum monococcum* (5,6), *Lolium multiflorum* (7), and *Zea mays* (8).

In this article, therefore, three types of approach are considered for direct gene transfer into plant protoplasts. The first is a method involving the use of chemical agents; the second relies on the use of electrical pulses; and the third employs a combination of chemical and electrical agents. All three strategies act to increase the permeability of the plasmalemma to DNA, which, being phosphorylated, is not lipophilic.

Once inside the protoplast, the DNA fragment(s) may either become integrated into the host DNA or may remain free in the cytoplasm and/or nucleus. We will therefore discuss methods for the detection of both stable transformation and of transient gene expression in cells in which foreign DNA is not necessarily integrated, but may nevertheless be expressed.

2. Materials

1. For the preparation of protoplasts, the reader is referred to Chapter 41 in this volume (9,10).
2. For the preparation of plasmid and carrier DNA, the reader is referred to ref. 11 and Vol. 2 of this series.
3. CPW13M medium: KH_2PO_4 (27.2 mg/L), KNO_3 (101.0 mg/L), $CaCl_2 \cdot 2H_2O$ (1480.0 mg/L). A 10x concentrated stock of KH_2PO_4, KNO_3, $CaCl_2 \cdot 2H_2O$, and $MgSO_4 \cdot 7H_2O$ combined is made up in double-distilled deionized water, and a 10,000x concentration stock of KI and $CuSO_4 \cdot 5H_2O$ is made up similarly. These solutions can be stored at 4°C

for several weeks, and made up to the correct concentration prior to use. Mannitol is added to 13% (w/v), and the pH is adjusted to 5.8.

4. CPW13M-PEG medium: CPW13 medium containing 45% (w/v) polyethyleneglycol 1500.

5. Multiwell culture plates, plastic, presterilized (Sterilin, Teddington, Middlesex).

6. Calcium chloride solution, $2M$.

7. HEPES buffered saline: NaCl (8.0 g/L), KCl (0.37 g/L), $Na_2H_2PO_4 \cdot 2H_2O$ (0.125 g/L), glucose (1.0 g/L), N-2-hydroxyethylpiperazine-N'-2-ethanesulfonic acid (5.0 g/L), dissolved in double-distilled deionized water, and the final pH is adjusted to 7.05.

8. $0.25M$ Tris-HCl, pH 7.8.

9. Glycine buffer, pH 10: 50 mM glycine, 50 mM $CaCl_2 \cdot 2H_2O$, 13% (w/v) mannitol, pH 10.0 double-distilled deionized water, and add $0.5M$ NaOH until the pH reaches 10.0.

10. Polyvinylalcohol, 20% (w/v).

11. Fluorescein diacetate: The powder and a stock solution of 5 mg/mL in acetone should be stored in the dark at 4°C and is stable for several months.

12. [^{14}C]-Chloramphenicol, 50 mCi/mmol: This should be stored at 4°C.

13. 4 mM Acetyl CoA: This should be stored at –20°C.

14. Chloramphenicol acetyltransferase: This enzyme should be stored at –20°C.

15. Silica gel TLC plates: 20 x 20-cm plastic backed plates with 0.2-mm thick silica gel, as supplied by Merck (Darmstadt, FRG).

16. Electroporation apparatus: Either the Zimmermann-type cell fusion apparatus (GCA Corporation, Chicago, USA), employing just the dc fusion pulse facility for inducing permeabilization of the protoplasts, or the purpose-built electroporator designed by Neumann (Dialog, Dusseldorf, FRG).

17. For ultraviolet microscopy, we use an Olympus BH-2 fluorescence microscope system. For viewing fluorescein fluorescence, we use a blue exciter filter (IF-490), a blue dichroic mirror [DM500 (0-515)], and a barrier filter 0.530.

3. Methods

Methods will be described for the direct transfer of genes to sugar beet protoplasts, but we expect that in principle such techniques can be applied to most species.

3.1. Chemical Agents to Induce Direct Gene Transfer

The plasmamembrane of the plant cell is the principal barrier between the cytoplasm and the external environment. It is selectively permeable to ions, nutrients, and other solutes, but is generally impermeable to highly charged macromolecules, including nucleic acids and water-soluble proteins. The permeability of the plasmamembrane can however, be manipulated by the use of chemical agents such as DMSO (12), toluene (13) and other solvents, antibiotics (e.g., 14), pH treatments (e.g., 15), osmotic shock (16), temperature treatments (17–20), and electrical treatments (see later; for review, see ref. 21). Such techniques can be used for introducing substrates for in situ enzyme assays, for inducing the release of cytoplasmic and vacuolar contents, and more recently, for allowing the uptake of DNA fragments by cultured plant protoplasts and animal cells for genetic transformation.

Two such chemical techniques are described for the transformation of protoplasts; (1) the use of polyethylene glycol and (2) the coprecipitation of DNA with calcium phosphate, which both allow the transfer of DNA across the plasmamembrane, but apparently by different mechanisms.

3.1.1. Polyethylene Glycol

This technique is based on the method of Krens et al. (22) and has been used by Lorz et al. (5) for direct gene transfer to protoplasts of *Triticum monococcum.* The precise mode of action of polyethylene glycol is unknown, but it appears to disrupt the stability of the plasmamembrane structure, perhaps by acting as a bridging molecule between phospholipid or protein components.

1. Isolate protoplasts from suspension cultures of sugar beet 3–5 d after subculture (23), wash free of enzyme solution (*see* Jones et al., this volume), and store in CPW13 medium at 4°C at a density of 5 x 10^5/mL for no longer than 12 h.

2. To 10^6 protoplasts (in 2 mL) add 20 μg of sterile plasmid DNA (in 20 μL) and 100 μg (in 100 μL) of sterile herring sperm carrier DNA. To this then add 1 mL of CPW13M-PEG medium. This produces a final concentration of 15% (w/v) PEG, which allows survival of >75% of the protoplasts, as determined by fluorescein diacetate staining (see below), and a very low level of fusion (less than 5–10%), but the appropriate concentration will vary for individual species and should therefore be determined for the experimental system in question.

3. Incubate the reaction mixture at room temperature for 30 min in a presterilized plastic culture vessel. For multiple samples we find that multiwell culture plates (Sterilin) are convenient.

4. Transfer the reaction mixture, in which the protoplasts are now highly aggregated, to a sterile glass centrifuge tube and dilute by the stepwise addition of 10 mL of cold (4°C) CPW13M, adding 2 mL at 5-min intervals. This procedure disperses the aggregates.

5. Sediment the protoplasts by centrifugation at 100g for 10 min at 4°C, discard the supernatant, resuspend the protoplasts in cold CPW13M as a washing medium, then centrifuge again at 100g for 10 min at 4°C.

6. To ensure further removal of PEG, resuspend the protoplasts in 10 mL of cold CPW13M, centrifuge at 100g for 10 min at 4°C, and finally resuspend in 5 mL of culture medium and plate out in multiwell plastic culture dishes at a

density of 5×10^{-4}/mL, 2 mL/well. Incubate the cultures at 25°C in the dark.

7.   Transient gene expression studies (*see below*) can be performed after 36–48 h, or the cultures can be maintained for the determination of the stable integration of the DNA. If the latter studies are to be carried out, the mannitol concentration of the protoplast culture medium is reduced in a stepwise fashion by the addition, twice, of 1 mL of culture medium lacking mannitol at 7-d intervals. The protoplast-derived cell colonies are then transferred to agar medium containing selective agents (*see below*). This procedure is performed after any permeabilization treatment if protoplasts are to be maintained in prolonged culture.

3.1.2. Calcium Phosphate Co-precipitation

This method is based on that of Hain et al. (*24*). The basis of the technique is the formation of a coprecipitate of the DNA fragments with calcium phosphate, which is formed by adding calcium chloride to solutions of DNA in phosphate buffer. On contact with protoplasts the coprecipitate is transferred across the plasmamembrane in a calcium-requiring process (*25*).

1.   Prepare protoplasts and store them for use at a density of 5×10^5/mL in CPW13M at 4°C as described above.

2.   For preparation of the $Ca_3(PO_4)_2$–DNA precipitate, dissolve 50 µg (in 50 µL) of plasmid DNA in 1 mL of HEPES-buffered saline, and add 0.062 mL of $2M$ $CaCl_2$ solution to produce a final concentration of 125 mM $CaCl_2$. Allow the coprecipitate to form for 20 min at room temperature.

3.   Then add 1-mL aliquots of the coprecipitate to 2-mL samples of the protoplast suspension (5×10^5/mL) in multiwell culture vessels, and incubated for 30 min at room temperature.

4.   In order to encourage the endocytotic uptake of the coprecipitate, the protoplast–coprecipitate complex is treated with polyvinylalcohol (PVA) and high pH calcium (*24*–

25). Add a solution of 20% (w/v) polyvinylalcohol in 13% (w/v) mannitol to the reaction mixture, to make a final concentration of 10% (w/v) PVA, and incubate at room temperature for 10 min.

5. Add 5 mL of 50 mM glycine buffer, pH 10, containing 50 mM $CaCl_2$ and 13% (w/v) mannitol, and incubate the reaction mixture in multiwell culture plates sealed with Nescofilm in a water bath at 30°C for 20 min.

6. Transfer the protoplast suspension to a sterile glass centrifuge tube and centrifuge at 100g for 10 min at 4°C.

7. Wash the protoplasts twice in cold CPW13M, as described above, resuspend in culture medium at a density of 5 x 10^4/mL, and culture in the dark at 25°C for either transient gene expression analysis or for studies of stable transformation.

3.2. Electrical Techniques

The application of an electrical pulse to a cell can result in an increased permeability of membranes to macromolecules (26). Above a threshold field strength, electrical pulses can result in membrane breakdown and cell death, but below this threshold there may occur a nonlethal transient increase in membrane permeability termed reversible electrical breakdown (27,28). The effects on the membrane are long-lived compared to the duration of the pulse (29), and it has been hypothesized that the transient formation or enlargement of aqueous pores occurs in the lipid bilayer (30–32). The use of such electrical techniques for the transfer of macromolecules including DNA is termed "electroporation" or "electropermeabilization" and has been used successfully for the transformation of mouse lymphoma cells (33) and lymphocytes (34), and of protoplasts of carrot, tobacco, maize, and sugar beet (8,35,36; K. Lindsey and M.G.K. Jones, submitted). We will describe a method for the electroporation of sugar beet protoplasts, employing the Zimmermann cell fusion apparatus (*see* Fig. 1).

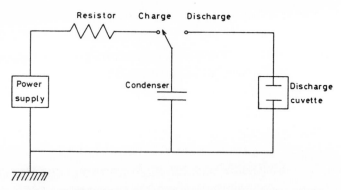

Fig. 1. Circuit diagram of simple equipment that can be used for electroporation (ref. 50, and Hauptmann et al., in press).

1. Isolate protoplasts and store at a density of 5×10^5/mL at 4°C in CPW13M, as described above.
2. For each pulse treatment, use 0.4 mL of the protoplast suspension (i.e., 2×10^5 protoplasts) in, for the Zimmermann apparatus, a lamellar electrode chamber (*see* Fish et al., this volume) of volume 1 mL (dimensions $2 \times 1 \times 0.5$ cm, mounted on a microscope slide), and add 10 µg (in 10 µL) of plasmid DNA, plus 50 µg (in 50 µL) of carrier DNA.
3. Sterilize the electrode by immersion in 90% ethanol for 10 min, allow it to dry in a laminar flow bench, introduce it into the reaction mixture, and apply up to 9 dc pulses of 10–99.9 µs duration, and of field strength 250–2500 V/cm. In our experience, the accumulation of membrane-impermeable marker compounds and transient gene expression within protoplasts of sugar beet is reduced if pulse number, field strength, and duration are high (23; and K. Lindsey and M.G.K. Jones, submitted). Different species will have different optimum conditions for electroporation. It is necessary to find a balance between maximum permeabilization and minimum cell death. R.M. Hauptmann et al. (in press) report that, using the Zimmermann cell fusion apparatus, DNA delivery was only successful in carrot protoplasts if multiple pulses of 10–100 each at 240 V dc and 99 µs were used.

4. Remove the electrode, transfer the protoplast solution to a sterile glass centrifuge tube and incubate on ice for 10 min, to keep the membrane pores open longer, so that DNA transfer from solution to protoplast can continue. (This step is optional, in our view.)

5. Give the protoplast solution a short heat shock by incubating the centrifuge tube in a 35°C water bath for 3 min (optional), and then maintain at room temperature (22–24°C) for a further 30–60 min, to allow the membrane pores to reseal, so trapping the DNA within protoplasts.

6. Add 4.5 mL of cold CPW13M to the reaction mixture, and sediment the protoplasts by centrifugation at 100g for 10 min at 4°C. Discard the supernatant, and further wash the protoplasts by two resuspensions and centrifugations in 10 mL cold CPW13M, as described above.

7. Finally resuspend the pelleted protoplasts in 2 mL of culture medium at a density of 5 x 10^4/mL, and culture at 25°C in the dark for the determination of transient gene expression or stable transformation.

3.3. Combined Electrical and Chemical Technique

Although the individual chemical and electrical techniques for membrane permeabilization have proved effective for direct gene transfer, it has been shown by Shillito et al. (37) that the efficiency of transformation can be significantly increased, for tobacco protoplasts at least, by employing a combination of electroporation, polyethylene glycol, and heat shock treatments. It was concluded that each of these factors was of equal importance in obtaining high efficiencies of transformation, and it has certainly been observed in our own laboratory that a combination of chemical and electrical permeabilization treatments is often more effective than either chemical or electrical treatment alone.

We will describe such a combined method for direct gene transfer based on that of Shillito et al. (37).

1. Isolate protoplasts and store at a density of 5 x 10^5/mL in CPW13M at 4°C as described above.
2. Transfer an aliquot of 0.2 mL of protoplast suspension (i.e., 10^5 protoplasts) to a sterile glass centrifuge or test tube, and apply a heat shock by incubating at 45°C for 5 min.
3. Cool the protoplasts on ice for 5 min and add 10 µg (in 10 µL) of plasmid DNA plus 50 µg (in 50 µL) of carrier DNA at room temperature.
4. Add 0.3 mL of a solution of CPW13M-PEG, containing 19% PEG 1500, to the protoplast suspension to make a final concentration of 8% (w/v) PEG, and incubate the mixture at room temperature for 10 min.
5. Transfer the sample to the sterile 1-mL electrode chamber for electroporation under the conditions described above. Transfer the sample back to a sterile centrifuge tube, incubate on ice for 10 min, and then incubate in a water bath at 30°C for 3 min and at room temperature for a further 30 min, to allow the pores to reanneal.
6. Finally wash the protoplasts stepwise in CPW13M as described above for simple PEG treatment, and resuspend in culture medium for the determination of transient gene expression stable transformation.

3.4. Protoplast Viability

Once permeabilized, it is important that the putatively transformed protoplasts remain viable. Viability can be assessed in a variety of different ways, and for the determination of whether stable transformation has occurred, the most important indication of viability is the ability of the protoplasts to regenerate cell walls, divide, and form cell aggregates and calluses upon which can be performed antibiotic selection, marker enzyme assays, DNA extraction and analysis, and, ultimately, plant regeneration. These procedures are time-consuming, however, and a number of rapid tests are available that indicate one facet or other of cell viability. One popular method, relevant here in that it indicates plasma membrane integrity, involves the use of fluorescein diacetate (*38*). All protoplasts will take up fluorescein diacetate, but intracellular esterase activity in viable protoplasts cleaves the diacetate moiety from the fluorescein, which, being membrane impermeable, is trapped within intact protoplasts or cells. When viewed under UV illumination, fluorescence is observed in those cells with an undamaged plasmamembrane.

1. Make up a stock solution of 5 mg/mL fluorescein diacetate in acetone, and, when required, a working solution is made up fresh by adding, dropwise, the stock solution to 5 mL of CPW13M until the solution develops the first permanent milky appearance.
2. Add one drop of this solution to a drop of protoplast supension on a microscope slide or, preferably, a hemocytometer, and view under UV illumination. Fluorescence develops after 3–5 min, and the proportion of intact protoplasts is easily determined.

3.5. DNA Integration and Expression

So far we have considered a variety of method for introducing DNA into plant protoplasts and a method for determining whether the plasmamembrane has reannealed successfully. The plating efficiency of the culture can now be determined as a further indication of viability (*see* Jones et al., this volume). In order to determine whether the DNA has been taken up, and is capable of being expressed, two general strategies can be adopted. The first, which has been well reviewed elsewhere and we will not consider in detail, relies on the prolonged culture of the transformed protoplasts to form colonies of callus that, ideally, can be manipulated to regenerate into whole plants. The precise procedures involved depend both on the DNA construct used and the host cell culture species. For tissue culture and regeneration procedures, the reader is directed to Jones et al., this volume (*9,10*). The determination of stable DNA integration and expression falls into two areas: a selection procedure, dependent upon the construct containing genes that encode marker enzymes, most commonly for antibiotic resistance; and analysis of the genomic DNA itself, usually by Southern blot analysis (39). Constructs have been used that contain genes for resistance to, in particular, the neomycin-type aminoglycosides neomycin, kanamycin (5,8,24,40), and G418 (7), and to hygromycin (41, 42), and the reader is referred to these articles and to Paskowski and Saul (43) for details of the procedures involved. Southern blot analysis involves the hybridization of restricted total genomic DNA, isolated from antibiotic-resistant callus or from the regenerated plant, with a nick-translated fragment of the plasmid DNA (see 43,44), and this provides the critical evidence of integration into the host.

This strategy, although providing a large amount of information on the fate of the DNA, requires much work, and should transformation be unsuccessful, it could be several weeks before this is realized. A rapid means of testing, first, whether the DNA has been taken up by the protoplasts, and, second, whether the regulatory sequences in the construct allow expression of the marker genes, can be performed, and does not necessaily require either stable integration of the DNA or division of the protoplasts. This is called "transient gene expression," and involves simply an assay of the activity of one or more enzymes encoded in the construct. The two most commonly used enzyme activity markers, of bacterial origin, are aminoglycoside phosphotransferase activity, encoded by the APH (3')II gene (see 43), and chloramphenicol acetyltransferase (CAT) activity, which has been used by Herrera-Estrella et al. (45) as a marker of transformation in cultured tumor cells derived from Agrobacterium-infected tobacco seedlings, and by Fromm et al. (36) as a marker of transient gene expression in electroporated protoplasts of carrot and maize, and is currently used in our laboratory. The method used by us is as follows, and is based on that of Gorman et al. (46).

1. After permeabilization in the presence of DNA, using any of the methods described above, incubate protoplasts for 36–48 h in culture medium in the dark at 25°C, and then sediment by centrifugation at 100g for 10 min at 4°C. The supernatant is discarded.
2. Resuspend the protoplasts in 100 µL of cold 0.25M Tris-HCl (pH 7.8), transfer them to an Eppendorf tube, add a pinch of sand, and disrupt them by hand using a glass homogenizer.
3. Heat samples to 60°C for 10 min. CAT enzyme is stable, but others such as proteases are broken down.
4. Pellet cell debris by spinning at maximum speed for 5 min in a microfuge at 4°C. Transfer the supernatant to a new Eppendorf tube and assay for CAT activity.
5. To the supernatant, add 0.15 µCi (in 0.75 µL) of [^{14}C]-chloramphenicol (50 mCi/mmol, Amersham Radiochemicals) and 20 µL of 4 mM acetyl CoA. Controls contain 0.01 U of CAT enzyme (a) in the presence of absence of cell extract and (b) extracts of nonpermeabilized protoplasts in both the presence and absence of extracellular DNA. Maintain all reagents except acetyl CoA at 37°C prior to use, and incubate the reaction mixture at 37°C for 2 h.
6. Stop the reaction by adding 1 mL of ethylacetate, and mix by vortexing for 30 s. The organic phase contains chloramphenicol and its acetylated forms.
7. Vacuum-dry the organic layer at room temperature, take it up in 50 µL ethylacetate, and load onto silica gel TLC plates. Separate the sample in a solvent system of chloroform/methanol (95/5), and autoradiograph for 48 h at room temperature. Chloramphenicol (unacetylated) has the lowest mobility in this system, followed by 1-acetyl chloramphenicol. 3-acetyl chloramphenicol, and then 1,3-diacetyl chloramphenicol, with the greatest mobility.

4. Notes

1. Storage conditions: Media for washing and culturing
    protoplasts and purified DNA can be stored, sterile, for
    several weeks if kept at 4°C, and almost indefinitely if kept
    at –20°C. Fluorescein diacetate powder and stock solution
    should be stored at 4°C in the dark.
2. Electroporation: Some laboratories (35–37) advocate a
    simplified electroporation medium (essentially phos-
    phate buffer plus salts), but we have found that culture
    medium plus mannitol has no deleterious effect on the
    permeabilization process.

    If a combination of electroporation and PEG treatments
    are to be used, Shillito et al. (37) have evidence that, to
    increase the frequency of transformation events, the PEG
    should be added to the protoplasts *after* the DNA.

    For tobacco protoplasts, the heat shock appears to
    improve protoplast viability after the treatments (J.
    Paszkowski, personal communication). A simple circuit
    diagram for equipment for electroporation is shown in
    Fig. 1.
3. DNA for transformation is sterilized by precipitation in
    ethanol and is resuspended after centrifugation in sterile
    distilled water (11).
4. The Dialog Elektroporator, as supplied, has an electrode
    chamber of volume 0.3 mL. Because the electrical pulse is
    supplied by discharge of capacitors, the duration of the
    pulse is determined by the choice of capacitor (C), and of
    the resistance (R) of the filled sample chamber. The resis-
    tance of the sample in the chamber can be measured, and
    adjusted to 1 kΩ using $MgCl_2$ (approximately 12.5 mM)
    (37). A suitable setting for electroporation of plant proto-
    plasts (at 1–2 kV/cm) is $C = 10$ nF and $R = 1$ kΩ, giving the
    half-life of pulse decay of 10 µs. R.M. Hauptmann et al. (in
    press) report that, using a capacitor discharge electropora-
    tor, the expression of CAT activity in carrot and cereal

protoplasts depended on the interaction between the voltage and the capacitance. In general, CAT activity increased with increasing total charge delivered, until protoplast viability was detrimentally affected. For carrot protoplasts, maximum CAT activity was obtained at 110 μf for 400 V, 500 μf for 510 V, and 1000 μf for 200 V.

5. As indicated above, the determination of transient gene expression by enzyme asssay is a rapid way to determine whether the foreign DNA has been taken up by the protoplasts and whether the regulatory elements of the construct are functional. It should be realized, however, that stably introduced DNA may be subject to developmental regulation (47–49), and so optimum conditions for transient gene expression systems may differ from those for integration and expression of foreign DNA.

References

1. Bevan, M.W. and Chilton, M-D. (1982) T-DNA of the Agrobacterium Ti and Ri plasmids. *Ann. Rev. Genet.* **16**, 357–384.

2. Hoykaas-van Slogteren, G.M.S., Hoykaas, P.J.J., and Schilperoort, R.A. (1984) Expression of Ti plasmid genes in monocotyledonous plants infected with *Agrobacterium tumefaciens*. *Nature* **311**, 763–764.

3. Hernalsteens, J.P., This Toong, L., Schell, J., and van Montagu, M. (1984) An *Agrobacterium*-transformed cell culture from the monocot *Asparagus officinalis*. *EMBO J.* **3**, 3039–3041.

4. Paszkowski, J., Shillito, R.D., Saul, M.W., Mandak, W., Hohn, T., Hohn, B., and Potrykus, I. (1984) Direct gene transfer to plants. *EMBO J.* **3**, 2717–2722.

5. Lorz, H., Baker, B., and Schell, J. (1985) Gene transfer to cereal cells mediated by protoplast transformation. *Mol. Gen. Genet.* **199**, 178–182.

6. Werr, W. and Lorz, H. (1986) Transient gene expression in a Gramineae cell line: A rapid procedure for studying plant promoters. *Mol. Gen. Genet.* **202**, 471–475.

7. Potrykus, I., Saul, M.W., Petruska, J., Paszkowski, J., and Shillito, R.D. (1985) Direct gene transfer to cells of a graminaceous monocot. *Mol. Gen. Genet.* **199**, 183–188.

8. Fromm, M.E., Taylor, L.P., and Walbot, V. (1986) Stable transformation of maize after gene transfer by electroporation. *Nature* **319**, 791–793.

9.   Vasil, I.E., ed. (1984) *Cell Culture and Somatic Cell Genetics of Plants* Academic, Orlando.
10. Evans, D.A., Sharp, W.R., Ammirato, P.V., and Yamada, Y., eds. (1983) *Handbook of Plant Cell Culture* Macmillan, New York, London.
11. Maniatis, T., Fritsch, E.G., and Sambrook, J. (1982) *Molecular Cloning: A Laboratory Manual* Cold Spring Harbor Laboratory, Cold Spring Harbor, New York.
12. Delmer, D.P. (1979) Dimethylsulfoxide as a potential tool for analysis of compoartmentation in living plant cells. *Plant Physiol.* **64**, 623–629.
13. Lerner, H.R., Ben-Bassat, C., Reinhold, L., and Poljakoff-Mayber, A. (1978) Induction of "pore" formation in plant cell membranes by toluene. *Plant Physiol.* **61**, 213–217.
14. Weimberg, R., Lerner, H.R., and Poljakoff-Mayber, A. (1983) Induction of solute release from *Nicotiana tabacum* tissue cell suspensions by polymixin and EDTA. *J. Exp. Bot.* **34**, 1333–1346.
15. McDaniels, C. (1983) Transport of Ions and Organic Molecules, in *Handbook of Plant Cell Culture* vol. 1, (Evans, D.A., Sharp, W.R., Ammirato, P.V., and Yamada, Y., eds.) Macmillan, New York, London.
16. Tanaka, H., Hirao, C., Semta, H., Tozawa, Y., and Ohmomo, S. (1985) Release of intracellularly stored 5'-phosphodiesterase with preserved plant cell viability. *Biotechnol. Bioeng.* **27**, 890–892.
17. Rhodes, D. and Stewart, G.R. (1974) A procedure for the *in vivo* determination of enzyme activity in higher plant tissue. *Planta* **118**, 133–144.
18. Yoshida, S. and Niki, T. (1979) Cell membrane permeability and respiratory activity in chilling-stressed callus. *Plant Cell Physiol.* **20**, 1237–1242.
19. Clarkson, D.T., Hall, K.C., and Roberts, J.K.M. (1980) Phospholipid composition and fatty acid desaturation inthe roots of rye during acclimatization of low temperature: Positional analysis of fatty acids. *Planta* **149**, 464–471.
20. Lockwood, G.B. (1984) Alkaloids of cell suspensions derived from four *Papaver* spp. and the effect of temperature stress. *Z. Pflanzenphysiol.* **114**, 361–363.
21. Felix, H.R. (1982) Permeabilized cells. *Anal. Biochem.* **120**, 211–234.
22. Krens, F.A., Molendijk, L., Wullems, G.J., and Schilperoort, R.A. (1982) *In vitro* transformation of plant protoplasts with Ti-plasmid DNA. *Nature* **296**, 72–74.
23. Lindsey, K. and Jones, M.G.K. (in press) The permeability of electroporated cells and protoplasts of sugar beet. *Planta.*
24. Hain, R., Stabel, P., Czernilofsky, A.P., Steinbiss, H.H., Herrera-Estrella, L., and Schell, J. (1985) Uptake, integration, expression and

genetic transmission of a selectable chimaeric gene by plant protoplasts. *Mol. Gen. Genet.* **199**, 161–168.

25. Graham, F.L., Van der Eb, A.J., and Heijneker, H.L. (1974) Size and location of the transforming region in human adenovirus type 5 DNA. *Nature* **251**, 687–691.

26. Neumann, E. and Rosenheck, K. (1972) Permeability changes induced by electric impulses in vesicular membranes. *J. Membrane Biol.* **10**, 279–290.

27. Zimmerman, U., Pilwat, G., and Riemann, F. (1972) Reversible dielectric breakdown of cell membranes by electrostatic fields. *Z. Naturforsch.* **29c**, 304–305.

28. Rosenheck, K., Lindner, P., and Pecht, I. (1975) Effect of electric fields on light-scattering and fluorescence of chromaffin granules. *J. Membrane Biol.* **20**, 1–12.

29. Lindner, P., Neumann, E., and Rosenheck, K. (1977) Kinetics of permeability changes induced by electric impulses in chromaffin granules. *J. Membrane Biol.* **32**, 231–254.

30. Sugar, I.P. and Neumann, E. (1984) Stochastic model for electric field-induced membrane pores: *Electroporation. Biophys. Chem.* **19**, 211–225.

31. Weaver, J.C., Powell, K.T., Mintzer, R.A., Ling, H., and Sloan, S.R. (1984) The electrical capacitance of bilayer membranes: The contribution of transient aqueous pores. *Bioelectrochem. Bioelectroenerg.* **12**, 393–404.

32. Weaver, J.C., Powell, K.T., Mintzer, R.A., Sloan, S.R., and Ling, H. (1984) The diffusive permeability of bilayer membranes: The contributions of transient aqueous pores. *Bioelectrochem. Bioelectroenerg.* **12**, 405–412.

33. Neumann, E., Schaefer-Ridder, M., Wang, Y., and Hofschneider, P.H. (1982) Gene transfer into mouse lyoma cells by electroporation in high electric fields. *EMBO J.* **1**, 841–845.

34. Potter, H., Weir, L., and Leder, P. (1984) Enhancer-dependent expression of human K immunoglobin genes introduced into mouse pre-B lymphocytes by electroporation. *Proc. Natl. Acad. Sci. USA* **81**, 7161–7165.

35. Langridge, W.H.R., Li, B.J., and Szalay, A.A. (1985). Electric field mediated stable transformation of carrot protoplasts with naked DNA. *Plant Cell Rep.* **4**, 355–359.

36. Fromm, M., Taylor, L.P., and Walbot, V. (1985) Expression of genes transferred into moncot and dicot plant cells by electroporation. *Proc. Natl. Acad. Sci. USA* **82**, 5824–5828.

37. Shillito, R.D., Saul, M.W., Paszkowski, J., Muller, M., and Potrykus, I. (1985) High efficiency direct gene transfer to plants. *Bio/Technol.* **3**, 1099–1103.

38. Widholm, J.M. (1972) The use of fluorescein diacetate and phenosafranine for determining viability of cultured plant cells. *Stain Technol.* **47**, 189–194.

39. Southern, E.M. (1975) Detection of specific sequences among DNA fragments separated by gel electrophoresis. *J. Mol. Biol.* **98**, 503–517.

40. Deshayes, A., Herrera-Estrella, L., and Caboche, M. (1985) Liposome-mediated transformation of tobacco mesophyll protoplasts by an *Escherichia coli* plasmid. *EMBO J.* **4**, 2731–2737.

41. Van den Elzen, P.J.M., Townsend, J., Lee, K.Y., and Bedbrook, J.R. (1985) A chimaeric hygromycin resistance gene as a selectable marker in plant cells. *Plant Mol. Biol.* **5**, 299–302.

42. Waldron, C., Murphy, E.B., Roberts, J.L., Gustafson, G.D., Armour, S.L., and Malcolm, S.K. (1985) Resistance to hygromycin B: A new marker for plant transformation studies. *Plant Mol. Biol.* **5**, 103–108.

43. Paszkowski, J. and Saul, M.W. (1986) Direct gene transfer to plants. *Meth. Enzymol.* **118**, 668–684.

44. Potrykus, I., Paszowski, J., Saul, M.W., Petruska, J., and Shillito, R.D. (1985) Molecular and general genetics of a hybrid foreign gene introduced into tobacco by direct gene transfer. *Mol. Gen. Genet.* **199**, 169–177.

45. Herrera-Estrella, L., Depicker, A., van Montagu, M., and Schell, J. (1983) Expression of chimaeric genes transferred into plant cells using a Ti plasmid-derived vector. *Nature* **303**, 209–213.

46. Gorman, C.M., Moffat, L.F., and Howard, B.H. (1982) Recombinant genomes which express chloramphenicol acetyltransferase in mammalian cells. *Mol. Cell Biol.* **2**, 1044–1051.

47. Burrell, M.M., Twell, D., Karp, A., and Ooms, G. (1985) Expression of shoot-inducing Ti T_L-DNA in differentiated tissues of potato (*Solanum tuberosum* cv. Maris Bard). *Plant Mol. Biol.* **5**, 213–222.

48. Burrell, M.M., Temple, S., and Ooms, G. (1986) Changes in translatable poly (A) RNA from differentiated potato tissues transformed with shoot-inducing Ti T_L-DNA of *Agrobacterium tumefaciens*. *Plant Mol. Biol.* **6**, 213–220.

49. Ooms, G., Twell, D., Bossen, M.E., Hoge, J.H.C., and Burrell, M.M. (1986) Developmental regulation of RI T_L-DNA gene expression in roots, shoots and tubers of transformed potato (*Solanum tuberosum* cv. Desirée). *Plant Mol. Biol.* **6**, 321–330.

50. Okada, K., Nagata, T., and Takebe, I. (1986) Introduction of functional RNA into plant protoplasts by electroporation. *Plant Cell Physiol.* **27**, 619–626.

Chapter 43

Isolation of Wheat Chloroplasts and Chloroplast DNA

Matthew Clement Jones

1. Introduction

Chloroplasts in common with other plastids and mitochondria have their own nucleic material, in the form of multiple copies of a circular DNA molecule, normally of 120–160 kbp.

Enrichment for chloroplasts in a leaf homogenate is the initial stage of extraction of chloroplast DNA. This is usually a combination of filtration and differential centrifugation. The chloroplast-enriched fraction will, however, normally remain significantly contaminated with DNA of nuclear and mitochondrial origin.

Normally chloroplast DNA has a base composition so similar to that of nuclear DNA that it may not be separated from it by density ultracentrifugation (1). The density change induced in supercoiled chloroplast DNA by the limited intercalation of ethidium bromide compared to that into nuclear and relaxed circular chloroplast DNA has been used for chloroplast DNA purification (2). Only relatively low yields may be obtained by this method, however, principally because of the damage to chloroplast DNA during the initial stages of isolation.

One common method to remove contaminating nucleic acid is the use of DNAse I. This method relies on the envelope of intact chloroplasts to protect their DNA. Prior to lysing the intact chloroplast to release the DNA, the DNAse I is rendered inactive by the chelation of Mg^{2+} ions using EDTA (3).

Wheat has high levels of endogenous nucleases. Because of this, the DNAse I enrichment procedure is omitted from the following protocol. The levels and types of DNAses present are likely to vary from species to species. A protocol for employing DNAse I is given in Note 2 in section 4. It is strongly recommended that the levels and types of nucleases present in the tissue from which chloroplast DNA is to be isolated is tested using a rapid assay technique (e.g., ref. 4). The endogenous DNAses of wheat include those with that remains active following the use of EDTA to chelate Mg^{2+} ions (4). Heat is used to inactivate all the DNAses of wheat immediately prior to the lysis of chloroplasts to release their DNA.

2. Materials

1. The isolation buffer employed is that of Leegood and Walker (5): 330 mM sorbitol, 20 mM HEPES-NaOH (pH 7.6), 10 mM EDTA, 5 mM $MgCl_2$, 10 mM $NaHCO_3$ is prepared in bulk to the correct the final volume and autoclaved (15 min at 121°C, 103 kPa). 0.1% (w/v) bovine serum albumin (BSA) is added, and the buffer is split into 135 mL lots quickly to reduce microbial contamination. The buffer should be stored frozen at –18°C. Some buffer is maintained as a liquid at 4°C. Before use the frozen buffer has to be brought to a semiliquid slurry. The correct consistency may be achieved by defrosting the frozen buffer at 4°C for 17–20 h. Just before use, 15 mL of 2% (w/v) sodium D-isoascorbate in liquid buffer is added to 135 mL of slurry. If using the DNAse protocol (Note 2 in section 4), a separate lot of the isolation buffer lacking EDTA will be required. It should be prepared as above, but not

frozen (store at 4°C) and does not require sodium D-iso-ascorbate.

2. Washing buffer is 400 mM sorbitol, 10 mM EDTA, 50 mM HEPES-KOH (pH 8.2), made up to final volume, auto-claved as above, and stored at 4°C. Immediately before use, the buffer is made 10 mM NaHCO$_3$, by adding 100 mM NaHCO$_3$ (freshly prepared in washing buffer). If following the protocol in Note 2 (section 4), an additional solution of washing buffer made to be 30 mM EDTA will be required.

3. A main filter is employed that consists of eight layers of muslin, the first two layers of which are separated by a layer of absorbent cotton wool about 5 mm thick and a prefilter of two layers of butter muslin. The main filter is "prewetted" before use with liquid isolation buffer.

4. A mixture of organic solvents is used to denature and precipitate proteins. This consists of a 1/1 (v/v) solution of phenol mix (6) and chloroform. Phenol mix is 300 g of phenol AR crystals (these should be white: pink crystals must not be used), 0.3 g of 8-hydroxyquinoline, 42 mL of *m*-cresol, and 90 mL of distilled water. Chloroform is toxic and a potential carcinogen. Dichloromethane (methylene chloride) is regarded as a safer substitute. *Gloves must be worn while handling these solutions.*

5. DNAse-free RNAse is prepared by incubating a solution of bovine pancreas RNAse A (100 μg/mL) in 200 mM so-dium acetate (pH 5.6), 200 mM NaCl at 80°C for 10 min. The solution, allowed to cool to room temperature, is then stored at –20°C in small aliquots. The final concentration of RNAse at the time of use is 50 μg/mL, (5–8 Kunitz unit/mL for most commercial RNAse enzymes).

6. Sodium saline citrate (SSC) is 0.15M NaCl, 0.015M Na$_3$ citrate. This buffer will be required as a one-tenth-strength, autoclaved solution. It is referred to as 0.1 x SSC.

7. Microcentrifuge tubes should be autoclaved to destroy any nuclease activities present. One-milliliter micropipets

are good for transferring chloroplast suspensions. In order to protect the chloroplasts from hydrostatic damage, the hole in the micropipet tip can be enlarged by slicing off the end (use a razor blade). Tips should be autoclaved.

3. Method

1. Wheat seedlings, from 7 to 10 d old, grown in a regime of 15 h light, 9 h dark, should be used. The dark period immediately preceding use should be extended to 30 min before the start of the preparation, to reduce starch grain build-up in chloroplasts. Seedling leaves should be cut into 5–15-mm lengths and washed in ice-cold distilled water before use. Steps 2–5 should be carried out at 4°C.

2. Homogenization of the prepared leaf in the isolation slurry will depend on the hardware you have available. For wheat and other fibrous leaves, the Polytron (an overhead power unit with partially enclosed rotating blades as a probe) is ideal. In this case, split 15 g of the leaf sections and the prepared slurry (150 mL) between two 100-mL measuring cylinders. Subject each cylinder to homogenization at setting 6 for 2 s. If there is no Polytron available (and as a preference for broad, soft-leaved plants), a Waring blender may be used. The preparation may be homogenized as a whole.

3. Squeeze the combined homogenate through the prefilter, allowing the fluid to pass directly into the main filter—the filtrate being collected in a vessel on ice. The filtrate should be green and contain mainly intact and slightly damaged chloroplasts. Chloroplast fragments and other cellular debris should be trapped by the cotton wool and muslin. The prefilter retains large leaf fragments and ice crystals.

4. The filtrate (split between several centrifuge tubes) should be spun at between 2300 and 7300g for a few seconds only. The force applied to the filtrate during acceleration and deceleration is sufficient to pellet chloroplasts.

5. A total of 25 mL washing buffer should be added to the pellets after discarding the supernatant. A smooth suspension is best achieved by initially resuspending the pellet without addition of buffer. The pooled suspension can be split between two tubes, and the centrifugation repeated. The pellets from the second spin, initially resuspended without addition of further buffer, can be pooled together with 1–2 mL of washing buffer. This centrifugation step removes more chloroplast fragments and replaces the isolation buffer that contained substances to protect chloroplasts during homogenization.

6. Before lysis of the chloroplasts, any nucleases in the suspension must be inactivated: this is done by heating. The suspension should be heated to 75°C for 10 min in a water bath in the presence of 10 mg pronase. The pronase (a proteolytic preparation, active at high temperatures) is the first stage of the deproteinization process. After cooling, 20 μL of sarkosyl [a commercial preparation of 30% (w/v) sodium lauryl sarcosinate in aqueous solution] is added to lyse the chloroplasts. If the chloroplasts are not pure at this stage, the final result will be affected (*see* Note 2 in section 4).

7. The lysate can be transferred to several 1.15-mL microcentrifuge tubes (each not more than half full). In order to continue the deproteinization process, an equal volume of phenol mix/chloroform (or dichloromethane) should be added to the lysate. This is mixed by inversion (producing an emulsion) for 1 min. The two phases will begin to separate as soon as mixing stops: complete separation requires centrifugation for another minute. The chlorophyll (green coloration) will partition into the lower layer. The upper, aqueous layer (but not any interfacial material) should be recovered and extracted again with the phenol mix solution as before.

8. The aqueous layer from the second extraction is made 0.9*M* with respect to potassium acetate (by addition of 3/

7 of the volume of 3*M* potassium acetate). Twice the total volume of cold (–18°C) absolute ethanol is added to precipitate nucleic acids. Nucleic acids should be precipitated at either –18°C for 2 h or –70°C for 30 min. Precipitates are collected by centrifugation at 4°C. The surface of the pellets are washed by adding 70% ethanol/30% sterile distilled water and removing the liquid without resuspending the pellet. This reduces any residual levels of phenol and potassium acetate. The pellet is dried using N_2 gas, using care not to dislodge the pellet or to overdry the nucleic acid. If the pellet becomes too dry, it will not resuspend easily.

9.   Pellets should be resuspended in 0.1x SSC, and an equal volume of DNAse-free RNAse A (100 μg/mL) is added. The mixture is incubated at 25°C for 60 min.

10. The solution is deproteinized with phenol mix/chloroform twice as before (step 7). The nucleic acid is precipitated as before (step 8). The dry DNA pellet is then resuspended in 0.1x SSC (usually about 40 μL). The DNA can be subjected to restriction endonuclease digestion and agarose gel electrophoretic analysis to determine its purity (*see* Vol. 2 of this series).

4. Notes

1.   The technique outlined above should be suitable for all young cereal tissue. For older cereal tissue, it may be necessary to produce protoplasts initially and isolate chloroplasts from them (5). Dicotyledenous plant tissue may be accommodated by adopting the modifications outlined in note 2. Alternatively, conventional methods are detailed in refs. 2 and 3, and novel methods in refs. 7–9.

2.   Many species (especially dicotyledons, unpublished results) may lack the high levels of endogenous nucleases found in wheat. When isolating chloroplast DNA from these tissues, DNAse I may be used to digest nonchloro-

plast DNA prior to chloroplast lysis. After the chloroplasts have been pelleted (step 4 in section 3), they should be resuspended in 25 mL of isolation buffer lacking EDTA, and DNAse I should be added to a concentration of 50 µL/mL. In the absence of EDTA, the Mg^{2+} ions allow DNAse I to act, but the enzyme cannot enter intact chloroplasts. After 1 h on ice, 25 mL of washing buffer (30 mM EDTA) is added. The EDTA will chelate the Mg^{2+} ions, and the DNAse I will become inactive. The chloroplast suspension is then spun as in step 5 in section 3. If it has proved necessary to use DNAse I, the heat-inactivation step is probably not necessary: proceed from step 7.

If you are not sure about using DNAse I, prepare a chloroplast DNA sample without using DNAse I. Subject this to PstI digestion and run a gel (*see* Vol. 2 of this series). Most chloroplast DNAs are distinguished by the lack of 5-methyl cytosine modifications. This makes them susceptible to PstI digestion, although contaminating DNA is not (7). The presence of contaminating DNA will be revealed by an undigested band (or smear) superimposed on the chloroplast DNA digestion pattern.

3. Pronase is a fairly crude proteolytic enzyme preparation from *Streptomyces griseus* and may contain contaminating nucleases. If the data sheet from your supplier does not claim complete absence of nuclease activity (including nicking activities), then the pronase should be pretreated. This may be done by preincubating the pronase in solution at 80°C for 30 min.

4. A 250-mL, wide-necked polypropylene bottle is the storage vessel of choice for the isolation buffer. It is robust, easily cleaned, and the wide neck allows the flow of semisolid slurry. I do not recommend the autoclaving of buffer in polypropyene bottle.

References

1.   Wells, R. and Ingle, J. (1970) The constancy of the buoyant density of chloroplast and mitochondrial deoxyribonucleic acid in a range of higher plants. *Plant Physiol.* **46**, 178–179.
2.   Vedel, F. and Quetier, F. (1978) Hydrolyse specifique de l'ADN chloroplastique et l'ADN mitochondrial des vegetaux superieurs par les enzymes de restriction. *Physiol. Veg.* **16**, 411–425.
3.   Kolodner, R. and Tewari, K.K. (1975) The molecular size and conformation of chloroplast DNA from higher plants. *Biochem. Biophys. Acta* **402**, 372–390.
4.   Jones, M.C. and Boffey, S.A (1984) Deoxyribonuclease activities of wheat seedlings. *FEBS Lett.* **174**, 215–218.
5.   Leegood, R.C. and Walker, D.A. (1979) Isolation of protoplasts and chloroplasts from flag leaves of *Triticum aestivum* L. *Plant Physiol.* **63**,1212–1214.
6.   Grierson, D. and Covey, S. (1976) The properties and function of rapidly-labelled nuclear DNA. *Planta* **130**, 317–321.
7.   Bowman, C.M. and Dyer, T.A. (1982) Purification and analysis of DNA from wheat chloroplasts isolated in non-aqueous media. *Anal. Biochem.* **122**, 108–118.
8.   Rhodes, P.R. and Kung, S.D. (1981) Chloroplast deoxyribonucleic acid isolation: Purity achieved without nuclease digestion. *Can. J. Biochem.* **59**, 911–915.
9.   Terri, T., Saura, A., and Lokki, J. (1984) Chloroplast DNA from *Calmagrostis* species by selective lysis of organelles. *Hereditas* **101**, 123–126.

Chapter 44

DNA Labeling by Isolated Wheat Chloroplasts

Matthew Clement Jones

1. Introduction

In wheat, growth is from a basal meristem: The cells of the young shoot form a developmental array from the tip (oldest) to the base (youngest) of the leaf. In cells from a specific part of the leaf (and so of a certain age), chloroplast DNA is found to be replicating. In seedlings up to 10 d old, chloroplast DNA replication is found in the region from 1 to 4 cm from the leaf base (1) If chloroplasts are isolated from this region of leaf and incubated in the presence of a radioactively labeled DNA precursor, radioactively labeled DNA can be produced in vitro.

Two methods of monitoring chloroplast DNA labeling may be used. The first is liquid scintillation counting of chloroplast suspension samples, suitably treated for the removal of nonincorporated label. The second is the isolation of DNA from chloroplasts that have been incubated in the presence of labeled precursor. The labeled DNA recovered from these chloroplasts may be subjected to gel electrophoresis and (depending upon the label employed) autoradiography or fluorography.

The choice of label depends upon a number of factors. Labeled deoxyribonucleotide triphosphates (dNTPs) are incorporated into chloroplast DNA, but this is only into DNAse I-permeable chloroplasts (2) and is light-independent. Furthermore, any contaminating bacteria may be able to take up, and incorporate into DNA, labeled dNTPs. Labeled nucleotide incorporation into chloroplast DNA is light-dependent, has an absolute requirement for chloroplast intactness, and therefore is a consequence of chloroplast DNA polymerase activity. The use of labeled nucleotides ([methyl-^{3}H]-thymidine is recommended) in the light (with a dark control) can account for any bacterial incorporation of label and is essential for studies on chloroplastic DNA polymerases; since in DNAse I-permeable chloroplasts, nonchloroplast DNA polymerases may have access to templates and precursors.

Because intactness is so important, particular care must be placed in the preparation of the chloroplasts. The most important steps are tissue growth, homogenization, and centrifugation.

There are fewer chloroplasts in the chloroplast DNA-replicating region than in more mature regions of the leaf. In order to maximize the incorporation of label into DNA, the chloroplasts selected should all be from the juvenile region of leaf. Loss of DNA will be encountered, however, during the processing of chloroplasts following incubation. The loss of label can be reduced by reducing the proportion of labeled DNA in the total DNA. This is done by adding mature chloroplasts before processing the incubated chloroplasts. The mature chloroplasts are not added before incubation, since this would reduce the light availability to the DNA-replicating chloroplasts. I have found it easiest to use a single tray of wheat seedlings, the juvenile part of the leaf forming the chloroplast preparation for labeling, and the mature part of the leaf forming the carrier preparation.

2. Materials

1. The materials required for labeled chloroplast DNA extraction are essentially the same as for two unlabeled chloroplast DNA preparations (Chapter 43). Additionally, an illuminated water bath and incubation vessels (one foil wrapped to prevent light entering) are required. I have used 25-mL "Quick-Fit" conical flasks. To visualize chloroplast DNA, which has been labeled, on an electrophoretic gel, a dark room equipped to process X-ray film and a –70°C freezer will be required.

2. If only liquid scintillation counted results are required, the carrier preparation is not needed. Sample tubes are required. Washing solutions: 5% TCA, 0.1M tetrasodium pyrophosphate ($Na_4P_2O_7$), and absolute ethanol, are needed. A solution of 20% TCA, 0.1M $Na_4P_2O_7$ will be required for initial precipitation of samples.

3. [Methyl-3H]-thymidine can be obtained as an aqueous solution with 2% ethanol (as an antibacterial agent). In this protocol, the solution is mixed with buffer before the addition of chloroplasts. This prevents osmotic damage to the chloroplasts.

4. For fluorography (Note 2 in section 4), a 3% (w/v) solution of 2,5-diphenyloxazole (PPO) in absolute alcohol is needed.

3. Method

Steps 2 to 5 of this protocol are essentially the same as the first four steps of the protocol outlined in Chapter 43.

1. Before starting to prepare the tissue, make sure that the illuminated water bath is at 20°C. Prepare the reaction vessels (one light and one dark) and place 2 mL of washing buffer and up to 100 µL (normally 100 µCi or 3.7 MBq at the highest specific activity) of [methyl-³H]-thymidine in the

dark vessel. Leave the vessel on ice while the tissue is prepared. Prepare sampling tubes by marking tubes appropriately and placing 1 mL of 20% TCA, $0.1M$ $Na_4P_2O_7$ in each.

2. Wheat seedlings, from 7 to 10 d old, grown in a regime of 15 h light, 9 h dark, should be used. The dark period immediately preceding use should be extended to 30 min before the start of the preparation, to reduce starch grain build-up in chloroplasts. The lowest 3 or 4 cm of seedling should be cut into 5–15 mm lengths and washed in ice-cold distilled water. Fifteen grams of this juvenile tissue should be prepared. The more mature parts of the leaf should be cut up also and 15 g stored on ice.

3. Split the juvenile tissue and one 150-mL lot of isolation buffer slurry between two 100-mL glass measuring cylinders. Subject each cylinder to homogenization using a Polytron for 2 s at setting 6.

4. Squeeze the combined homogenate through the prefilter, allowing the fluid to pass directly into the main filter—the filtrate being collected in a vessel on ice. As an economy measure, the main filter can be saved and used for the carrier preparation. Since there will be large leaf fragments in the prefilter, however, a fresh prefilter must be used.

5. The filtrate (split between several centrifuge tubes) should be spun to about $7000g$ for 30 s only. The deceleration phase should be as brief as possible: Special breaks can be fitted to some centrifuges.

6. A total of 25 mL of washing buffer should be added to the pellets after discarding the superanatant. A smooth suspension is best achieved by initially resuspending the pellet without addition of further buffer. The pooled suspension can be split between two tubes and the centrifugation repeated. The pellets from the second spin should be resuspended without addition of further buffer.

7. Add the suspension to the dark incubation vessel, mix by swirling, and transfer half the solution to the light vessel (it

is a good idea to take a sample for processing as in step 13, before spliting up the suspension). Place the vessels in the illuminated water bath.

8. If you want to prepare DNA from the labeled chloroplasts, the carrier preparation should be made while the incubation is taking place (steps 9–12). Otherwise, samples should be taken from the vessels at specific times and processed as in step 13.

9. The reserved, mature tissue should be homogenized and filtered as for the juvenile preparation (using a fresh prefilter, but keeping the same main filter as before). To maximize pellet size, a relatively long, slow spin (e.g., 2500g for 30 s) should be employed. The supernatant is removed, and the pellets should be resuspended initially without addition of further buffer. Thirty milliliters of washing buffer should be added to wash the chloroplasts. Split the suspension between two tubes and employ the same type of spin again. The supernatant is removed and the pellets are superficially washed with 1 mL of washing buffer each. Pellets are resuspended, without addition, and equal volumes are added to the incubated preparations.

10. The light and dark preparations (each combined with a proportion of the carrier preparation) should be heated to 75°C for 10 min in a water bath in the presence of 10 mg of pronase each. The heating is done to inactivate nucleases and the pronase is the first stage of the deproteinization process. After cooling, 20 µL of sarkosyl (30% aqueous solution of sodium lauryl sarcosinate) is added to both light and dark preparations to lyse the chloroplasts.

11. Deproteinization, RNAse treatment, and DNA precipitation of the incubated chloroplast suspensions are carried out as for the unlabeled preparation detailed in Chapter 43, steps 7–10. *It must be remembered, however, that all waste is potentially radioactive and should be disposed of accordingly.*

12. The labeled DNA can be digested with a restriction endonuclease, and the resulting digest run on an agarose gel.

This should give a unique pattern of bands, which should also show up on X-ray film exposed to the gel. These bands demonstrate the purity of the preparation and that only chloroplast DNA has been labeled (*see* Fig. 1). Because tritium is a weak beta emitter, fluorography (rather than autoradiography) must be used to visualize the radioactivity. This involves replacing the water in the gel with alcohol, impregnating the gel matrix with PPO, and precipitating the PPO *in situ* by rehydrating the gel (*see* Vol. 1 in this series). The gel is then dried and X-ray film is exposed to it: *see* Note 2 in seciton 4.

13. As samples of the incubated chloroplast suspensions are taken, they should be precipitated by addition to 20% TCA, $0.1M$ $Na_4P_2O_7$, and chilling. They can be processed in two ways. Glass fiber filters are commonly employed (3). As an alternative, cycles of microcentrifugation of the precipitate, followed by resuspension (using an ultrasonic bath) in washing solutions may be employed. In both cases, washing solutions should be 5% TCA, $0.1M$ $Na_4P_2O_7$, followed by absolute ethanol. The $Na_4P_2O_7$ is used to prevent nonspecific binding of unincorporated label. When dry, glass fiber filters can be added to scintillation fluid for counting. Scintillation fluid can be added directly to microcentrifuge tubes so that they may be used as insert vials for liquid scintillation counting. In the latter case, counting efficiency may be slightly compromised. Although the microcentrifuge tube method has advantages in terms of economy of reagents, efficiency of washing, and time involved, the glass fiber filter method may be more efficient at precipitate retention and is probably more satisfactory from a safety point of view.

4. Notes

1. The absence of detergent from the plastic- and glassware associated with the preparation and incubation of juvenile

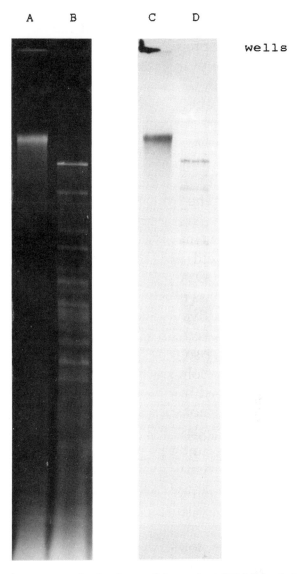

Fig. 1. Agarose gel of wheat chloroplast DNA isolated by this method (lanes A and B). The fluorograph (lanes C and D) shows that the chloroplast DNA has been labeled. The fluorograph shows some shrinkage compared with the gel. The mark at the top of the gel is from the radioactive ink showing the position of the well. Lanes A and C are uncut wheat chloroplast DNA. Lanes B and D are wheat chloroplast DNA digested with 10 U of EcoRI restriction endonuclease.

chloroplasts is very important. Any detergent present could lead to the premature lysis of chloroplasts: Lysed chloroplasts cannot incorporate labeled thymidine into their DNA (2).

2. Agarose gels should be photographed (ideally so that a negative is produced, from which a full-size photograph can be made) and cut down to minimal size to economize on reagents. Dehydration is achieved by subjecting the gel to a series of washes in alcohol. Each wash in the sequence should last at least 30 min. A shaking platform will improve results. The last stage should last 45 min and employ fresh absolute alcohol. Complete dehydration normally takes three changes of alcohol (including the absolute alcohol step). The gel is then impregnated with PPO as a 3% (w/v) solution in absolute alcohol, by soaking for at least 3 h. In order to precipitate the PPO out into the matrix of the gel structure, the gel is rehydrated in distilled water for 1 h. The gel (which is then an opaque white color) is dried down, under vacuum, using commercial gel-drying apparatus and a vacuum pump, with a moisture trap to protect the pump. When fully dry (normally about 3 h), the gel is marked out using radioactive ink (normal fountain pen ink mixed with an aqueous solution of a tritiated isotope) to identify it and mark the positions of the wells. The dried gel is then wrapped in Saran wrap and exposed to X-ray film in the normal way. The alcohol and PPO solutions can be used again, but fresh alcohol must be used for the last step, and the PPO solution (which should be stored in a dark bottle) has a finite life (the initial sign of exhaustion is a reduction in the white color in the rehydrated gel).

For maximum sensitivity, the X-ray film should be pre-exposed, and the exposure should take place at $-70°C$ (4). An X-ray film is darkened by exposure to photons (emitted by PPO in response to ^3H decay). Up to a certain stability threshold, however, the reaction is reversible.

Unless sufficient photons are incident upon a grain in the film to reach the stability threshold within a given time, the grain will not darken during development and no image will be observed. By pre-exposing the film (to sufficient light to give it an absorbence of 0.15 at 540 nm), enough photons will have reached the grains for them to pass over the stability threshold. The grains are made of silver atoms, which revert to silver ions if stability is not reached. Exposure at −70°C increases the half life of the silver atoms, increasing sensitivity.

References

1. Boffey, S. A., Ellis, J. P., Sellden, G., and Leech, R. M. (1979) Chloroplast division and DNA synthesis in light-grown wheat leaves. *Plant Physiol.* **64**, 502–505.
2. Bohnert, H. J., Schmitt, J. M., and Herrmann, R. G. (1974) Structural and functinal aspects of the plastome: III DNA- and RNA-synthesis in isolated chloroplasts. *Proc. Int. Symp. Plant Cell Differentiat. Portugaliae Acta Biologica* **14**, 71–90.
3. Bollum, F. J. (1968) Filter paper disk techniques for assaying radioacitive macromolecules. *Meth. Enzymol.* **12B**, 169–173.
4. Lasky, R. A. (1984) *Radioisotope Detection by Fluorography and Intensifying Screens* Amersham International, Amersham, England.

Index